Water Activity: Theory and Applications to Food

ift Basic Symposium Series

Edited by
INSTITUTE OF FOOD TECHNOLOGISTS
221 N. LaSalle St.
Chicago, Illinois

Foodborne Microorganisms and Their Toxins: Developing Methodology *edited by Merle D. Pierson and Norman J. Stern*

Water Activity: Theory and Applications to Food *edited by Louis B. Rockland and Larry R. Beuchat*

Water Activity: Theory and Applications to Food

edited by

Louis B. Rockland
Food Science Research Center
Chapman College
Orange, California

Larry R. Beuchat
Department of Food Science and Technology
University of Georgia
Agricultural Experiment Station
Experiment, Georgia

Marcel Dekker, Inc. New York and Basel

Library of Congress Cataloging-in-Publication Data

Water activity.

(IFT basic symposium series)
Proceedings of the tenth basic symposium held in Dallas, Tex., June 13-14, 1986, sponsored by the Institute of Food Technologists and the International Union of Food Science and Technology.
Includes bibliographies and index.
1. Food--Water activity--Congresses. 2. Food--Moisture--Congresses. I. Rockland, Louis B. II. Beuchat, Larry R. III. Institute of Food Technologists. IV. International Union of Food Science and Technology. V. Series
TX553.W3W36 1987 664 87-8967
ISBN 0-8247-7759-X

MARCEL DEKKER, INC.
270 Madison Avenue, New York, New York 10016

Current printing (last digit):
10 9 8 7 6 5 4 3 2 1

PRINTED IN THE UNITED STATES OF AMERICA

Preface

This book presents the proceedings of the Tenth Basic Symposium sponsored jointly by the Institute of Food Technologists (IFT) and the International Union of Food Science and Technology (IUoFST). It was held at the Grenelefe Hotel, Dallas, Texas, on June 13–14, 1986, prior to the 46th Annual IFT Meeting.

The symposium was organized for the purpose of disseminating new information on relationships between water activity (a_w) and the quality and stability of food products. It featured internationally recognized authorities who have pioneered this important new technology.

The symposium provided an update on new developments in an expanding literature on a_w and its influence on the physical and chemical properties of food products. This new discipline has been hindered by ingrained classical concepts of the role of water on the quality and stability of food products, and by the subtle differential influences of tightly bound, partially bound, and free water. During the past few years it has been shown that the properties of foods can be related more critically and explicitly to a_w. The varied configurations of moisture sorption isotherms illustrate diagrammatically that a_w is nonlinearly related to total moisture content, and that the sigmoid shape of isotherms varies because of the different types and proportions of food product constituents. Chemical and physical changes that occur during processing and storage can be related directly to the equilibrium relative humidity, expressed as a_w, of a food product. As new generations of students are exposed to a_w concepts, applications of these principles should facilitate significant improvements in the processing and stabilization of foods. Increased applications of a_w concepts would also be facilitated by development of rugged, stable, rapid, and precise instruments for estimating a_w in small containers.

Water activity principles can be applied directly to many common industrial problems, such as dehydration, the development of intermediate moisture foods, stabilization of flavor, color, and texture, and elimination of microbiological problems.

The symposium program was approved and supported by the IFT Basic Symposium Committee, which included Drs. R. V. Josephson (Chairman), P. G. Crandall, D. T. Gordon, J. A. Maga, M. D. Pierson, J. M. Regenstein, and R. A. Scanlan. Charles J. Bates, 1985–86 IFT President and an enthusiastic supporter, welcomed participants and expressed his gratitude to the speakers who came from distant places and foreign lands. Calvert L. Wiley, IFT Executive Director, John B. Klis, Director of Publications, Anna May Schenck, IFT Associate Scientific Editor, and other IFT staff members provided the support and coordination required to facilitate a nearly flawless symposium.

Primary credit for the success of this symposium belongs to the speakers. They performed professionally, meeting all the necessary deadlines for submission of abstracts and manuscripts to result in the rapid and timely publication of this volume.

Louis B. Rockland
Larry R. Beuchat

Introduction

Critical interest in the influence of water activity (a_w) on food product, quality, and stability began unceremoniously about 30 years ago. It was promoted by empirical observations inconsistent with a direct relationship between total moisture and product stability.

Microbiologists were among the first to recognize that a_w, rather than total moisture content, governed the growth, survival rate, death, sporulation and toxin production by diverse microorganisms. Scott (1936, 1953, 1957) and Mossel (1949, 1951, 1955), who conducted the classical work in this area, did not consider applications of their observations to dry products having low a_w. As early as 1949, Karon and Hillery noted that at any given a_w, the moisture content of various tissues of peanuts were very different, and that despite these differences, all of the tissues were at moisture equilibrium. In classical papers, Lea et al. (1943) and Bryce and Pearce (1946) reported an optimum moisture content for dried milk powder. Rockland (1957) reported an optimum moisture content for shelled walnuts and related this to an optimum a_w range. These observations were in contrast to the generally accepted belief that stability increased consistently with decreasing moisture content. It was also suggested that the optimum a_w was related to the manner in which the moisture was bound by individual chemical moieties (Pauling, 1945) and that this could be delineated using a mathematical treatment of moisture sorption data. This provided the basis for the Local Isotherm Concept and the utilization of moisture sorption data for derivation of Stability Isotherms (Rockland, 1957, 1969). On the basis of the Local Isotherm Concept, which delineated, at least qualitatively, moisture binding properties and hence stability, it was also proposed that changes in temperature would displace moisture sorption isotherms and result in predictable changes in quality and stability. Previously Bull (1944) and subsequently others have subscribed to the premise that the Brunauer-Emmet-Teller (BET) equation (1938) might be useful in providing a theoretical basis for differentiating moisture binding characteristics. Of classical importance was the thermodynamic treatment of moisture sorption phenomena by McLaren and Rowen (1951). During the past 20 years, there has been an avalanche of equations proposed to describe moisture sorption isotherms and to provide theoretical justification for moisture sorption phenomena (Van den Berg and Bruin, 1981).

During recent years, emphasis has been placed upon the relationships between moisture sorption characteristics, i.e., physical and chemical changes, as well as microbiological influences upon the stability and quality of food and other natural products (see figure, facing page). The primary objective of the 1986 Basic Symposium was to disseminate new knowledge and to explore some basic principles relating to the influences of a_w on food quality.

Louis B. Rockland

REFERENCES

Brunauer, S., Emmett, P. H., and Teller, E. 1938. *J. Am. Chem. Soc.* 60:309.

Bryce, W. A. and Pearce, J. A. 1946. *Can. J. Res.* 24:61.

Bull, H. B. 1944. *J. Am. Chem. Soc.* 66:1499.

Karon, M. L. and Hillery, R. E. 1949. *J. Am. Oil Chem. Soc.* 26:16.

Lea, C. H., Moran, T., and Smith, J. A. B. 1943. *J. Dairy Sci.* 13:164.

McLaren, A. D. and Rowen, J. W. 1951. *J. Polymer Sci.* 7:289.

Mossel, D. A. A. 1951. *Antonie von Leeuwenhoek* 17:146.

Mossel, D. A. A. and Kuijk, H. J. L. 1955. *Food Res.* 20:415.

Mossel, D. A. A. and Westerdijk, J. 1949. *Antonie von Leewenhoek* 15:190.

Pauling, L. 1945. *J. Am. Chem. Soc.* 67:555.

Rockland, L. B. 1957. *Food Res.* 22:604.

Rockland, L. B. 1969. *Food Technol.* 23:11.

Scott, W. J. 1936. *J. Council Sci. Ind. Res.* 9:177.

Scott, W. J. 1953. *Aust. J. Biol. Sci.* 6:549.

Scott, W. J. 1957. *Adv. Food Res.* 7:83.

Van den Berg, C. and Bruin, S. 1981. In *Water Activity: Influences on Food Quality* (L. B. Rockland and G. F. Stewart, eds.), p. 1. Academic Press, New York.

WATER ACTIVITY — STABILITY DIAGRAM

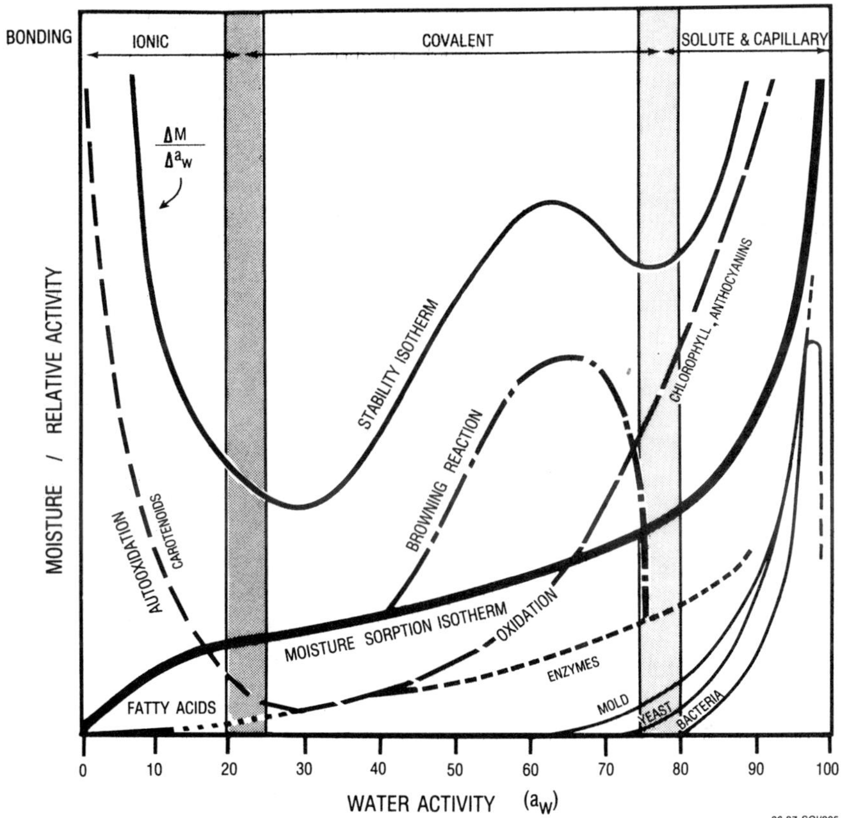

86-87 SCI/005

Contributors

Larry R. Beuchat, Ph.D. Department of Food Science and Technology, University of Georgia, Agricultural Experiment Station, Experiment, Georgia

David P. Bone Quaker Fellow, Technology & New Business R&D, The Quaker Oats Company, Barrington, Illinois

Malcolm C. Bourne, Ph.D. New York State Agricultural Experiment Station and Institute of Food Science, Cornell University, Geneva, New York

Ailsa D. Hocking, Ph.D. Division of Food Research, Commonwealth Scientific and Industrial Research Organisation, North Ryde, New South Wales, Australia

Melvin R. Johnston, Ph.D.* Division of Food Technology, Center for Food Safety and Applied Nutrition, FDA/DHHS, Washington, D.C.

John G. Kapsalis, Ph.D. Biological Sciences Division, Science and Advanced Technology Directorate, U.S. Army Research, Development & Engineering Center, Natick, Massachusetts

Lothar Leistner, Ph.D. Institute for Microbiology, Toxicology and Histology, Federal Centre for Meat Research, Kulmbach, Federal Republic of Germany

Marc Le Maguer, Ph.D. Food Science Department, The University of Alberta, Edmonton, Alberta, Canada

Lawrence M. Lenovich, Ph.D. Microbiology Research, Research and Development, Hershey Foods Corporation, Hershey, Pennsylvania

Henry K. Leung† Department of Food Science and Human Nutrition, Washington State University, Pullman, Washington

---

**Current affiliation:* Canned food consultant, New Braunfels, Texas

†Current affiliation: Campbell Soup Company, Camden, New Jersey

Rong C. Lin, Ph.D. Plant and Protein Technology Branch, Division of Food Technology, Food and Drug Administration, Washington, D.C.

John I. Pitt, Ph.D. Division of Food Research, Commonwealth Scientific and Industrial Research Organisation, North Ryde, New South Wales, Australia

Shelly J. Richardson, Ph.D. Division of Foods and Nutrition, University of Illinois, Urbana, Illinois

Walter E. L. Spiess, Ph.D. Federal Research Centre for Nutrition, Karlsruhe, Federal Republic of Germany

Marvin P. Steinberg, Ph.D. Department of Food Science, University of Illinois, Urbana, Illinois

J. Antonio Torres, Ph.D. Department of Food Science and Technology, Oregon State University, Corvallis, Oregon

John A. Troller, Ph.D. Procter & Gamble Co., Winton Hill Technical Center, Cincinnati, Ohio

Joachim H. von Elbe, Ph.D. Department of Food Science, University of Wisconsin–Madison, Madison, Wisconsin

Walter Wolf, Ph.D. Federal Research Centre for Nutrition, Karlsruhe, Federal Republic of Germany

Contents

Water Activity: Theory and Applications to Food

1

Mechanics and Influence of Water Binding on Water Activity

Marc Le Maguer

The University of Alberta
Edmonton, Alberta, Canada

INTRODUCTION

Water, as the main component of food and biological materials, plays a predominant role in determining their shape, structure, and physical and chemical properties. It also is a major control component in mass transfer, chemical reactions, and the activity of microorganisms. The food technologist involved in product development, process design, or production is confronted daily with the effect, desirable or undesirable, of water on the food material. In either case, a better understanding of the water relations with the food would certainly be of help. Most of the unit operations used in food processing have as a goal, in one way or another, either the removal of water to stabilize the material, as in drying and concentration, its transformation into a nonactive component in freezing, and its immobilization in gels, structured foods, and low and intermediate moisture foods. The main and essential way in which the immobilization of the water is measured is through the consideration of a_w and its relationship to moisture content. Based on the thermodynamic concept of water chemical potential in solutions, a_w has served and still is serving as an index of how successful we are at controlling water behavior in food systems. It is also the parameter that controls the driving force in water removal operations and is therefore essential for design purposes. Its effect on many of the other properties important to food manufacturing and preservation will be reviewed in the subsequent chapters in this volume. What all this indicates, however, is that we need to be able to deal with a_w quantitatively if we want to effectively control our food processing.

This review is concerned with understanding water and a_w as they relate to the other components of the food material, and will subsequently describe an approach to the modeling of the complex material that food represents.

BASIC STRUCTURE OF THE FOOD MATERIAL

In all of the operations involved in food processing, it is rare that we deal with a pure, well-characterized component (with the possible exception of sucrose). For the sake of simplification and clarity the following categories are proposed.

Cellular Material of Plant or Animal Origin

The process used in the transformation of plant and animal tissue will preserve to a great degree the initial cellular structure of the material. Blanching, conditioning in syrup or brine, freezing, and drying of fruits and vegetables and some meat products would fall in this category. The material can be considered made of a matrix, usually water insoluble, cell membranes, intact, partially or totally destroyed, and a complex aqueous solution of sugars, amino acids, proteins, salts, aromatic compounds, and lipids. The cell membranes may or may not play a role in the water relation depending on whether or not they have been damaged during processing. This last consideration is important in the removal of water from high-moisture fresh fruits or vegetables.

Mixture Water-Soluble and Water-Insoluble Food Components

No structure of the initial cellular material remains. Usually a mixing step of the ingredients with water is involved and subsequent processing may or may not induce structure. Texturization, filtration, wet mechanical separation, membrane separation, and concentration are typical processing operations involved in this category of products. Here again the resulting material is made of a complex aqueous solution, as defined previously, and contains structured or unstructured insoluble components which may have an effect on the water.

Aqueous Solutions of Soluble Food Components

This last category includes the complex aqueous solutions mentioned previously. It may also cover the area of emulsions and gels. Most of the solutes involved—salts, sugars, and proteins—have limited solubility in water, and therefore phase changes will occur in these solutions depending on the processing steps involved. Crystallization may or may not occur, and metastable states resulting from supersaturation or supercooling can occur and are dealt with by extrapolation of the properties of the concentrated solution. This is a very important area since in most foods the properties of the solutions will control to a great extent the a_w down to values of 0.3–0.5.

From the above classification it should be clear that, in general, food materials can be considered made of a complex aqueous solution, which may or may not contain insoluble components or membranes. The key to describing the water relations in these materials will depend on the availability of a reasonable model that reflects our current understanding of the interaction of water with soluble components and solid surfaces.

WATER INTERACTION WITH SOLUBLE AND INSOLUBLE COMPONENTS

Structure of Bulk Water

A comprehensive review of the notions concerning the structure of bulk water is not possible here; however, good and extensive literature reviews are available (Wood, 1979). Theories to explain the essential features of liquid water fall into two general classes: continuum and mixture models. Mixture models generally regard water to be a mixture of species of which one is ice-like in character. The range of low-density species used has included ice, pentagonal dodecahedra, and a range of other polyhedral structures as described by Speedy (1984).

Continuum theories owe their origin to Bernal and Fowler (1933), who postulated that each molecule is hydrogen bonded to four others with continuous variability in the hydrogen-bound angles, lengths, and energies. Two types of hydrogen bonds are usually postulated (Fig. 1.1) as discussed by Jeffrey (1982). This model has subsequently been developed by Pople (1951) and Sceats and Rice (1981). Finally, in the most recent development geometri-

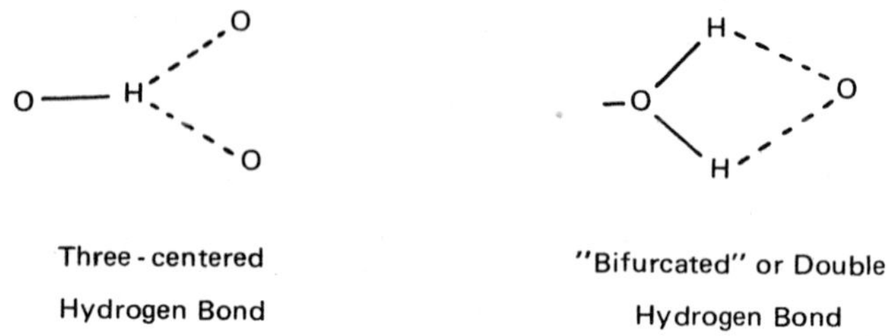

FIG. 1.1 Typical hydrogen bonds.

cal cooperativity is introduced by Stillinger (1981) to explain some of the behavior of the water in the supercooled state and its anomalies. These models can account for most of the properties of water as a liquid and serve as a support to explain the behavior of other substances when introduced in water.

Effect of Solutes on the Structure of Water

Both changes in temperature and the introduction of a solute perturb the water structure and alter the nature and extent of hydrogen bonding. Very often these effects have been reported in terms of a "structure making" and "structure breaking," showing empirically the ability of the solute to perturb the solvent water structure. In this perspective it is convenient to follow Frank and Wen (1957) in subdividing the solvent into regions labelled A, B, and C. The A region is the immediate vicinity of the solute, whereas region C is bulk water. The B region interfaces regions A and C. The net structuration of a solution is the result of A region and B region solvent perturbation.

Effect of ions. Ionic state will invariably introduce electrostrictive structure in the solvent A region due to the strong ion–water binding interactions. Water in the B region must interface two structurally incompatible structures, the A region and the hydrogen-bonded C region. Consequently it is destructured compared even to the C region.

Effect of apolar molecules. In hydrophobic hydration, the solute induces a clathrate-like cage (Fig. 1.2) structure in the solvent A region, a structure with increased ice-like character and stronger intermolecular hydrogen bonds than in bulk water. Consequently the B region shows increased ice-like character as it interfaces the highly structured water clathrate with bulk water (Mezei and Beveridge, 1981).

Effect of hydrophilic molecules. Polyhydroxy compounds occur in many of our food materials, e.g., sugars, cellulose, starch, and other high molecular

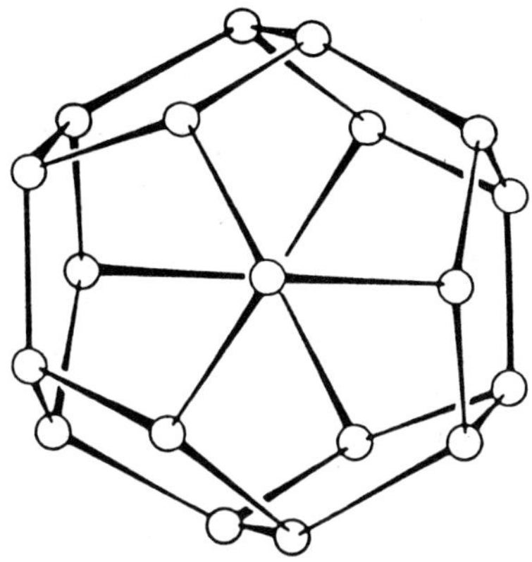

FIG. 1.2 Clathrate hydrate case geometry. Circles represent oxygen atoms and lines represent hydrogen bonds.

weight polymers. Franks (1983) indicates that in aqueous solutions the OH groups, particularly in an equational position, are almost indistinguishable from water molecules. This in turn leads to the observed effect that water will prefer the molecule in the case of sugars which have more equational OH. One of the other important effects related to the similarity of the OH groups on the polyhydroxy compounds and the water is in inducing supersaturation in these solutions (Franks, 1982), which is of great consequence in drying and aroma retention (Flink, 1983).

Effect of Solid Surfaces on the Structure of Water

The notion that water may be structurally modified by solid surfaces is not new and has been reviewed in particular by Drost-Hansen (1982). More recently Etzler (1983) has presented a model for water near interfaces. This water, also called vicinal water, will be regarded as water that is structurally modified. Estimates of the range of vicinal structuring run from about 5 to 200 water molecule diameters. It was postulated that vicinal water differs from bulk water in that it has a greater fraction of 4-hydrogen bonded "ice-like" molecules (Etzler, 1983). As a result, the model predicts, as is observed experimentally (Peschel and Adlfinger, 1970), that viscosity increases near the interfaces and the self-diffusion coefficient of water is reduced significantly (Hazelwood, 1979). Israelachvili and Pashley (1982) studied the effect of electrolytes on the hydration forces at the surfaces of solids. Upon binding to negatively charged surfaces, cations are only partially dehydrated and the remaining hydration shells give rise to repulsive hydration forces well in excess of ten water diameters. One of the postulated effects is that in very dilute electrolyte solutions surfaces can adhere but addition of a small amount of NaCl (1–10 mM), displaces the bound H^+ by Na^+ and the surfaces separate under the action of the hydration forces. At higher Na^+ concentrations the surfaces tend to get closer again but remain well separated by about 40–60 Å of water. Changing the pH will affect this separation. All these findings apply to macromolecules in solution with a possible use in texturization and stabilization of colloid and micellar systems.

Location of Water on Carbohydrates and Proteins

In many food products a level of dehydration has been reached, such that very low moisture contents are achieved. In these cases it is important to know where the water is located to be able to better control desired properties, for example, texture.

Water on carbohydrates. At very low moisture content internal hydrogen bonding occurs in the molecules of carbohydrates. Such molecules then organize themselves in a crystal lattice by including as many hydrogen bonds as possible (Jeffrey, 1982), offering to the outside a hydrophobic interface. This configuration is quite stable, using the cooperative effect to strengthen the structure, and is based on the well-established fact (Del Bene and Pople, 1970) that a chain of —O–H—O–H—O–H—O–H– hydrogen bonds has a greater bond energy than the sum of the individual bonds. Furthermore it is to be noted that the anomeric hydroxyls in reducing sugars are stronger than normal hydrogen-bond donors (Jeffrey et al., 1977).

As a result of this preferred organization very few water molecules are included in the crystal structure, but at intermediate moisture level the water could act as bridges between the different parts of the molecule, particularly in high molecular weight compounds.

Water on proteins. In these structures the hydrogen bonding is dominated by the Zwitterion bonds. The NH_3^+ invariably donates three bonds, whereas each oxygen of the carboxylate group accepts at least two and frequently three hydrogen bonds. In amino acids and small peptides, the water molecules, whenever present in the crystal structure, tend to have approximately tetrahedral environments (Fawcet et al., 1975). In the larger cyclic peptides, it is possible to distinguish between three distinct roles of the water molecules (Karle and Duesler, 1977). Some water molecules seem to be an integral part of the peptide in that they reside in the interior of the molecule and stabilize the conformation. Water molecules that are hydrogen bonded to the carboxyl oxygens which are directed to the outer surface of the molecule represent "bound" water; the remaining water molecules are bound to each other. When these molecules hydrate (Finney, 1977), the water molecules divide themselves into internal water molecules tightly bound in the protein interior and more mobile water molecules situated at the protein surface. These are usually associated with the charged and uncharged hydrophilic side-chain groups at the surface as well as more distant water molecules which form a transitional layer merging into the bulk water (Vinogradov, 1980). Some of the external water molecules bridge across to neighboring molecules in the crystal and others are found in active sites and along channels into the protein. Internal clusters of water molecules hydrogen bond to external water and thereby provide communication between the protein interior and the surrounding aqueous environment.

Summary

The hydrogen bonding of water is well documented and computer modeling describes quite well the properties of liquid water and dilute solutions from the supercooled to 100°C. From this emerges the picture of a tight hydrogen-bonded structure which can bend to accommodate different types of molecules or surfaces in its interior. What is necessary now is to try to put all of these descriptions and facts into a common framework for practical calculations.

MODELING COMPLEX SOLUTIONS

The best way to approach the prediction of the behavior of complex solutions is through the use of the excess Gibbs energy function to represent thermodynamic properties. This allows for the description of the deviation from ideality of the solution as a whole (Prausnitz, 1969). Such a model can then be used to predict phase changes, a_w, and estimate partial equilibrium properties.

UNIQUAC and UNIFAC

A summary of the attempts made at representing fluid phase equilibria is presented by Renon (1985) and Fredenslund and Rasmussen (1985). The two models thus retained allow for a description of electrolyte and nonelectrolyte solutions. The essential equations are presented in Appendix I for UNIFAC and Appendix II for UNIQUAC. The following simply illustrates the results that these models give on simple solutions containing sugar, ethanol, and sodium chloride.

Glucose-water solutions. The parametric values presented in Tables 1.1 and 1.2 were used with the UNIFAC equations (Appendix I). Results are presented on Fig. 1.3 for the prediction of a_w with concentration. In Fig. 1.4 is the predicted freezing curve. Figures 1.5 and 1.6 represent the experimental and predicted excess enthalpies for water and glucose, and Figs. 1.7 and 1.8 show the experimental and predicted behavior of the excess entropy—this is where the model has the most difficulty in its prediction capability. However, all in all it does offer a good representation of the solution's characteristics given that it has no adjustable parameter.

TABLE 1.1 Area and Volume Parameters of Functional Groups in a Binary Solution of Glucose and Water

Components	Functional groups	Number of groups (ν_k)	Volume parameter (R_k)	Area parameter (Q_k)
	CH_2	1	0.6744	0.540
Glucose	CH	4	0.4469	0.228
	OH	5	1.0000	1.097
	CHO	1	0.9980	0.948
Water	H_2O	1	0.9200	1.400

TABLE 1.2 Group-Interaction Parameters (a_{mn})

Groups	CH_2	CH	OH	H_2O	CHO
CH_2	0.0	0.0	986.5	1318.0	677.0
CH	0.0	0.0	986.5	1318.0	677.0
OH	156.4	156.4	0.0	−169.6	441.8
H_2O	300.0	300.0	−6.1	0.0	−257.3
CHO	505.7	505.7	−404.8	232.7	0.0

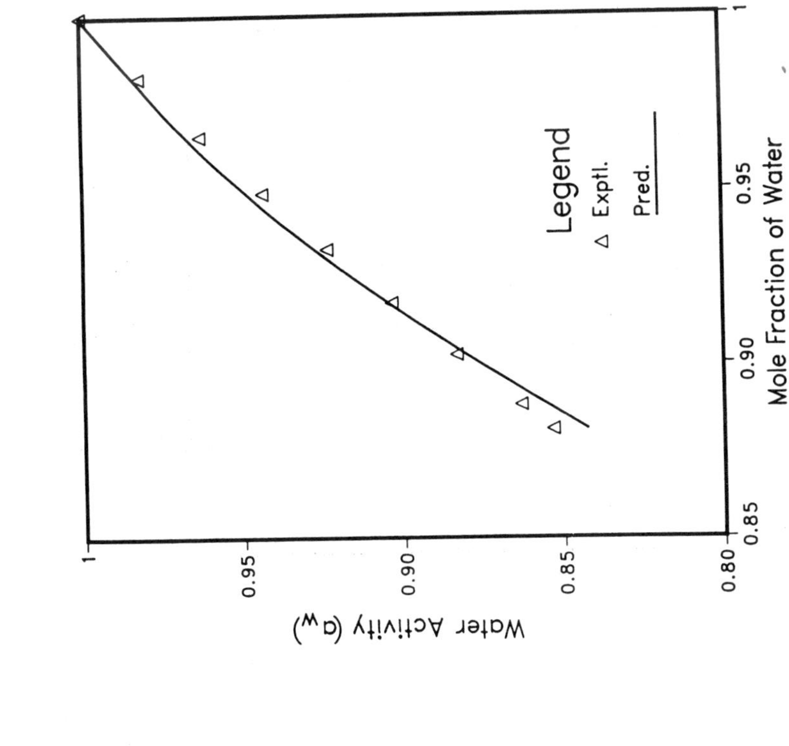

FIG. 1.3 Water activity at 25°C in glucose–water solutions.

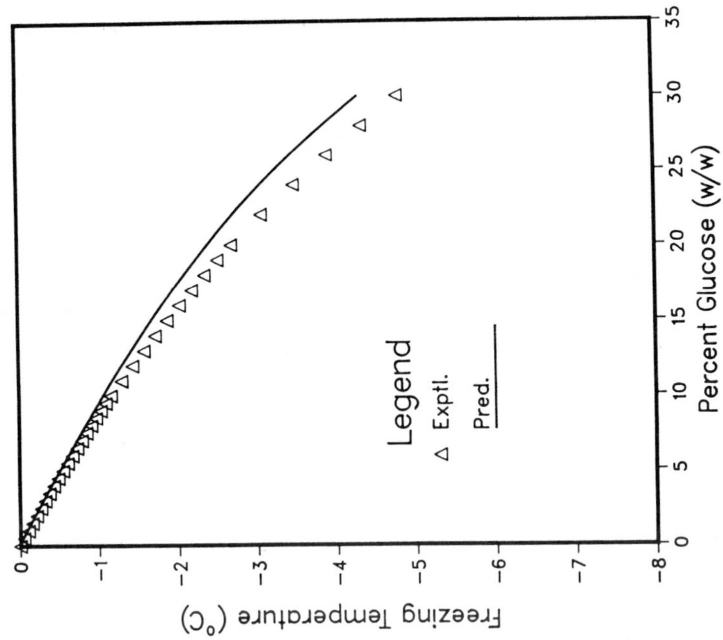

FIG. 1.4 Freezing curve of glucose–water solutions.

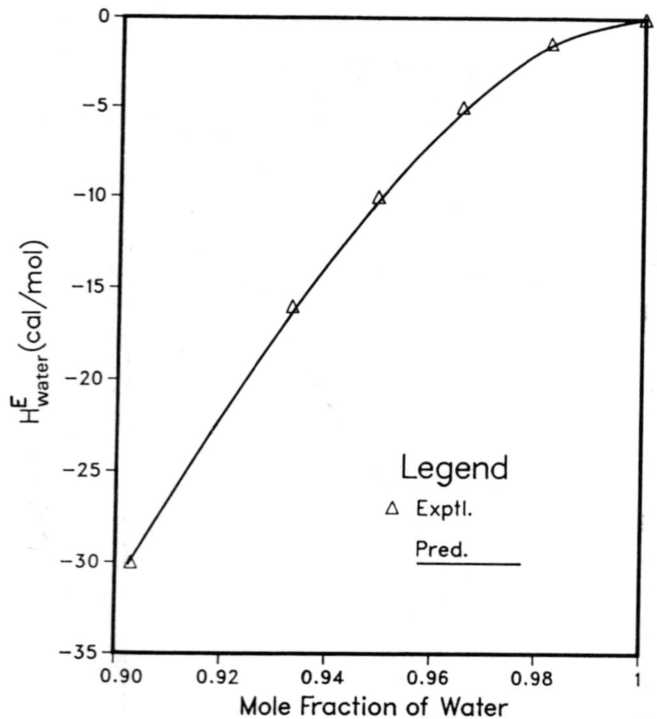

FIG. 1.5 Water excess partial molar enthalpy.

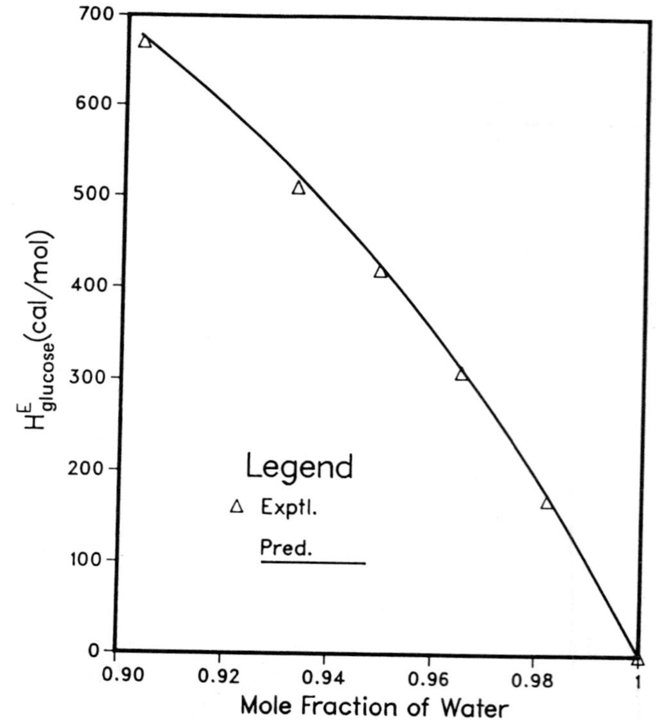

FIG. 1.6 Glucose excess partial molar enthalpy.

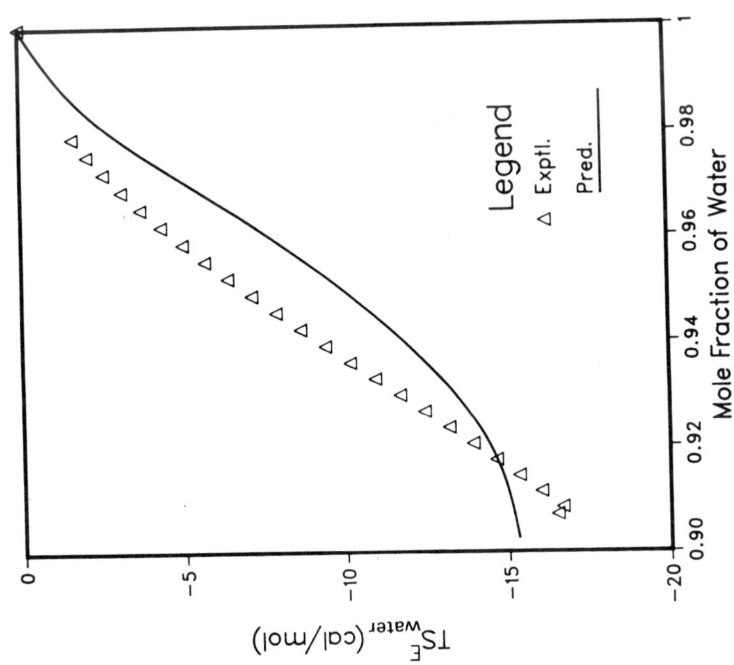

FIG. 1.7 Water partial molar entropy contribution.

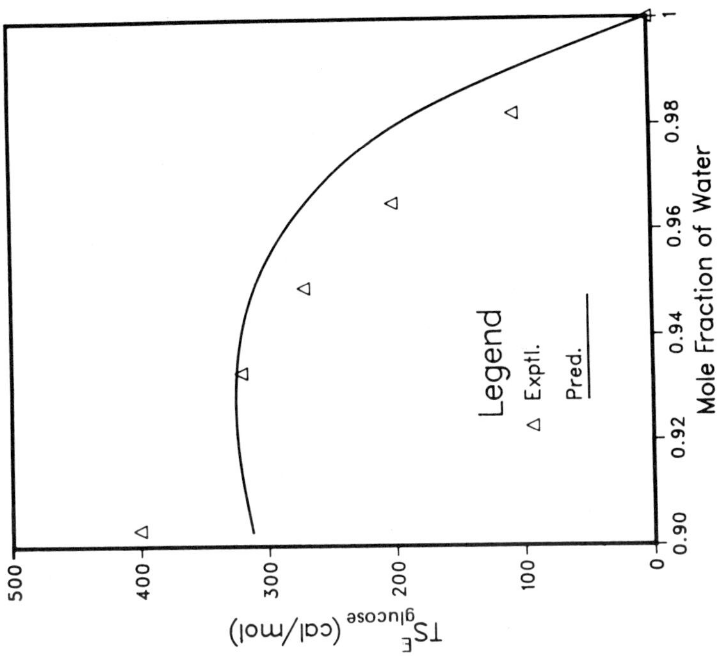

FIG. 1.8 Glucose partial molar entropy contribution.

Glucose-water sorption. This is a more severe test of the model's capabilities. The basic unit cell used to describe the "solution" is presented in Fig. 1.9. Using this unit cell, which takes into account the observations made previously on carbohydrate structure, the data of Smith et al. (1981) were used and the results are shown in Fig. 1.10. It was found, however, that the number

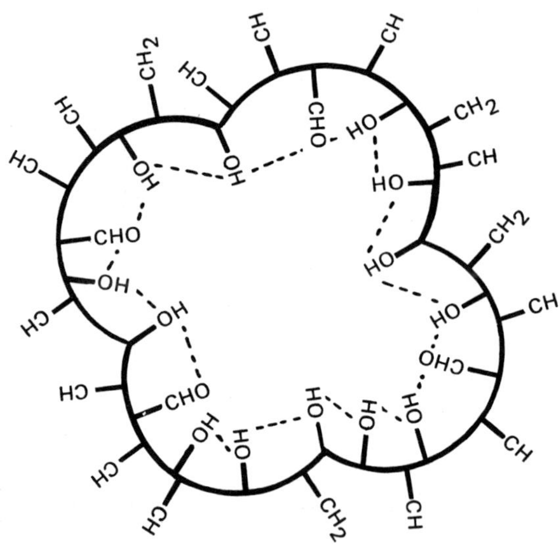

FIG. 1.9 Pictorial representation of an unit cell of a glucose-crystal. There are four molecules of glucose per unit cell (density = 1.56g/cm³). The packing arrangement is dictated mainly by H-bond. Molecules are held together by a complete system of strong hydrogen bonds and all OH groups are H-bonded, presenting a hydrophobic surface.

TABLE 1.3 Group Composition of Glucose Crystal–Water Mixture

Functional groups	Number of groups
CH_2	4
CH	16
OH	f(T, H_2O content)
H_2O	1
CHO	0

of OH groups internally bounded was a function of temperature and the amount of water that had penetrated the structure as per the functionality presented in the last equation in Appendix I.

It is remarkable how well the data can be represented given the severe assumptions under which the model is operating.

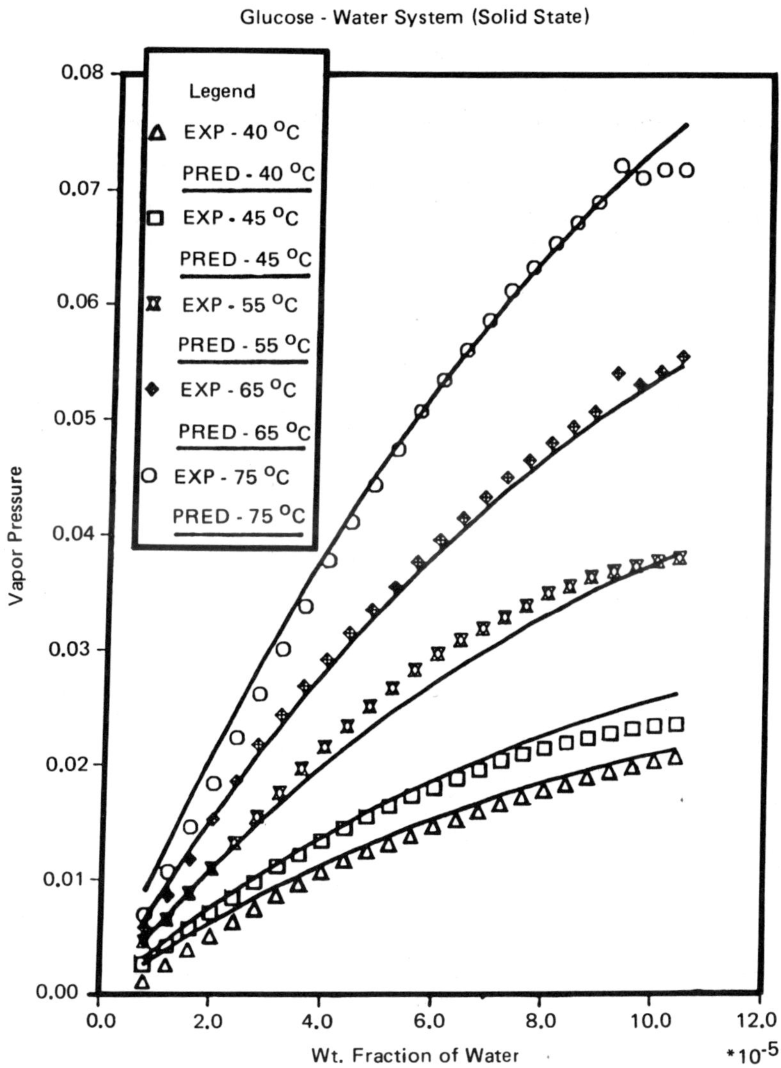

FIG. 1.10 Water vapor pressure (atm) vs. equilibrium moisture content (kg/kg) on glucose crystal.

Sucrose-water solution. Here, the UNIQUAC model was used and proved to be effective, as shown in Fig. 1.11.

Aroma-water solutions. These results were obtained by Le Maguer (1981) for terpene–water solutions and represent the application of the UNIQUAC equation to these limited solubility systems. In Fig. 1.12 the experimental solubilities of piperitone, carvone, and pulegone are compared to the predicted

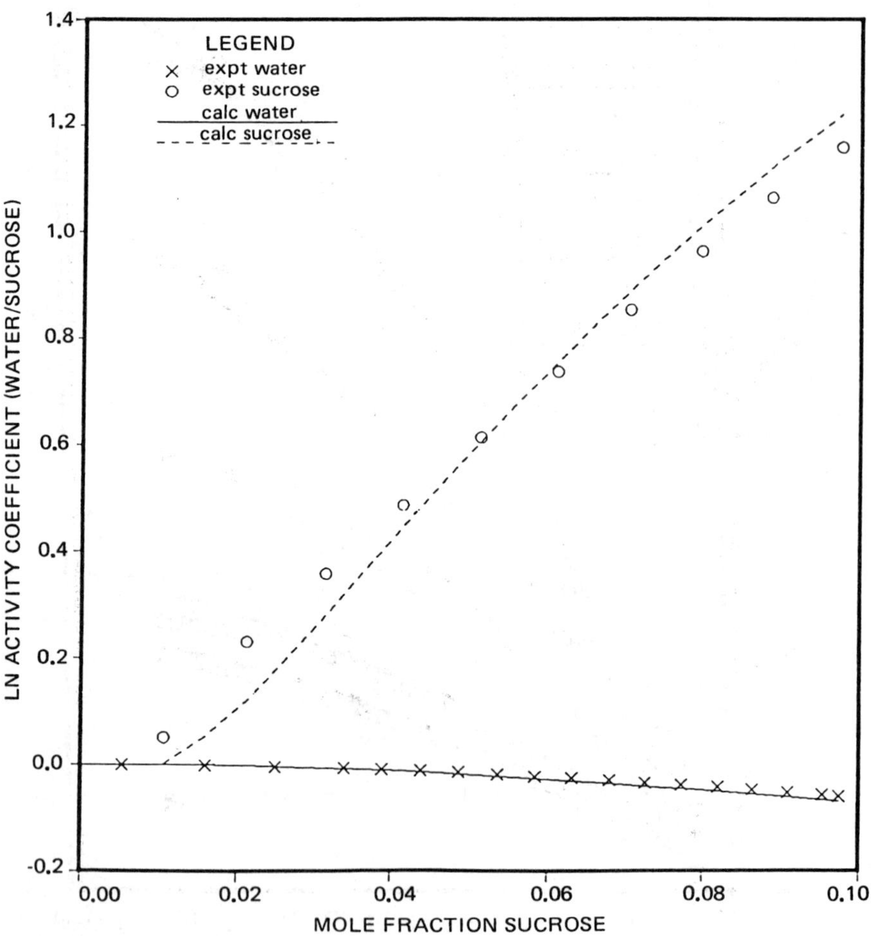

FIG. 1.11 Extended UNIQUAC equation: activities of sucrose and water. (*From Robinson and Stokes, 1965.*)

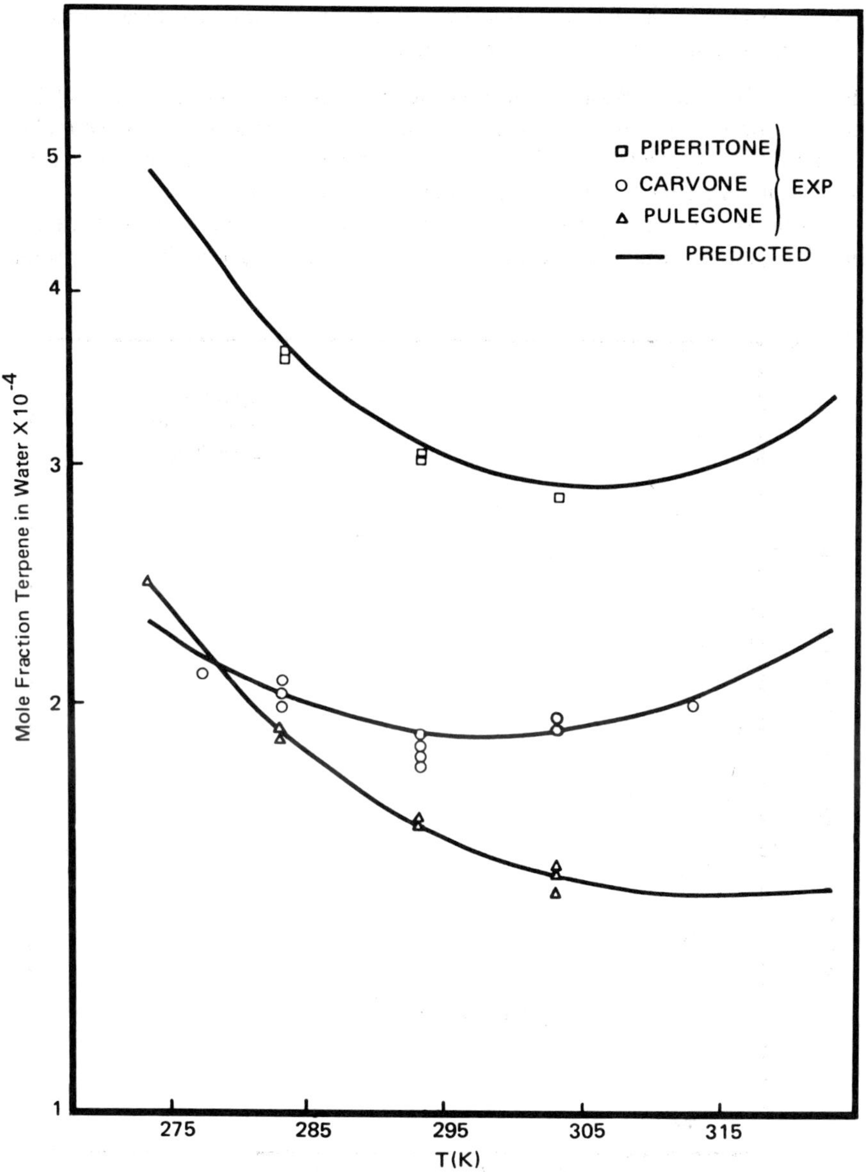

FIG. 1.12 Solubility of terpenes in water.

values. Again the model with only two adjustable parameters gives an excellent representation of the solution properties.

Sodium chloride–ethanol–water solution. Using a modified UNIQUAC model, the electrostatic contributions have been introduced as presented in Appendix II. Prediction of the activity coefficient for sodium chloride (Fig. 1.13) and water and ethanol (Fig. 1.14) is very satisfactory.

As illustrated above, it is clear that the functional group approach offered by UNIQUAC is capable of a good representation of carbohydrate and aroma

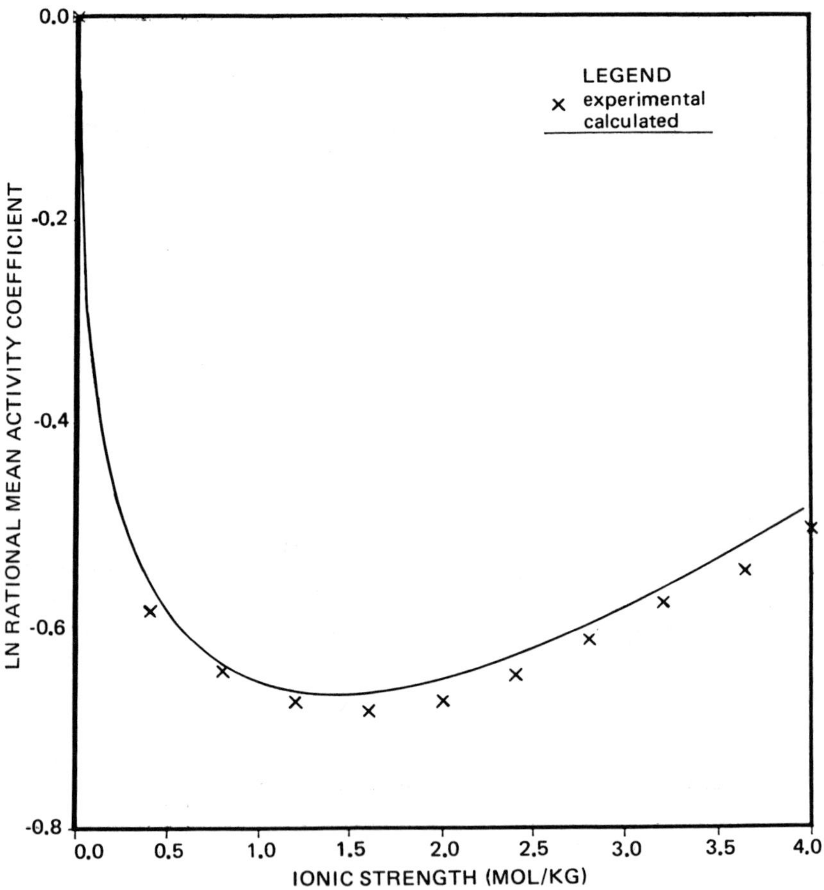

FIG. 1.13 Extended UNIQUAC equation: activity of NaCl. Results for system at constant solvent composition 10 mol% ethanol (salt-free basis) at 298.15 K. (*From Smirnov and Ovchinnikova, 1981.*)

solutions with a minimum of adjustable parameters. The introduction of the Debeye-Huckel and Bromsted contributions in the Gibbs free energy function allow for the introduction of the ionic components always present in the food material. It should be emphasized at this point that the functional group approach is only dependent on the functional groups present in the solution and not on their arrangement within individual molecules. It offers the possibility to reduce considerably the quantity of information necessary to de-

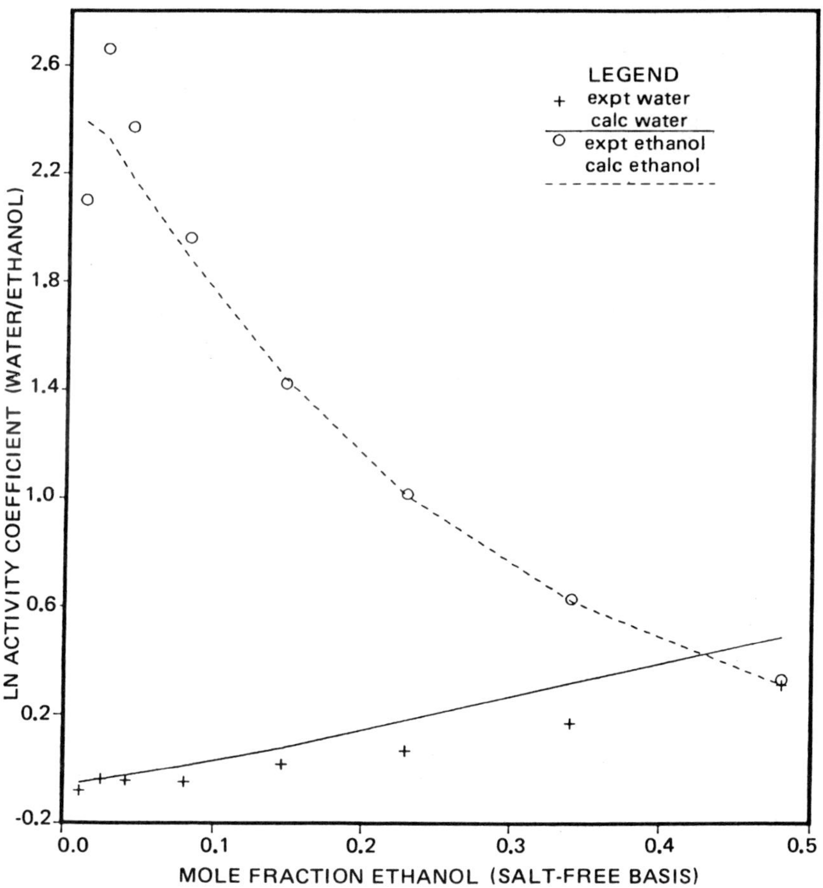

FIG. 1.14 Extended UNIQUAC equation: activities of water and ethanol. Solvent activity coefficients vs. mole fraction ethanol solutions saturated with NaCl at 760 mm Hg. (*From Furter, 1958.*)

scribe complex solutions, since it has been noted that with about 42 functional groups all molecules can be represented.

CELLULAR SYSTEMS

The water chemical potential is useful to describe cellular systems at high moisture content since it can contain terms that represent pressure effect, or through the partitioning of the Gibbs free energy separate the sorptional force field from the pressure and solution effects. In the equation presented in Appendix III, these three contributions are made explicit. This is a standard method used in thermodynamics, which has been used by plant physiologists to describe water relations in plant material (Nobel, 1974) and applied to some fruits (Rotstein and Cornish, 1978). The remarkable feature as illustrated on Fig. 1.15 for apple is the fact that the sorption term in foods containing

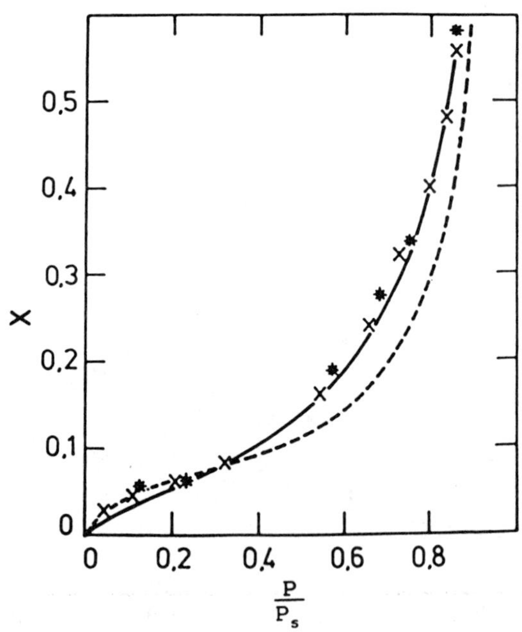

FIG. 1.15 Apple isotherm prediction. BET equation (broken line), equation in Appendix III (continuous line), and experimental (*, X). (*From Rotstein and Cornish, 1978.*)

complex solutions becomes significant only at very low a_w, usually below 0.3. This offers, in combination with the previous predictive model for solutions, a powerful tool for the estimation of the desorption branch of cellular materials.

CONCLUSION

The field of water relations is a very complex and challenging area as much for the researcher as for the technologist in the field. The current state of knowledge concerning interactions of water and solutes and surfaces has been reviewed and it can be concluded that much insight has been gained from the exploration at the molecular level of the water–water water–solute interactions. Also, the development of powerful thermodynamic models, based on functional groups and the partitioning of the excess Gibbs free energy, opens the door to dealing with food material in a much more efficient and productive way. A few examples have been presented, illustrating how these tools may be used, but an enormous amount of work still remains before food materials can be satisfactorily described. However, the foundation is there, waiting to be used.

APPENDIX I

Activity

$$a_i = x_i \gamma_i$$

UNIFAC-FV Model

$$\ln\gamma_i = \underset{\text{(Combinatorial)}}{\ln\gamma_i^c} + \underset{\text{(Residual)}}{\ln\gamma_i^R} + \underset{\text{(Free-volume)}}{\ln\gamma_i^{FV}}$$

Combinatorial Part

$$\ln\gamma_i^c = \ln\frac{\phi_i}{x_i} + \frac{z}{2} q_i \ln\frac{\theta_i}{\phi_i} + l_i - \frac{\phi_i}{x_i}\sum_j x_j l_j$$

Where,

$$\phi_i = \frac{r_i x_i}{\sum_j r_j x_j} = \text{component volume fraction}$$

$$\theta_i = \frac{q_i x_i}{\sum_j q_j x_j} = \text{component area fraction}$$

$$l_i = \frac{z}{2}(r_i - q_i) - (r_i - 1)$$

$$x_i = \text{mole fraction of component } i$$

$$z = \text{coordination number, set equal to 10}$$

Residual Part

$$\ln\gamma_i^R = \sum_k \nu_k^{(i)} [\ln\Gamma_k - \ln\Gamma_k^{(i)}]$$

The group activity coefficient is:

$$\ln\Gamma_k = Q_k\left[1 - \ln\left\{\sum_m \theta_m \psi_{mk}\right\} - \sum_m \left\{\frac{\theta_m \psi_{km}}{\sum_n \theta_n \psi_{nm}}\right\}\right]$$

Where,

$$\theta_m = \frac{Q_m X_m}{\sum_n Q_n X_n} = \text{area fraction of group } m$$

$$X_m = \frac{\Sigma_n x_n \nu_{mn}}{\sum_n x_n \sum_m \nu_{mn}} = \text{mole fraction of group } m$$

$$x_n = \text{mole fraction of component } n$$

$$\psi_{mn} = \exp\left[-\frac{U_{mn} - U_{nn}}{RT}\right] = \exp\left[-\frac{a_{mn}}{T}\right]$$

$$= \text{group interaction parameter}$$

Free-volume Part

$$\ln\gamma_i^{FV} = 3c_i \ln\left[\frac{(\tilde{v}_i^{1/3} - 1)}{(\tilde{v}_m^{1/3} - 1)}\right] - c_i\left[\left\{\frac{\tilde{v}_i}{\tilde{v}_m} - 1\right\}\left\{1 - \frac{1}{\tilde{v}_i^{1/3}}\right\}^{-1}\right]$$

Where,

$$\tilde{v}_i = \frac{v_i}{15.17 r_i} = \text{reduced volume of component } i$$

$$\tilde{v}_m = \frac{\Sigma_i v_i x_i}{15.17 b \sum_i r_i x_i} = \text{reduced volume of mixture}$$

v_i = molar volume of component i

x_i = mole fraction of component i

b = proportionality constant of the order of unity

$3c_i$ = number of external degrees of freedom per molecule of component i

Group Volume and Area Parameters

$$r_i = \sum_k \nu_k^{(i)} r_k = \text{component volume parameter}$$

$$q_i = \sum_k \nu_k^{(i)} Q_k = \text{component area parameter}$$

$$R_k = \frac{V_{wk}}{15.17} = \text{group volume parameter}$$

$$Q_k = \frac{A_{wk}}{2.5 \times 10^9} = \text{group area parameter}$$

V_{wk} and A_{wk} = van der Waals volume and area of group k

15.17 and 2.5×10^9 are normalization factors determined by volume and surface area of a CH_2 unit in polyethylene.

Availability of OH Group

$$\nu_{OH} = \exp\left[A_o + \frac{B_o}{T} + \left\{\frac{SW}{T}\right\}\right]$$

Where,

W = weight fraction of water

T = temperature in K

A_o, B_o, S = optimized constants for the glucose–water system

APPENDIX II

Extended UNIQUAC Equation (Christensen et al., 1983)

$$G^{E++} = G^{E++}{}_{UNI} + G^{E++}{}_{D-H} + G^{E++}{}_{B-G}$$

UNIQUAC Equation (Abrams and Prausnitz, 1975; Maurer and Prausnitz, 1978)

$$\frac{G^E}{RT}\text{(Combinatorial)} = \Sigma_i \frac{\phi_i}{X_i} + \frac{Z}{2}\ \Sigma_i q_i X_i \ln \frac{\theta_i}{\phi_i}$$

$$\frac{G^E}{RT}\text{(Residual)} = -\Sigma_i q_i X_i \ln(\Sigma_j \theta_j \tau_{ji})$$

Where

$$\tau_{ji} = \exp\left(-\frac{[U_{ji} - U_{ii}]}{T}\right)$$

Debye-Huckel Contribution (Robinson and Stokes, 1965)

$$\frac{G^{E++}{}_{D-H}}{RT} = -\Sigma_k X_k M_k \frac{4A}{b^3}\left(\ln(1 + b\sqrt{I}) - b\sqrt{I} + \frac{b^2 I}{2}\right)$$

Where

$$A = c\,\frac{d^{1/2}}{(DT)^{3/2}}$$

Bronsted-Guggenheim Contribution (Guggenheim, 1935)

$$\frac{G^{E++}{}_{D-G}}{RT} = \Sigma_k X_k M_k \Sigma_c \Sigma_a \frac{\beta_{ca}}{T} m_c m_a$$

Activity Coefficients

Solvent

$$\ln\gamma_{solvent} = \ln\gamma^s{}_{UNI} + M_s \frac{2A}{b^3}\left(1 + b\sqrt{I} - \frac{1}{1 + b\sqrt{I}} - 2\ln(1 + b\sqrt{I})\right)$$

$$- M_s \Sigma_c \Sigma_a \frac{\beta_{ca}}{T} m_c m_a$$

Electrolyte (1:1)

$$\ln\gamma_{\pm} = \{(\ln\gamma^{c}{}_{UNI} - \ln\gamma^{c,\infty}{}_{UNI})(\ln\gamma^{a}{}_{UNI} - \ln\gamma^{a,\infty}{}_{UNI})\}^{1/2} - Z\frac{A\sqrt{I}}{1 + b\sqrt{I}} + \frac{1}{2}\left(\frac{\Sigma_a\beta_{ca}}{T}m_a + \frac{\Sigma_c\beta_{ca}}{T}m_c\right)$$

APPENDIX III

$$RT\ln\left\{\frac{P}{P_s}\right\} = \overline{V}_w\epsilon\left\{\frac{X - X_o}{X_o}\right\} + RT\ln a_w + \overline{V}_w\psi_m$$

$$a_w = x_w\gamma_w$$

$$x_w = \frac{X}{X + \sum_j W_j(M_w/M_j)}$$

$$\gamma_w = f(x_w)$$

REFERENCES

Abrams, D. S. and Prausnitz, J. M. 1975. *AIChE J.* 21:116.

Bernal, J. D. and Fowler, R. H. 1933. *J. Chem. Phys.* 1: 515.

Christensen, C., Sander, B., Fredenslund, A. A., and Rasmussen, P. 1983. *Fluid Phase Equilibria* 13:297.

Del Bene, J. and Pople, J. A. 1970. Theory of molecular interactions. I. Molecular orbital studies of water polymers using a minimal Slater-type basis. *J. Chem. Phys.* 52: 4858.

Drost-Hansen, W. 1982. The occurrence of extent of vicinal water. In *Biophysics of Water.* Franks, F. and Mathias, S. F. (Ed.), p. 163. John Wiley and Sons Ltd., New York.

Etzler, F. M. 1983. A statistical thermodynamic model for water solid interfaces. *J. Colloid Interface Sci.* 92: 43.

Fawcet, J. K., Camerman, N., and Camerman, A. 1975. *Acta Cryst.* B 31: 658.

Finney, J. L. 1977. *Phil. Trans. Roy Soc.* (London) B 278: 3.

Flink, J. M. 1983. Structure and structure transitions in dried carbohydrate materials. In *Physical Properties of Foods.* Peleg, M. and Bagley, E. B. (Ed.), p. 473. Avi Publishing Co., Westport, Ct.

Frank, H. S. and Wen, W. Y. 1957. *Discuss. Faraday Soc.* 24: 133.

Franks, F. 1983. Solute-water interactions. Do polyhydroxy compounds alter the properties of water? *Cryobiology* 20: 335.

Franks, F. 1982. The properties of aqueous solutions at subzero temperatures. In *Water, A Comprehensive Treatise*. Franks, F. (Ed.), Vol. 7, p. 215. Plenum, New York.

Fredenslund, A. and Rasmussen, P. 1985. From UNIFAC to SUPERFAC and back? *Fluid Phase Equilibria* 24: 115.

Furter, W. F. 1958. PhD. Thesis, University of Toronto, Toronto, Ontario, Canada.

Guggenheim, E. A. 1935. *Phil. Mag.* 19:588.

Hazelwood, C. F. 1979. In *Cell-Associated Water*. Drost-Hansen, W. and Clegg, J. S. (Ed.). Academic Press, New York.

Israelachvili, J. N. and Pashley, R. H. 1982. Double-layer, Van der Waals and hydration forces between surfaces in electrolyte solutions. In *Biophysics of Water*. Franks, F. and Mathias, S. F. (Ed.). John Wiley and Sons Ltd., New York.

Jeffrey, G. A. 1982. Hydrogen bonding in amino acids and carbohydrates. In *Molecular Structure and Biological Activity*. Griffin, J. F. and Duax, W. L. (Ed.), p. 135. Elsevier Science Publishing Co., Inc.

Jeffrey, G. A., Gress, M. E., and Takagi, S. 1977. Some experimental observations on H—O hydrogen bond lengths in carbohydrate crystal structures. *J. Amer. Chem. Soc.* 99: 609.

Karle, I. L. and Duesler, E. 1977. *Proc. Natl. Acad. Sci.* (U.S.) 74: 2602.

Le Maguer, M. 1981. A thermodynamic model for terpene-water solutions. In *Water Activity: Influences on Food Quality*. Rockland, L. B. and Stewart, G. F. (Ed.), p. 347. Academic Press, New York.

Maurer, G. and Prausnitz, J. M. 1978. *Fluid Phase Equilibria* 2:91.

Mezei, M. and Beveridge, D. L. 1981. Theoretical studies of hydrogen bonding in liquid water and dilute aqueous solutions. *J. Chem. Phys.* 74: 622.

Nobel, P. S. 1974. In *Introduction to Biophysical Plant Physiology*. W. H. Freeman & Co., San Francisco.

Peschel, G. and Adlfinger, K. H. 1970. *J. Colloid Interface Sci.* 34: 505.

Pople, J. A. 1951. *Proc. R. Soc.* (London) Ser. A, 205: 163.

Prausnitz, J. M. 1969. In *Molecular Thermodynamics of Fluid-Phase Equilibria*. Prentice-Hall, Englewood Cliffs, NJ.

Renon, H. 1985. NRTL: An empirical equation or an inspiring model for fluid mixture properties? *Fluid Phase Equilibria* 24: 87.

Robinson, R. A. and Stokes, R. H. 1965. *Electrolyte Solutions,* 2nd Edition (revised), Butterworths, London.

Rotstein, E. and Cornish, A. R. H. 1978. Prediction of the sorptional equilibrium relationship for the drying of foodstuffs. *AIChE J.* 24: 956.

Sceats, M. G. and Rice, S. A. 1981. *J. Phys. Chem.* 85: 1108.

Smirnov, V. D. and Ovchinnikova, V. D. 1981. *Khimiia İ Khimcheskaia Technologiia* 24:440.

Smith, D. S., Mannheim, C. H., and Gilbert, S. G. 1981. Water sorption isotherms of sucrose and glucose by inverse gas chromatography. *J. Food Sci.* 46: 1051.

Speedy, R. J. 1984. Self-replicating structures in water. *J. Phys. Chem.* 88: 3364.

Stillinger, F. H. 1981. In *Waters in Polymers.* Rowland, S. P. (Ed.), p. 11. ACS Symposium Ser. No. 127. American Chemical Society, Washington, DC.

Vinogradov, S. N. 1980. Structural aspects of hydrogen bonding in amino acids, peptides, proteins, and model systems. In *Molecular Interactions.*

Ratajczak, H. and Orville-Thomas, W. J. (Ed.), Vol. 1, p. 179. John Wiley and Sons Ltd., New York.

Wood, D. W. 1979. Computer simulation of water and aqueous solutions. In *Water, A Comprehensive Treatise.* Franks, F. (Ed.), Vol. 6, p. 279. Plenum Press, New York.

2

Influence of Water Activity on Chemical Reactivity

Henry K. Leung*

Washington State University
Pullman, Washington

INTRODUCTION

As early as 1957, Salwin recognized that maximum storage stability of many dehydrated foods occurs at moisture contents close to the BET monolayer values (Brunauer et al., 1938) corresponding to a_w 0.2–0.4 (Salwin, 1959). He suggested that the water molecules covering the active sites of the dry solids form a protective film against oxygen. In 1970, Labuza presented a comprehensive review on the influence of water on chemical reactions in foods. Since then, extensive studies have been conducted in this area. More recent research has been reviewed in two articles (Labuza, 1980; Rockland and Nishi, 1980) and three books on properties of water in foods (Duckworth, 1975; Rockland and Stewart, 1981; Simato and Multon, 1985).

Water may influence chemical reactivity in different ways. It may act as a

**Current affiliation*: Campbell Soup Company, Camden, New Jersey reactant, such as in the case of sucrose hydrolysis. As a solvent, water may exert a dilution effect on the substrates, thereby decreasing the reaction rate. Water may also change the mobility of the reactants by affecting the viscosity of the food systems. Water may form hydrogen bonds or complexes with the reacting species. For example, lipid oxidation rate may be affected by hydration of trace metal catalysts or hydrogen bonding of hydroperoxides with water. Structure of a solid matrix may change substantially with changes in moisture content, thus indirectly influencing reaction rates. In addition, water influences protein conformation and transition of amorphous-crystalline states of sugar and starch.

Due to the vast amount of information available in this subject area, no attempt will be made to give an in-depth review here. Instead, an overview and update of the influence of a_w on some major deteriorative reactions in foods are presented. The discussion will include nonenzymatic browning, lipid oxidation, degradation of vitamins, enzymatic reactions, protein denaturation, starch gelatinization, and starch retrogradation.

NONENZYMATIC BROWNING

The Maillard reaction has been reviewed recently by Mauron (1981) and Baltes (1982). In the early stage of the reaction, the carbonyl group of the reducing sugar reacts with the free amino group of the amino acid to form the Schiff base and then the *N*-substituted glycosylamine. Glycosylamines are converted to the 1-amino-1-deoxy-2-ketose by the Amadori rearrangement. The early Maillard reactions forming Amadori compounds do not cause browning but do reduce nutritive value (Mauron, 1981).

In the advanced Maillard reaction, there are five pathways. The first two pathways start from the 1-2 enol or 2-3 enol forms of the Amadori product, yielding various flavor compounds. The third pathway is the Strecker degradation, which involves oxidative degradation of amino acids by the dicarbonyls produced in the first two pathways. The fouth pathway involves transamination of the Schiff base. The fifth pathway starts with a second substitution of the amino-deoxy-ketose. The final step of the advanced Maillard reaction is the formation of many heterocyclic compounds such as pyrazines and pyrroles.

Brown melanoidin pigments are produced in the final stage of the Maillard reaction. The pigments are formed by polymerization of the reactive compounds produced during the advanced Maillard reactions, such as unsaturated carbonyl compounds and furfural. The polymers have a molecular weight greater than 1,000 and are relatively inert (Mauron, 1981).

The effect of water on enzymatic reaction has been reviewed by Eichner

(1975) and Labuza and Saltmarch (1981a). Water may accelerate browning by imparting mobility to the substrates. On the other hand, an increase in water content may decrease browning rate by diluting the reactive species. The mobility factor predominates in the low a_w range, whereas the dilution factor predominates in the high a_w range. Therefore, browning rate generally increases with increasing a_w at low moisture content, reaches a maximum at a_w of 0.4–0.8, and decreases with a further increase in a_w. Besides the mobility and dilution factors, water may affect nonenzymatic browning by inhibiting or enhancing some of the intermediate reactions (Labuza and Saltmarch, 1981a).

Model Systems

Losses of lysine and other essential amino acids in foods during the early stage of Maillard browning have a direct influence on protein quality. Because of its highly reactive epsilon-amino group, lysine has been used extensively as an index of browning in food systems (Friedman, 1982).

Using a model food system containing protein, sugar, oil, salt, cellulose and water, Wolf et al. (1981) demonstrated that losses of free lysine and methionine were highly dependent on a_w, protein, and sugar. Thermal degradation of both amino acids followed first-order reaction kinetics. Reaction rates at 65 and 115°C decreased with increasing a_w (0.33–0.98). Leahy and Warthessen (1983) observed a more rapid decrease of lysine, tryptophan, and threonine at higher a_w in model systems heated at 95°C. The retention of tryptophan was greater than lysine at a_w 0.75, but lysine retention was greater than that of tryptophan at a_w 0.22 (Fig. 2.1). At higher a_w, the Maillard reaction predominates and a rapid loss of lysine occurs. At lower a_w, browning proceeds at a slower rate and reactions involving the indole ring of tryptophan become significant.

Glucose utilization in a model system consisting of glucose, monosodium glutamate, corn starch, and lipid during nonenzymatic browning was investigated by Kamman and Labuza (1985). The rates of glucose utilization at a_w 0.81 were greater than at a_w 0.41. Lipid accelerated the reaction rates at a_w 0.41 but had virtually no effect at a_w 0.81. Liquid oil is more effective than shortening in increasing the degradation rate of glucose. The activation energy at a_w 0.41 is twice that at a_w 0.81 (88 KJ/mol). These results illustrate the effects of water and oil on the mobility of reactants and the reaction rate. Liquid oil has little influence on the rate of glucose utilization at high a_w since enough water is present to impart mobility to the reactants.

Using a model system consisting of lysine, glucose, sodium chloride, and phosphate buffer, Cerrutti et al. (1985) showed that water had little or no effect on the rate of glucose loss at a_w 0.90–0.95. On the other hand, the reaction rate

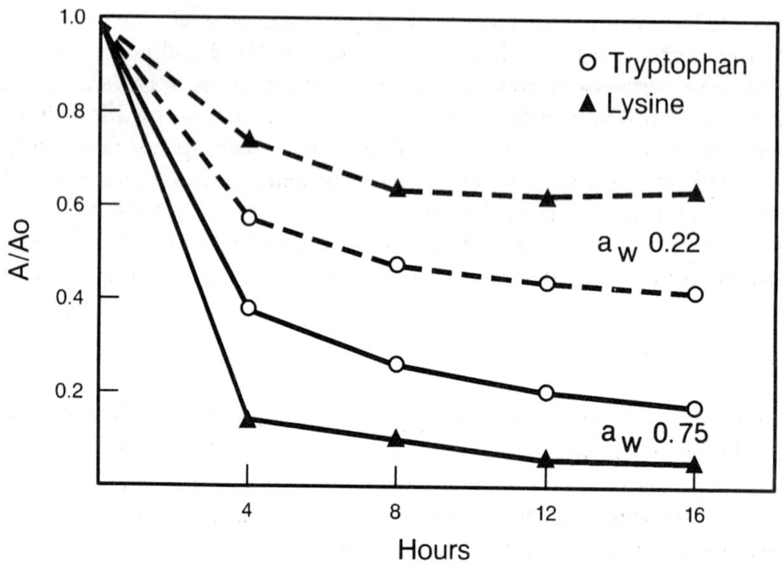

FIG. 2.1 Effect of a_w and heating time at 95°C on retention of lysine and tryptophan in a model system. (*Adapted from Leahy and Warthesen, 1983.*)

was highly dependent on temperature and pH. Similar behavior was observed for accumulation of 5-hydroxy-methyl-furfural, fluorescent compounds, and brown pigments (Petriella et al., 1985; Cerrutti et al., 1985).

The effect of a_w on nonenzymatic browning in a water–glycerol–sorbate–glycine model system at pH 4 was studied by Seow and Cheah (1985). As shown in Table 2.1, the zero-order reaction rate constant and the activation energy of browning decrease with increasing a_w. Glycerol was found to react with sorbate or glycine to form brown pigments at a_w 0.80. The glycerol–glycine system showed higher activation energy and lower reaction rates than the glycerol–sorbate system. The role of glycerol on nonenzymatic browning warrants further investigation.

Food Systems

The formation of 5-hydroxy-methyl-furfural (5-HMF), an intermediate product during nonenzymatic browning, in dehydrated apples during storage was observed to follow a zero-order reaction kinetics at different moisture contents (Resnik and Chirife, 1979). The effect of moisture content or a_w on

TABLE 2.1 Effect of a_w on Kinetic Parameters of Nonenzymatic Browning in a Water–Glycerol–Sorbate–Glycine System

a_w (40°C)	$K \times 10^2 (\Delta A_{420}/\text{day})$ 40°C	50°C	60°C	Ea (KJ/mol)
0.55	3.76	13.14	31.9	93
0.65	3.23	8.30	25.1	89
0.71	2.81	6.75	23.7	92
0.80	2.57	6.64	15.1	77
0.90	2.54	5.81	11.4	65

Source: Seow and Cheah (1985).

5-HMF accumulation was complex and did not follow a simple pattern. The results are different from those using browning or loss of reactants as indicators. Activation energy of the reaction decreased from 172 KJ/mol at zero moisture to 121 KJ/mol at 83% moisture (dry basis). In a computer simulation study of storage stability in intermediate moisture apples at fluctuating temperatures, nonenzymatic browning and ascorbic acid degradation rates were significantly greater at a_w 0.84 than at a_w 0.62 (Singh et al., 1984).

The influence of a_w on nonenzymatic browning of apple juice concentrate during storage was investigated by Toribio and Lozano (1984) and Toribio et al. (1984). As shown in Fig. 2.2, browning rate reached a maximum at a_w 0.53–0.55. This confirms the importance of the dilution effect of water and reactant mobility on browning (Labuza and Saltmarch, 1981a).

Two recent Japanese studies were conducted on nonenzymatic browning as affected by a_w. In one study, significant browning of skim milk powder was found to occur after a month of storage at a_w below 0.23. Losses of total lysine and methionine in skim milk powder increased with increasing a_w (0.23–0.80). Losses of tryptophan and arginine were about 10% after one month of storage at 40°C and a_w 0.80, but the losses were slight at a_w 0.57 (Okamoto and Hayashi, 1985). In another study by Homma and Fujimaki (1982), browning of kori-tofu during storage at various a_w values was followed by changes in color, carbonyl value, and CO_2 production (Fig. 2.3). Browning, as indicated by color changes, increased with increasing a_w (0–0.95). It was suggested that CO_2 was formed mainly through lipid oxidation at low a_w, and through browning and oxidation at high a_w. Carbonyl compounds produced from lipid oxidation were utilized in

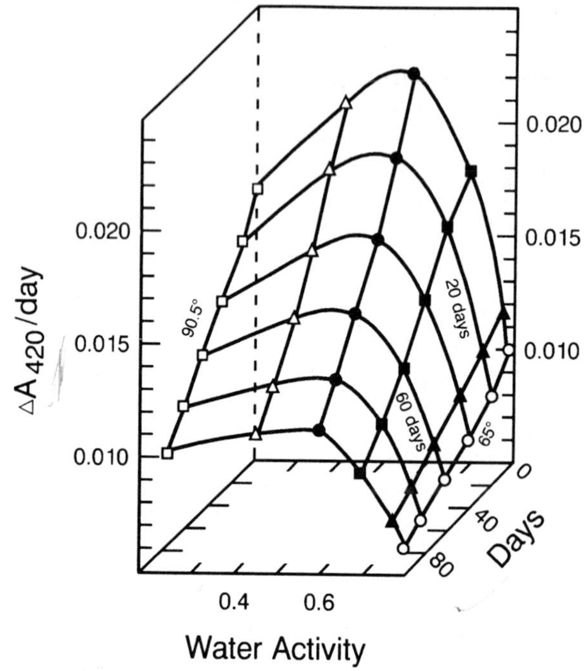

FIG. 2.2 Rate of nonenzymatic browning in apple juice concentrate as a function of a_w and storage time at 37°C. (*Adapted from Toribio et al., 1984.*)

Maillard reactions at high a_w (0.75–0.95). The carbonyl value remained high at low a_w (0–0.30) because of the relatively slow rate of browning.

Kinetics of browning and protein quality loss in sweet whey powders stored at various a_w values (0.33–0.65) and under constant or fluctuating temperature conditions was studied by Labuza and Saltmarch (1981b). The maximum rate of browning, loss of available lysine and loss of relative nutritive value (RNV) by *Tetrahymena* assay occurred at a_w 0.44. The activation energies were 126–142 KJ/mol, 84–105 KJ/mol, and 96–136 KJ/mol for browning, loss of available lysine, and loss of RNV, respectively.

A later study on fish flour showed that the rate of loss of available lysine was greater when flour was stored in a sealed system (foil pouch) than in an open system (Kaanane and Labuza, 1985). The reaction rate constants for the closed system are greater than those for the open system at 25, 38, and 45°C (Table 2.2). In a closed system, moisture content of the sample is held constant and its a_w increases with increasing temperature. Also, volatile products trapped inside the package may catalyze the reaction. It was suggested that free radicals

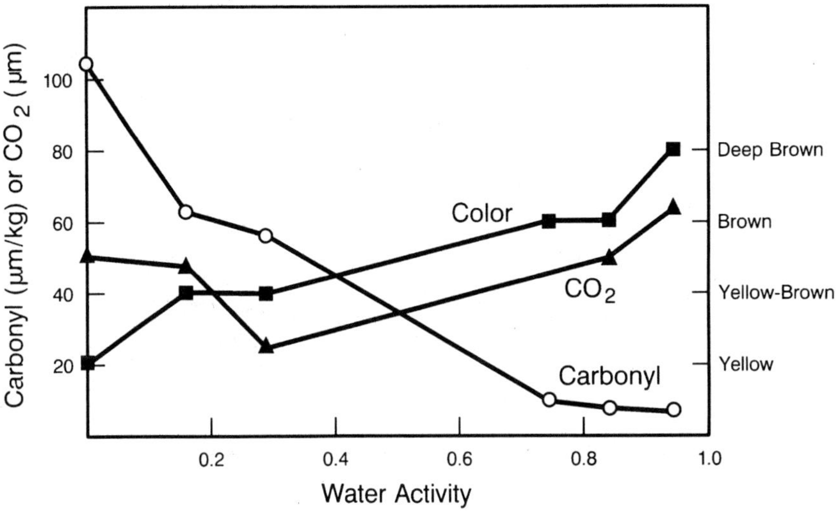

FIG. 2.3 Effect of a_w on browning, carbonyl value, and CO_2 production in kori-tofu stored at 50°C for 10 months. (*Adapted from Homma and Fujimaki, 1982.*)

TABLE 2.2 Kinetic Parameters for Loss of Available Lysine in Fish Flour Stored in Open and Closed Systems at Different a_w

		25°C		38°C		45°C	
Nominal a_w	System	Actual a_w	$K(wk^{-1})$	Actual a_w	$K(wk^{-1})$	Actual a_w	$K(wk^{-1})$
0.33	open	0.33	9.8	0.33	17.5	0.33	22.5
0.44	open	0.44	9.9	0.44	22.5	0.44	26.3
0.65	open	0.65	19.6	0.65	33.1	0.65	39.4
0.33	closed	0.33	15.7	0.44	17.6	0.51	27.2
0.44	closed	0.44	19.5	0.54	22.8	0.60	33.8
0.65	closed	0.65	21.4	0.70	36.3	0.72	43.0

Source: Kaanane and Labuza (1985).

or compounds formed by lipid oxidation may contribute to loss of available lysine since the reaction rate was significant even at low a_w. Activation energy for the open system decreased with increasing a_w, but showed a slight increase with increasing a_w for the closed system (Fig. 2.4). This suggests that the reaction mechanisms may be different in the two systems.

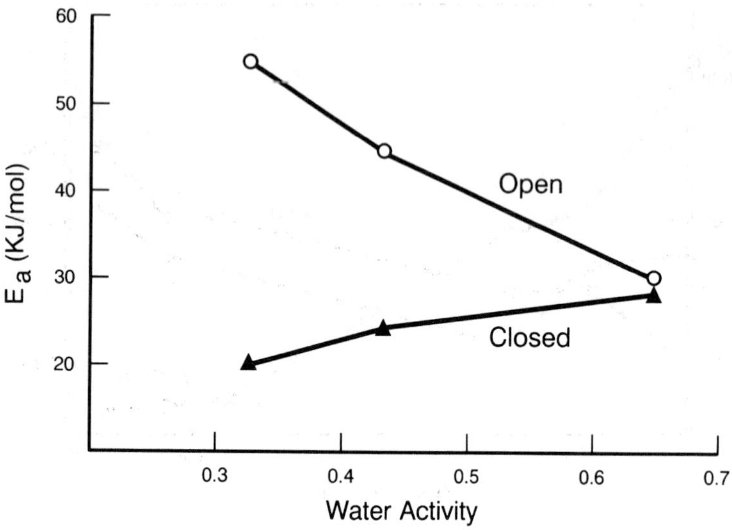

FIG. 2.4 Activation energy for loss of available lysine in fish flour stored at 25–45°C in open and closed systems. (*Adapted from Kaanane and Labuza, 1985.*)

Maillard Reaction During Drying

Formation of Amadori compounds during drying of carrots was investigated by Eichner et al. (1985). The accumulation of the browning intermediates increased substantially when moisture content decreased below about 20%. Also, the reaction rate was highly temperature dependent. When carrots were dried to 7% moisture, a drying temperature of 110°C resulted in unacceptable quality, while no perceptible change in sensory quality was observed at 60°C. Therefore, a reduction in drying temperature is suggested during the later stage of drying to minimize nonenzymatic browning.

To further study the effect of water content and temperature on formation of Amadori compounds, freeze-dried carrot powder was equilibrated to different a_w values and subjected to 30 min of heating at 60, 75, and 90°C. At 90°C, the reaction rate increased with decreasing a_w, whereas at lower temperatures, the reaction rate leveled off or decreased as a_w was reduced below about 0.3 (Fig. 2.5). It was suggested that at high temperature and low a_w, mobility of reactants is sufficient to allow the reaction to proceed at a rapid rate; at lower temperatures, both the dilution and mobility effects of water are important factors influencing the reaction rate (Eichner et al., 1985).

Activation energy for the formation of Amadori compounds decreases with increasing a_w and levels off at a_w of about 0.50 (Fig. 2.6). Therefore, the

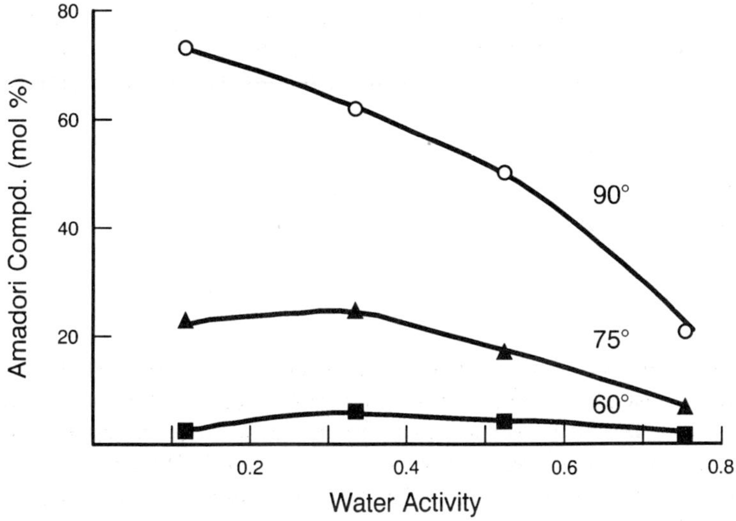

FIG. 2.5 Formation of Amadori compounds in freeze-dried carrot powder heated for 30 min at different temperatures and a_w. (*Adapted from Eichner et al., 1985.*)

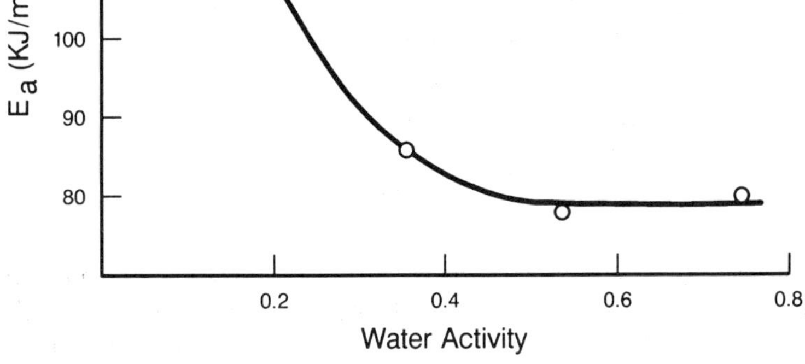

FIG. 2.6 Activation energy for formation of Amadori compounds in freeze-dried carrots as a function of a_w. (*Adapted from Eichner et al., 1985.*)

reaction rate is highly temperature dependent at low a_w and becomes much less dependent on temperature at a_w greater than 0.50. This confirms the importance of low air temperature during the final stage of drying (Eichner et al., 1985).

LIPID OXIDATION

Mechanism

The mechanism of lipid oxidation has been reviewed recently by Frankel (1984). Free radical autoxidation of unsaturated lipids is characterized by four major steps: initiation, propagation, branching, and termination. Initiation takes place by loss of a hydrogen radical due to heat, light, or trace metals. In propagation, the lipid free radical reacts with oxygen to form peroxy free radicals, which in turn react with more lipid molecules to form hydroperoxides. In the branching process, free radicals increase geometrically from decomposition of hydroperoxide. Termination involves the elimination of free radicals by addition of two free radicals or transfer of the radical to a compound to form a stable radical. The four processes can be summarized as follows:

Initiation $RH \rightarrow R\bullet + H\bullet$

Propagation $R\bullet + O_2 \rightarrow ROO\bullet$

$ROO\bullet + RH \rightarrow ROOH + R\bullet$

Branching $ROOH \rightarrow RO\bullet + \bullet OH \xrightarrow{2RH} 2R\bullet + ROH + H_2O$

(monomolecular decomposition)

$2ROOH \rightarrow ROO\bullet + RO\bullet + H_2O$

(bimolecular decomposition)

Termination $ROO\bullet + ROO\bullet \rightarrow ROOR + O_2$

$R\bullet + R\bullet \rightarrow R - R$

$R\bullet + ROO\bullet \rightarrow ROOR$

A number of products are produced during autoxidation. Decomposition of hydroperoxides results in formation of aldehydes, ketones, alcohols, hydrocarbons, and other products. Hydroperoxides may react with oxygen to form secondary products such as epoxyhydroperoxides which decompose to form volatile breakdown products. In addition, hydroperoxides and their products may react with proteins, enzymes, and membranes (Frankel, 1984).

Effect of Water

The influence of a_w on lipid oxidation has been studied extensively and reviewed by Labuza (1975) and Karel (1980). Most of the studies in foods utilized model systems consisting of methyl linoleate, microcrystalline cellulose, water, and sometimes trace metals and antioxidants. The mixture was freeze-dried and adjusted to the desired a_w by equilibrating with saturated salt slurries. Oxidation was measured by oxygen uptake using a Warburg manometer (Maloney et al., 1966; Chou and Labuza, 1974).

Figure 2.7 shows the general effect of a_w on lipid oxidation on food and model systems (Labuza, 1975). At a_w below the monolayer value, oxidation rate decreases with increasing a_w. The rate reaches a minimum around the monolayer value and increases with a further increase in a_w. The "antioxidant effect" of water at low a_w has been attributed to bonding of hydroperoxides and hydration of metal catalysts, whereas the "pro-oxidant effect" of water at higher a_w is due to the increased mobility of reactants (Heidelbaugh and Karel, 1970). Karel (1980) stated that water may influence lipid oxidation by influencing the concentrations of initiating radicals present, the degree of contact and mobility of reactants, and the relative importance of radical transfer versus recombination reactions.

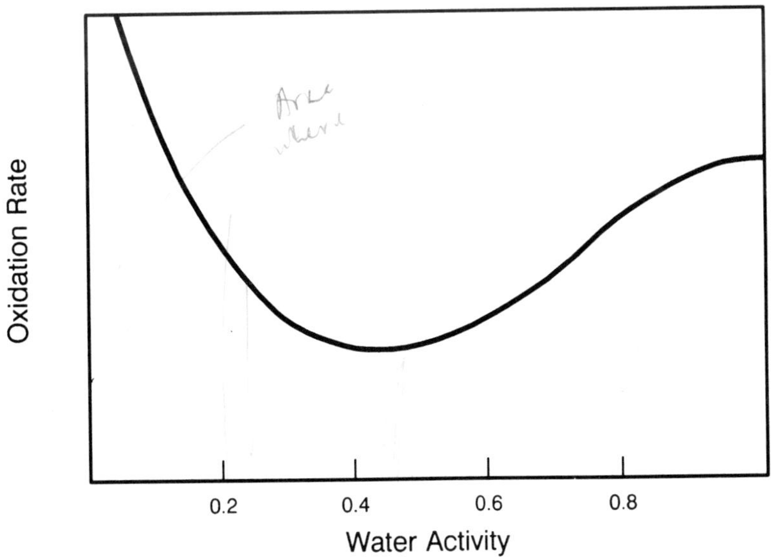

FIG. 2.7 Rate of lipid oxidation as influenced by a_w in food.

TABLE 2.3 Oxygen Uptake of Freeze-Dried Emulsions Exposed to Different a_w and Oxygen for 90 hr

	O_2 Uptake (μL/g)	
Water activity	Linoleic acid (17.4%) and maltodextrin	Linoleic acid (20%) and maltose
0	0	0
0.43	0	190
0.75	0	200
0.93	100	280

Source: Karel (1980).

The importance of water in controlling food structure was pointed out as a major factor in lipid oxidation by Karel (1980). Addition of water to a freeze-dried emulsion may cause collapse of the metastable structure. When this occurs, the encapsulated lipid flows from the matrix interior to the surfaces. When exposed to air, surface lipid is readily oxidized, whereas encapsulated lipid is protected from oxygen. As a_w increases to a certain critical value, the protective matrix collapses and the encapsulated lipid is distributed on the surface and oxidized. This encapsulation effect is demonstrated in Table 2.3, which shows the oxidation of linoleic acid in freeze-dried emulsions containing maltodextrin or maltose at different a_w. It was noted (Karel, 1980) that the surface lipid was not distributed uniformly but rather as "puddles" associated with specific morphological features of the surface.

In a recent study on lipid oxidation of model and milk systems at reduced oxygen pressure, oxidation of methyl linoleate was shown to occur much more rapidly at a_w 0.55 than in the dry state (Kacyn et al., 1983). These researchers suggested that release of the encapsulated lipid by humidification contributed to the increased oxidation rate at higher a_w.

Several reactions initiated by lipid oxidation have been shown to be affected by a_w. These include crosslinking of proteins, enzyme inactivation by lipid peroxidation products, protein scission and degradation of amino acids, production of free radicals in proteins, and generation of fluorescent pigments (Karel, 1980).

Browning Intermediates

The antioxidative effect of Maillard reaction products has been observed by many investigators (Lingnert and Eriksson, 1980; Eichner, 1980; Homma and

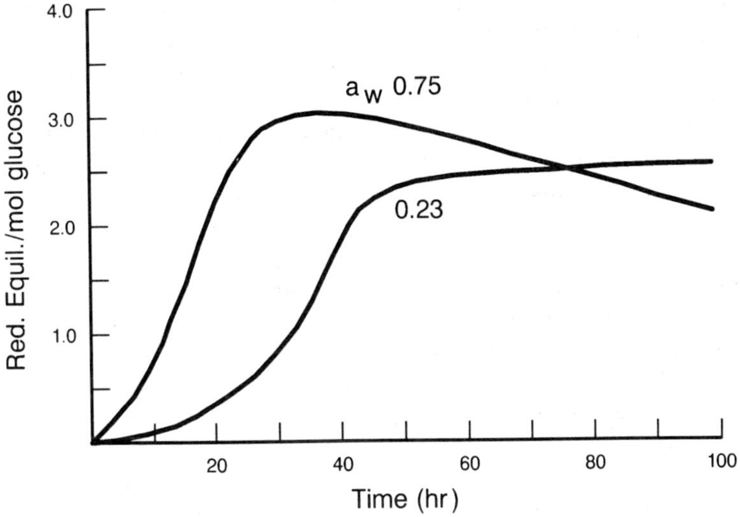

FIG. 2.8 Increase in reducing power of a glucose–lysine–cellulose model system as a function of heating time at 40°C and a_w 0.23 and 0.75. (*Adapted from Eichner, 1980.*)

Fujimaki, 1982). Figure 2.8 shows the increase in reducing power of a glucose–lysine–cellulose model system at 40°C and a_w 0.23 and 0.75. Formation of reducing browning intermediates, which are primarily Amadori compounds, accelerates with increasing a_w (Eichner, 1980). Oxygen uptake of methyl linoleate was shown to decrease in a model system containing browning intermediates or products. Colorless reducing intermediates formed at lower a_w are more effective in retarding lipid oxidation than browning products formed at higher a_w. The presence of Maillard reaction intermediates results in a marked decrease in hexanal formation from hydroperoxide decomposition (Eichner, 1980).

A New Model System

As pointed out by Karel (1980), the freeze-dried emulsion model system contains encapsulated and surface lipids that are not uniformly distributed on the matrix surface. The oxidation rate of the lipid is dependent on its accessibility to oxygen. Thus, oxygen diffusion is difficult to control in a model system consisting of freeze-dried lipid, cellulose, and water. Also, the surface area of lipid in the system cannot be readily estimated. To eliminate the above problems, a new model system for studying the effect of a_w in lipid oxidation was devised by Kahl (1986).

The new model system consisted of a thin lipid film formed on a glass coverlip where surface area and thickness of the film can be easily controlled. The coverlip was placed inside a modified Warburg reaction flask with a pedestal and a reservoir for saturated salt slurry. A sidearm port was added for oxygen purging. Because of the large surface area and small film thickness, oxygen is readily accessible and the reaction rate is not dependent on oxygen diffusion.

Using the new model system, Kahl (1986) observed that the rate of oxygen uptake is linear with time. This is different from the curvilinear relationship reported in previous studies (Heidelbaugh et al., 1970; Chou and Labuza, 1974). The linear oxidation rate observed in this study was attributed to the elimination of oxygen diffusion as a factor. Using tritiated water, an increase in water binding of methyl linoleate during oxidation was observed. This suggests that water is taken up by some of the oxidation products.

Using this model system, the rate of lipid oxidation as a function of water activity was determined (Fig. 2.9). In the absence of any catalyst, the reaction rate shows a maximum at a_w of about 0.3–0.4. Kahl (1986) suggested that the initial addition of water to the lipid system increased the oxidation rate through

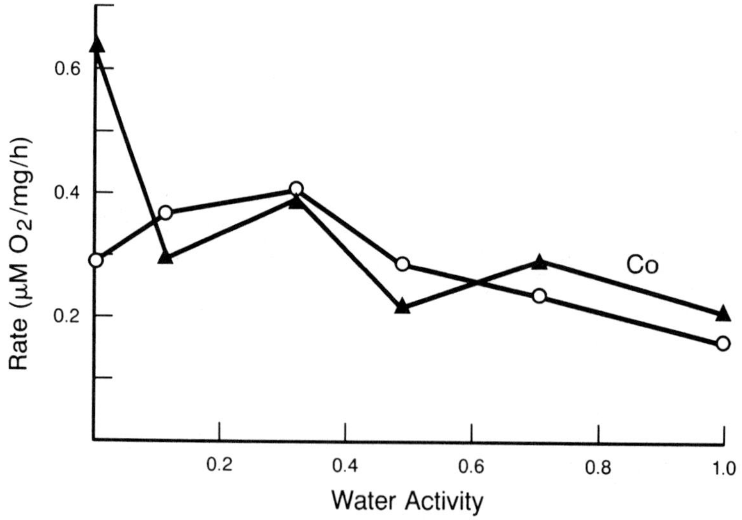

FIG. 2.9 Rate of oxidation of methyl linoleate film as a function of a_w in the presence or absence of cobaltous laurate. (*Adapted from Kahl, 1986.*)

solvation and stabilization of the propagation state. The addition of more water above a_w 0.3–0.4 solvated the peroxy radical and decreased its reactivity by steric hindrance. When 1 ppm cobaltous laurate was added, a maximum oxidation rate was observed in the dry state. The rates were significantly different between the two systems only at a_w 0 and a_w greater than 0.70. The catalytic effect of cobalt was cancelled by the increased water content at a_w above 0.11. It was hypothesized that water may interfere with the formation of metal–hydroperoxide complex.

The results obtained with the lipid film model system are quite different from those of the emulsion model and food systems (Figs. 2.7 and 2.9). The lipid film system does not show an increased oxidation rate at high a_w, although reactant mobility increases with addition of water. The reactants are probably uniformly distributed on the lipid layer which is well exposed to oxygen. Thus reactant mobility may have little or no effect on the rate of oxidation in the new model system. Interestingly, the cobalt-catalyzed lipid layer system shows a high oxidation rate in the dry state, similar to the behavior of most food systems.

Although the new model system is different from most food systems, it does provide a tool for identifying some of the effects of water on lipid oxidation. Further studies in this area may provide additional information about the antioxidant and pro-oxidant effects of water on oxidation.

VITAMIN DEGRADATION

Little additional work has been reported on a_w and vitamin degradation in food systems since Kirk reviewed this subject in 1981. Some of the previous results using a dry model food system containing corn starch, soy protein isolate, fat, sugar, and salt are summarized in Table 2.4. With only one exception, the reaction rates of vitamins A, B_1, B_2, and C increase with increasing a_w (0.24–0.65). The B vitamins are more stable than vitamins A and C at various a_w values. Widicus et al. (1980) showed that the degradation of α-tocopherol in a model system containing no fat increased with increasing a_w (0.10–0.65), storage temperature (20–37°C), and molar ratio of oxygen: α-tocopherol (15:1–1450:1). The changes in first-order reaction rate constant of α-tocopherol stored in TDT cans (minimum oxygen) and 303 cans as a function of a_w are shown in Fig. 2.10. The reaction rates decreased when oxygen was restricted. The activation energies (37–54 KJ/mol) are not significantly different at different a_w or oxygen:tocopherol molar ratios. The results indicate that the degradation mechanism for α-tocopherol is similar regardless of storage conditions.

TABLE 2.4 Effect of a_w on the Degradation Rates of Some Vitamins in Model Food Systems

		K(10^{-3} day^{-1})[b]		
Vitamin[a]	T (°C)	a_w 0.24	a_w 0.40	a_w 0.65
A	30	2.9	6.5	32
A	37	7.0	7.6	46
A	45	20	59	23
B_1	45	0.9	6.8	8.7
B_2	37	1.9	2.6	5.0
C	20	9.5	12.8	14.4
C	30	18	31	113
C	37	50	70	157

[a]Vitamin A was stored in 303 cans; the other vitamins were stored in TDT cans.
[b]First-order rate constant.
Source: Kirk (1981).

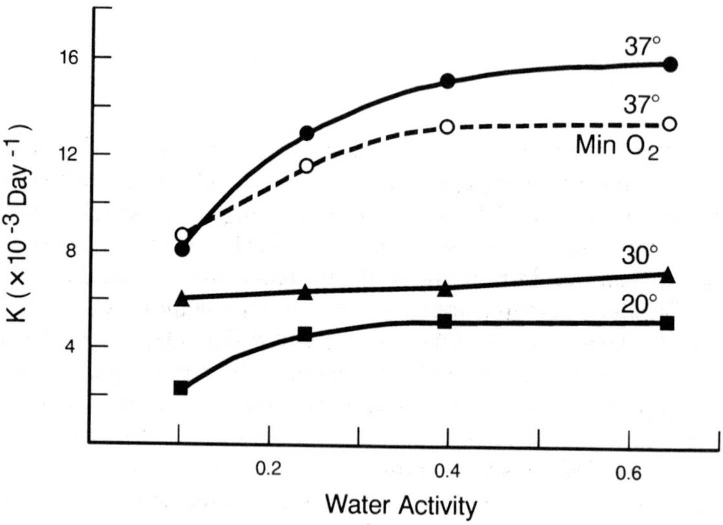

FIG. 2.10 Effects of a_w and temperature on the first order reaction rate constants of α-tocopherol stored in 303 cans or TDT cans (minimum oxygen). (*Adapted from Widicus et al., 1980.*)

Storage stability of ascorbic acid in a model food system was studied as a function of a_w (Dennison and Kirk, 1982). At a_w below 0.40, addition of transition metals (iron, copper, and zinc) to the system had no catalytic effect on ascorbic acid destruction. However, the reaction rate increased by two- to fourfold at a_w 0.65 when trace metals were added. The increased rate of ascorbic acid degradation at high a_w was attributed to greater mobility of the metal ions.

Ascorbic acid degradation in dehydrated sweet potato was shown by Haralampu and Karel (1983) to vary exponentially with a_w (Fig. 2.11). The following exponential function describes the ascorbic acid deterioration well at low a_w:

$$K = b_1 \exp (b_2 a)$$

where K is the first order reaction rate constant, a is the a_w, and b_1 and b_2 are constants. The model seems to be general, but model parameters may vary for different systems.

A recent study on the kinetics of riboflavin photodegradation as affected by a_w was conducted by Furuya et al. (1984). A simple first-order reaction was observed in liquid systems, while a two-phase mechanism was observed in dry

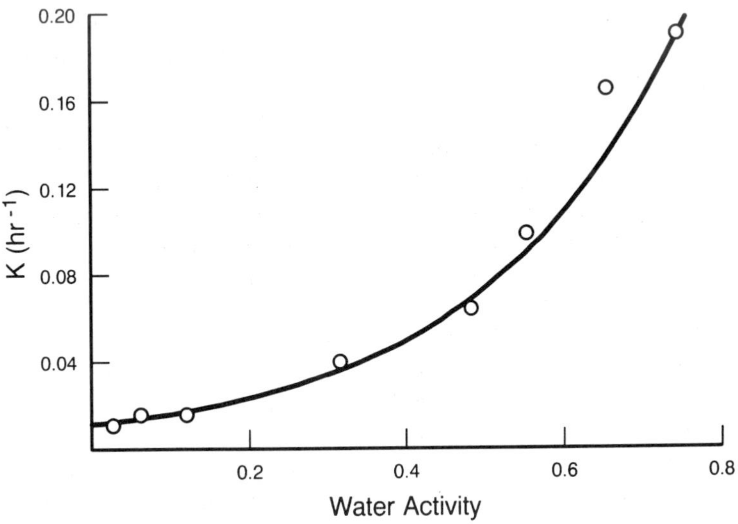

FIG. 2.11 Degradation of ascorbic acid in sweet potatoes as influenced by a_w. *(Adapted from Haralampu and Karel, 1983.)*

food systems. Riboflavin retention in macaroni exposed to light (75 ft-c) at room temperature for 14 days was 32% at a_w 0.11–0.43, 28% at a_w 0.52, 24% at a_w 0.66, and 17% at a_w 0.75.

ENZYME ACTIVITY

The influence of a_w on enzyme activity has been reviewed by Blain (1962), Acker (1969), Potthast (1978), Schwimmer (1980), and more recently by Drapron (1985). In food systems, enzymes may be naturally present in the raw materials, they may be secreted from microorganisms, or they may be added to serve a certain function. As pointed out by Potthast (1978), enzymatic reactions in biological systems and in model systems are different in that enzymes and substrates are generally separated in biological materials, but are usually in close contact in model systems. With almost no exception, enzyme activity increases with increasing a_w or increased substract mobility. Substrates of high molecular weight (e.g., starch, protein) are less mobile than low-molecular-weight substrates, such as glucose, and generally have a higher a_w threshold for enzyme activity. For a liquid oil or nonaqueous substrate, enzyme activity may be observed even at very low a_w. In such cases water is not needed to provide mobility of substrates.

The minimum a_w values required for several enzymatic reactions in food and model systems are given in Table 2.5. Note that lipase is active even at a_w as low as 0.025 when olive oil, trilaurin, and triolein were used as substrates in a model system. Similarly, enzymatic oxidation of sunflower seed oil catalyzed by lipoxygenase occurs at a_w as low as 0.05 (Fig. 2.12). This is due to the mobility of liquid oil in the system. Lipoxygenase activity increases almost linearly with a_w because of the increased substrate mobility (Fig. 2.12).

In lipolysis of oat oil, the increase in free fatty acids increases with increasing a_w. If the enzyme–substrate mixture with a certain a_w, having reached its hydrolysis limit, is raised to a higher a_w, a new hydrolysis limit is attained (Fig. 2.13). This new hydrolysis limit is the same as that for the high a_w (Acker, 1969).

Following lipolysis of a mixture of defatted wheat bran and olive oil at 30°C and a_w below 0.8 for different substrate concentration, the kinetic parameters V_m and K_m were obtained (Drapron, 1985). In the low moisture range, the K_m values decreased with increasing a_w and reached a minimum at a_w of about 0.3–0.4 (Fig. 2.14). The minimum K_m corresponds to maximum affinity of lipase for olive oil. At a_w above 0.4, excess water would impede formation of the enzyme–substrate complex because of substrate hydrophobicity. The V_m values increased with increasing a_w, reflecting the role of water in substrate deacylation.

TABLE 2.5 Minimum a_w Values for Enzymatic Reactions in Selected Food and Model Systems

Product/Substrate	Enzyme	T(°C)	a_w Threshold
Grains	Phytases	23	0.90
Wheat germ	Glycoside-hydrolases	20	0.20
Rye flour	Amylases Proteases	30	0.75
Macaroni	Phospholipases	25–30	0.45
Wheat flour dough	Proteases	35	0.96
Bread	Amylases Proteases	30	0.36
Casein	Trypsin	30	0.50
Starch	Amylases	37	0.40/0.75
Galactose	Galactosidase	30	0.40–0.60
Olive Oil	Lipase	5–40°C	0.025
Triolein, Trilaurin	Phospholipases	30	0.45
Glucose	Glucose oxidase	30	0.40
Linoleic acid	Lipoxygenase	25	0.50/0.70

Source: Drapron (1985).

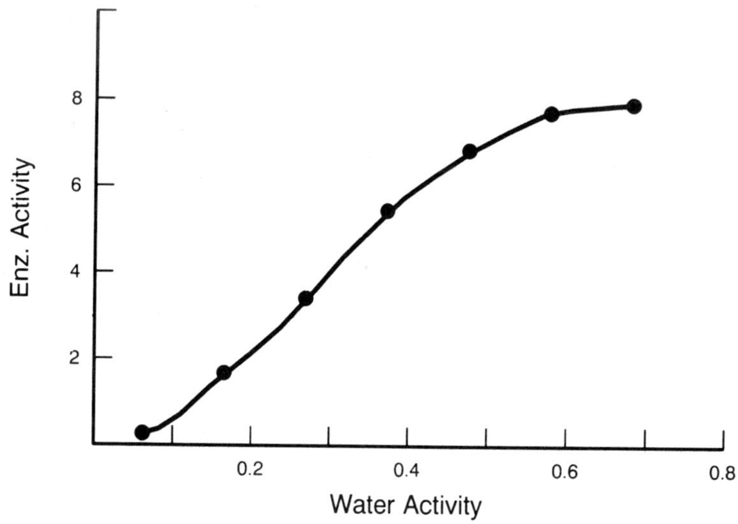

FIG. 2.12 Effect of a_w on enzymatic oxidation of sunflower seed oil by lipoxygenase at 25°C. (*Adapted from Brockmann and Acker, 1977.*)

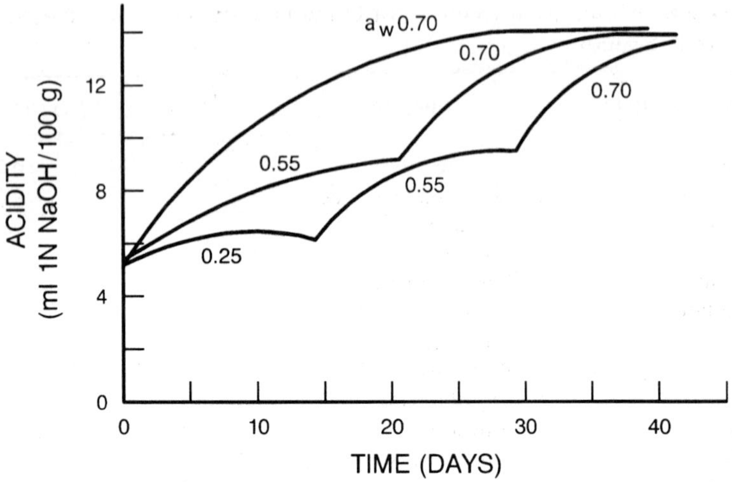

FIG. 2.13 Oxidation of glucose in a cellulose–glucose–glucose oxidase model system stored at 25°C and different a_w. Water activities of all systems were changed to 0.70 after 10 days. (*Adapted from Acker, 1969.*)

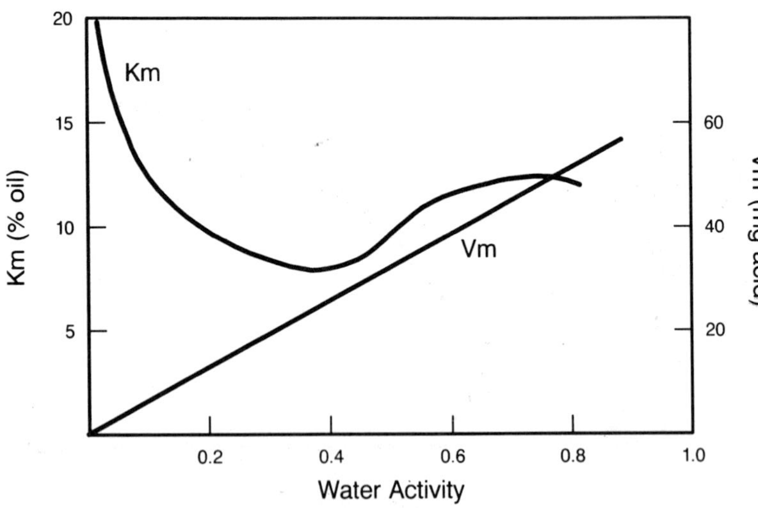

FIG. 2.14 Changes in V_m and K_m for lipase in wheat bran at 30°C as a function of a_w. K_m = % olive oil, V_m = mg fatty acids formed after 72 hr. (*Adapted from Drapron, 1985.*)

Activities of some enzymes in freeze-dried muscle have been shown to decrease with storage time (Potthast, 1978). Inactivation of ATPase activity, which is accompanied by the denaturation of myosin, increased with increasing a_w. Activity of glycogen-splitting enzymes in freeze-dried muscle also decreased with increasing time and a_w during storage. The glycolytic enzyme activity was not detectable after 100 days of storage at a_w 0.65. The enzyme interaction could be due to chemical interaction of proteins with carbohydrates such as the Maillard reaction (Potthast, 1978).

Enzyme kinetics at low temperature and reduced a_w is the subject of a review by Fennema (1978). At subfreezing temperatures, a_w is defined as the vapor pressure of ice divided by the vapor pressure of the supercooled water at the same temperature. According to Fennema, a_w of a frozen food is related to temperature by the following equation:

$$a_w = K \exp(\Delta H/RT)$$

where K is a constant, ΔH is the latent heat of fusion for ice, R is the gas constant, and T is the absolute temperature. Therefore, the relationship between a_w and rates of enzymatic reactions at subfreezing temperatures can be obtained by converting temperature to a_w if the temperature dependence of the reaction is known. Table 2.6 shows the effect of subfreezing temperature and a_w on hydrolysis of lipids in haddock. As expected, the reaction rate generally decreases with decreasing temperature and a_w. Besides the temperature factor, increase in viscosity of the partially frozen system and enzyme denaturation may also contribute to the reduction in reaction rate.

TABLE 2.6 Effect of Subfreezing Temperature and a_w on Lipid Hydrolysis in Haddock

Temp (°C)	a_w	K(day^{-1})	Fatty acids (mg/100g)
−7	0.93	0.067	300
−14	0.87	0.044	200
−20	0.82	0.012	227
−29	0.75	0.014	103

Source: Fennema (1978).

PROTEIN DENATURATION AND STARCH RETROGRADATION

Just as water activity has an important influence on enzyme activity, denaturation of most food proteins is affected by moisture content of the system. As an example, the denaturation temperatures of legumin and vicilin in fava bean as a function of water content are shown in Fig. 2.15. The denaturation temperature of protein was determined by differential scanning calorimeter (DSC). The denaturation temperature decreased sharply with increasing water content to about 0.9 g water/g protein, and gradually leveled off with further addition of water. It has been suggested that at low moisture content, or low a_w, there is insufficient water in the vicinity of the protein to bring about the thermal transition (Hagerdal and Martens, 1976). Similar observations have also been made in enzyme inactivation (Acker, 1969).

Moisture content or a_w has been shown to influence the gelatinization temperature and retrogradation rate of starch. Gelatinization temperature of starch generally increases with decreasing moisture content (Donovan, 1979; Eliasson, 1980). Two to three endothermic peaks are usually observed in water–starch mixtures during heating in a DSC. In general, the temperature of the first transition does not change appreciably with water content. However,

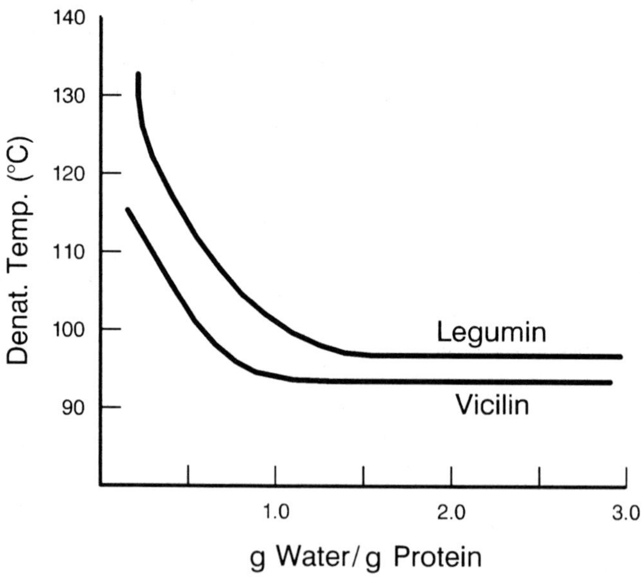

FIG. 2.15 Effect of water content on denaturation temperature of legumin and vicilin from fava beans. (*Adapted from Arntfield et al., 1985.*)

the second and third transitions shift toward higher temperatures as water content of the system decreases. Eliasson (1980) observed a linear relationship between the enthalpy of the first endosperm and water content up to 70°C moisture. Extrapolation of the enthalpy to zero gives the minimum water content necessary for starch gelatinization. Therefore, starch gelatinization is highly dependent on the availability of water in the system.

The relationship between starch retrogradation and water content is more complex; the reaction rate is slow at very high and very low moisture contents (Longton and Le Grys, 1981). In the case of bread, the rate of starch retrogradation as detected by DSC increases with decreasing water content. As shown in Fig. 2.16, breads containing 10% sorbitol (a_w 0.95) or 20% sorbitol (a_w 0.93) show a more rapid rate of starch retrogradation than regular bread (a_w 0.97) during the first week of storage (Leung, 1986). Increasing water content of bread was shown to result in a slower rate of starch retrogradation (Mackey, 1985). Also, the retrogradation rate of potato starch gel has been shown to increase with decreasing water content (Mackey and Leung, 1986). Therefore, the changes of starch from the amorphous state to the crystalline state and vice versa are closely associated with the availability of water in the system.

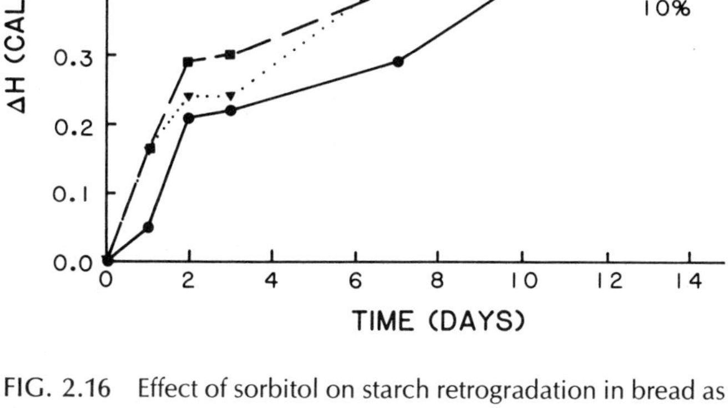

FIG. 2.16 Effect of sorbitol on starch retrogradation in bread as detected by differential scanning calorimetry during storage at 30°C. (*Adapted from Leung, 1986.*)

SUMMARY

A brief overview and update of the effect of a_w on lipid oxidation, nonenzymatic browning, vitamin degradation, enzyme activity, and protein denaturation, as well as starch gelatinization and retrogradation was presented. One of the most important effects of water on chemical reactivity in foods is its ability to mobilize and act as a solvent. Chemical reaction rates generally accelerate with increasing a_w due to increased reactant mobility. The major exception to this rule is lipid oxidation. At low a_w, water acts as an antioxidant by complexing with hydroperoxides and/or metal catalysts. As a result, the minimum oxidation rate of many food systems occurs at an a_w range of 0.2–0.4. In the intermediate or high moisture range, chemical reaction rates may decrease with increasing a_w because of the dilution effect of excess water on reactant concentrations. Water is usually required for enzymatic reactions to take place except when the substrate is a nonaqueous liquid, such as oil, which is mobile at the reaction temperature. The availability of water also plays an important role in protein denaturation, starch gelatinization, and starch retrogradation.

REFERENCES

Acker, L. W. 1969. Water activity and enzyme activity. *Food Technol.* 23: 1257.

Arntfield, S. D., Murray, E. D., and Ismond, M. A. H. 1985. The influence of processing parameters on food protein functionality. III. Effect of moisture content on the thermal stability of fababean protein. *Can. Inst. Food Sci. Technol. J.* 18: 226.

Baltes, W. 1982. Chemical changes in food by the Maillard reaction. *Food Chem.* 9: 59.

Blain, J. A. 1962. Moisture levels and enzyme activity. In *Recent Advances in Food Science. Vol. II. Processing.* Hawthorn, J. and Leitch, J. M. (Ed.), p. 41. Butterworths, London.

Brockmann, R. and Acker, L. 1977. Lipoxygenase activity and water activity in systems of low water content. *Ann. Technol. Agric.* 26: 167.

Brunauer, S., Emmett, P. H., and Teller, E. 1938. Adsorption of gases in multimolecular layers. *J. Am. Chem. Soc.* 60: 309.

Cerrutti, P., Resnik, S. L., Seldes, A., and Ferro-Fontan, C. 1985. Kinetics of deteriorative reactions in model food systems of high water activity: glucose loss, 5-hydroxymethyl furfural accumulation and fluorescence development due to nonenzymatic browning. *J. Food Sci.* 50: 627.

Chou, H. E. and Labuza, T. P. 1974. Antioxidant effectiveness in intermediate moisture content model system. *J. Food Sci.* 39: 479.

Dennison, D. B. and Kirk, J. R. 1982. Effect of trace mineral fortification on the storage stability of ascorbic acid in a dehydrated model food system. *J. Food Sci.* 47: 1198.

Donovan, J. W. 1979. Phase transitions of the starch-water system. *Biopolymers* 18: 263.

Drapron, R. 1985. Enzyme activity as a function of water activity. In *Properties of Water in Foods.* Simato, D. and Multon, J. L. (Ed.). Martinus Nijhoff Publishers, Dordrecht, The Netherlands.

Duckworth, R. B. 1975. *Water Relations of Foods.* Academic Press, New York.

Eichner, K. 1975. The influence of water content on non-enzymic browning reactions in dehydrated foods and model systems and the inhibition of fat oxidation by browning intermediates. In *Water Relations of Foods.* Duckworth, R. B. (Ed.), p. 417. Academic Press, New York.

Eichner, K. 1980. Antioxidative effect of Maillard reaction intermediates. In *Autoxidation in Food and Biological Systems.* Simic, M. G. and Karel, M. (Ed.), p. 367. Plenum Press, London.

Eichner, K., Laible, R., and Wolf, W. 1985. The influence of water content and temperature on the formation of Maillard reaction intermediates during drying of plant products. In *Properties of Water in Foods.* Simato, D. and Multon, J. L. (Ed.). Martinus Nijhoff Publishers, Dordrecht, The Netherlands.

Eliasson, A. C. 1980. Effect of water content on the gelatinization of wheat starch. *Starch/Starke* 32: 270.

Fennema, O. 1978. Enzyme kinetics at low temperature and reduced water activity. In *Dry Biological Systems.* Crowe, A. H. and Clegg, J. S. (Ed.). Academic Press, New York.

Frankel, E. N. 1984. Lipid oxidation: Mechanisms, products and biological significance. *J. Amer. Oil Chem. Soc.* 61: 1908.

Friedman, M. 1982. Chemically reactive and unreactive lysine as an index of browning. *Diabetes* 31(6): 5.

Furuya, E. M., Warthesen, J. J., and Labuza, T. P. 1984. Effects of water activity, light intensity and physical structure of food on the kinetics of riboflavin photodegradation. *J. Food Sci.* 49: 525.

Hagerdah, B. and Martens, H. 1976. Influence of water contents on the stability of myoglobin to heat treatment. *J. Food Sci.* 41: 933.

Haralampu, S. G. and Karel, M. 1983. Kinetic models for moisture dependence of ascorbic acid and β-carotene degradation in dehydrated sweet potato. *J. Food Sci.* 48: 1872.

Heidelbaugh, N. and Karel, M. 1970. Effects of water binding agents on oxidation of methyl linoleate. *J. Am. Oil Chem. Soc.* 47: 539.

Homma, S. and Fujimaki, M. 1982. Effect of water activity on lipid oxidation and browning of kori-tofu. *Agric. Biol. Chem.* 46: 301.

Kaanane, A. and Labuza, T. P. 1985. Change in available lysine loss reaction rate in fish flour due to an a_w change induced by a temperature shift. *J. Food Sci.* 50: 582.

Kacyn, L. J., Saguy, I., and Karel, M. 1983. Kinetics of oxidation of dehydrated food at low oxygen pressures. *J. Food Proc. Pres.* 7: 161.

Kahl, J. 1986. Effect of water and cobalt on methyl linoleate autoxidation. M.S. thesis, Washington State Univ., Pullman, WA.

Kamman, J. F. and Labuza, T. P. 1985. A comparison of the effect of oil versus plasticized vegetable shortening on rates of glucose utilization in nonenzymatic browning. *J. Food Proc. Pres.* 9: 217.

√ Karel, M. 1980. Lipid oxidation, secondary reactions, and water activity of foods. In *Autoxidation in Food and Biological Systems.* Simic, M. G. and Karel, M. (Ed.). Plenum Press, London.

Kirk, J. R. 1981. Influence of water activity on stability of vitamins in dehydrated foods. In *Water Activity: Influences on Food Quality.* Rockland, L. B. and Stewart, G. F. (Ed.), p. 631. Academic Press, New York.

Labuza, T. P. 1970. Properties of water as related to the keeping quality of foods, p. 618. *Proceedings of the Third International Congress of Food Science & Technology,* Washington, DC.

Labuza, T. P. 1975. Oxidative changes in foods at low and intermediate moisture levels. In *Water Relations of Foods.* Duckworth, R. B. (Ed.), p. 455. Academic Press, New York.

Labuza, T. P. 1980. The effect of water activity on reaction kinetics of food deterioration. *Food Technol.* 34(4): 36.

Labuza, T. P. and Saltmarch, M. 1981a. The nonenzymatic browning reaction as affected by water in foods. In *Water Activity: Influences on Food Quality.* Rockland, L. B. and Stewart, G. F. (Ed.), p. 605. Academic Press, New York.

Labuza, T. P. and Saltmarch, M. 1981b. Kinetics of browning and protein quality loss in whey powders during steady state and nonsteady state storage conditions. *J. Food Sci.* 47: 92.

Leahy, M. M. and Warthesen, J. J. 1983. The influence of Maillard browning and other factors on the stability of free tryptophan. *J. Food Proc. Pres.* 1: 25.

Leung, H. K. 1986. Bread quality: The effects of water binding ingredients. *Activities Report of the R&D Associates.* Schenck, A. M. (Ed.) 38(1): 43.

Lingnert, H. and Eriksson, C. E. 1980. Antioxidative Maillard reaction products. I. Products from sugars and free amino acids. *J. Food Proc. Pres.* 4: 161.

Longton, J. and LeGrys, G. A. 1981. Differential scanning calorimetric studies on the crystallinity of aging in wheat starch gels. *Starch/Starke* 33: 410.

Mackey, K. L. 1985. Staling of potato bread and retrogradation of starch gels. M.S. thesis, Washington State Univ., Pullman, WA.

Mackey, K. L. and Leung, H. K. 1986. Retrogradation of wheat, potato, and wheat-potato starch gels. *J. Food Sci.* (in press).

Maloney, J. F., Labuza, T. P., Wallace, D. H., and Karel, M. 1966. Autoxidation of methyl linoleate in a freeze-dried model system. I. Effect of water on the antocatalyzed oxidation. *J. Food Sci.* 31: 878.

Mauron, J. 1981. The Maillard reaction in food: A critical review from the nutritional standpoint. *Progr. Fd. Nutr. Sci.* 5: 5.

Okamoto, M. and Hayashi, R. 1985. Chemical and nutritional changes of milk powder proteins under various water activities. *Agric. Biol. Chem.* 49: 1683.

Petriella, C., Resnik, S. L., Lozano, R. D., and Chirife, J. 1985. Kinetics of deteriorative reactions in model food systems of high water activity: Color changes due to nonenzymatic browning. *J. Food Sci.* 50: 622.

Potthast, K. 1978. Influence of water activity on enzymic activity in biological systems. In *Dry Biological Systems.* Crowe, J. H. and Clegg, J. S. (Ed.), p. 323. Academic Press, New York.

Resnik, S. and Chirife, J. 1979. Effect of moisture content and temperature on some aspects of nonenzymatic browning in dehydrated apple. *J. Food Sci.* 44: 601.

Rockland, L. B. and Nishi, S. K. 1980. Influence of water activity on food product quality and stability. *Food Technol.* 34(4): 42.

Rockland, L. B. and Stewart, G. F. 1981. *Water Activity: Influences on Food Quality.* Academic Press, New York.

Salwin, H. 1959. Defining minimum moisture contents for dehydrated foods. *Food Technol.* 13: 594.

Seow, C. C. and Cheah, P. B. 1985. Reactivity of sorbic acid and glycerol in nonenzymatic browning in liquid intermediate moisture model systems. *Food Chem.* 18: 71.

Schwimmer, S. 1980. Influence of water activity on enzyme reactivity and stability. *Food Technol.* 34(5): 64.

Simato, D. and Multon, J. L. 1985. *Properties of Water in Foods.* Martinus Nijhoff Publishers, Dordrecht, The Netherlands.

Singh, R. K., Lund, D. B., and Buelow, F. H. 1984. Computer simulation of storage stability in intermediate moisture apples. *J. Food Sci.* 49: 759.

Toribio, J. L. and Lozano, J. E. 1984. Nonenzymatic browning in apple juice concentrate during storage. *J. Food Sci.* 49: 889.

Toribio, J. L., Nuñes, R. V., and Lozano, J. E. 1984. Influence of water activity on the nonenzymatic browning of apple juice concentrate during storage. *J. Food Sci.* 49: 1630.

Widicus, W. A., Kirk, J. R., and Gregory, J. F. 1980. Storage stability of α-tocopherol in a dehydrated model food system containing no fat. *J. Food Sci.* 45: 1015.

Wolf, J. C., Thompson, D. R., Warthesen, J. J., and Reineccius, G. A. 1981. Relative importance of food composition in free lysine and methionine losses during elevated temperature processing. *J. Food Sci.* 46: 1074.

3

Influence of Water Activity on Pigment Stability in Food Products

Joachim H. von Elbe

University of Wisconsin-Madison
Madison, Wisconsin

INTRODUCTION

The stability of pigments in foods is influenced by several environmental factors. Depending upon the pigment class under study, stability during processing can be a function of the presence or absence of oxygen, light, oxidizing or reducing substances, heavy metals, a_w, pH, and temperature. Recognizing the importance of water in many of these reactions, it is not surprising that a_w is included among the primary factors affecting pigment stability and/or color of a product. The literature dealing with the effect of a_w on pigment stability is limited, to a large extent, because the degradation reactions of a number of pigments are poorly understood. The major classes of pigments responsible for color in plant tissues are chlorophylls, carotenoids, anthocyanins, and betalaines.

CAROTENOIDS

Stability of Pigments

The carotenoids, in part because of their importance in nutrition, have received the most attention. They are subject to changes due to heating and both enzymatic and nonenzymatic oxidation, all of which are influenced by water. In nature, the majority of carotenoids exist in the stable *trans*-configuration, although *cis*-isomers occur in some plants (Weedon, 1971). Processing and storage of carotenoid-containing foods under acid conditions, elevated temperatures, and/or light favors isomerization (Singleton et al., 1961). Carotenoids are relatively heat-stable in foods devoid of oxygen. Several investigations have reported changes in carotenoid content during blanching, cooking, or heat sterilization. It is generally agreed that the total carotenoid content is unchanged (Eskin, 1979; Walter and Giesbrecht, 1972), and that color changes that occur during these processes can be attributed to the isomerization of *trans*-carotenoids to the less intensely colored *cis* form. In some fruits and vegetables, increases in total carotenoid content have been observed after heat sterilization (Weckel et al., 1962; Penalaks and Murray, 1970; Paulus and Saguy, 1980; Edwards and Lee, 1986). The increase has been attributed to the leaching of solids into the packing medium; no difference in total pigment content occurred when the leaching of solids was taken into account in the calculations.

Changes upon heating. Loss of carotenoids has been reduced by blanching of vegetable tissue, suggesting enzymatic destruction. Fresh peas, for example, when blended and held for 2 hr before extraction of carotenoids, showed a 68% decrease in total carotenoids (Edwards and Lee, 1986). Bleaching of carotenoids has been associated with activity of lipoxygenase, an enzyme whose presence has been demonstrated in many chlorophyll-containing vegetables (Rhee and Watts, 1966). The role of the enzyme can be demonstrated by extracting the tissue with a monohydric alcohol, such as methanol, which inhibits lipoxygenase (Mitsuda et al., 1967). Therefore, the extraction of plant tissue, with methanol compared to acetone, should yield a higher total carotenoid content. This was demonstrated in a study reported by Edwards and Lee (1986). Fresh carrots, which are known to have little or no lipoxygenase activity, showed little or no difference in carotenoid content when the tissue was extracted with acetone or methanol. However, green peas, which have a high lipoxygenase activity, showed a significant difference in carotenoid content (Table 3.1).

Oxidation. Similarly nonenzymatic oxidation of carotenoids with concurrent color loss is of major concern in dehydrated foods. Carotenoids in carrot

TABLE 3.1 Difference in Carotenoid Content of Fresh Carrots and Green Peas as Influenced by Extraction Solvent Used

Carotenoids (μg/g)[a]	Carrots		Peas	
	Acetone	Methanol	Acetone	Methanol
Total	1331	1358	14.60	22.41
Hydrocarbon	1246	1284	3.23	6.23
Monohydroxy	6.9	6.7	0.65	0.89
Polyhydroxy	14.00	12.60	6.73	10.40

[a]Dry weight basis.
Adapted from Edwards and Lee (1986).

powder were fully retained for six months when stored under nitrogen, compared to a 21% loss when stored in air (Mackinney et al., 1958). The importance of only small amounts of residual oxygen on β-carotene loss in model systems simulating dehydrated foods was demonstrated by Goldman et al. (1983). These researchers defined shelf life of the pigment as the time required for 50% pigment degradation. Applying this definition, levels of 1, 2, 10, 15, and 20.9% oxygen produced β-carotene shelf-life values of 37, 25, 10, 7, and 5 days at 35°C, respectively. Control samples stored under a nitrogen atmosphere showed only a 12% loss of β-carotene after 60 days of storage at 35°C. The authors further illustrated that in addition to reducing the oxygen concentration, the shelf life of the pigments would be extended by the addition of either antioxidants or the adjustment of the a_w of the samples. Samples with an a_w of 0.84 and 0.32 had shelf-life values of 45 and 30 days.

Protective effect of water (a_w). These data on the stability of carotenoids as a function of oxidation or heat exposure suggested a protective role of water and that a_w is an important factor in determining the stability of carotenoids in foods. This was demonstrated in a study by Martinez and Labuza (1968), who suggested that in freeze-dried salmon there is a direct relationship between a_w and carotenoid loss; the oxidation followed a free radical mechanism and, as in the case of lipid oxidation, water influenced the oxidation mode of the carotenoid loss. Oxygen absorption in the lipid fraction in freeze-dried salmon demonstrated that the major effect of humidification was a decrease in the overall rate of oxidation after the initial phase. The effect of moisture content on the oxidative deterioration of the pigment astaxanthine was similar to that

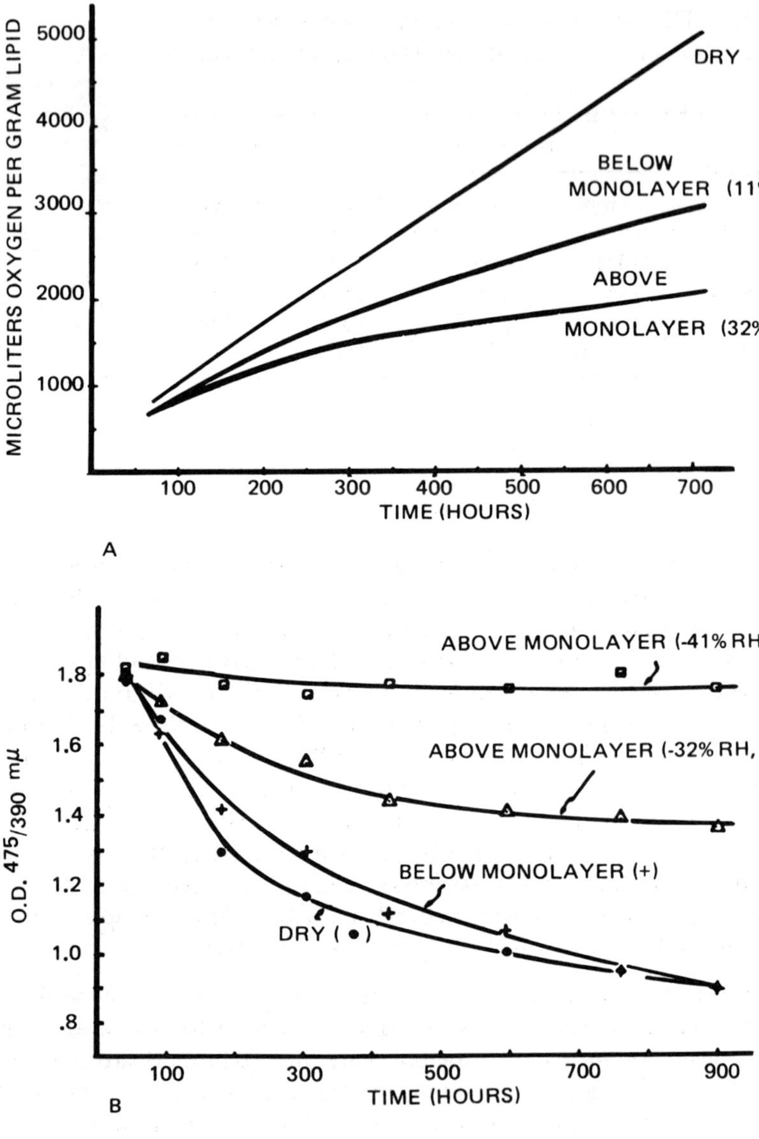

FIG. 3.1 Oxygen absorption (A) and astaxanthine degradation (B) in freeze-dried salmon. (*Adapted from Martinez and Labuza, 1968.*)

observed for the oxidative deterioration of lipids. Values of a_w above the monolayer (up to a_w 0.41) gave almost complete protection (Fig. 3.1).

Astaxanthine, like other carotenoids, is also sensitive to oxidation. If the oxygen concentration is not limiting, degradation of the pigment follows first-order reaction kinetics. Values for the rate constant and half-life values for freeze-dried salmon stored at 37°C are given in Table 3.2. Several mechanisms may be involved in the protective effect of water (Labuza et al., 1970):

1. Water on the surface of foods hydrogen bonds to the hydroperoxides produced during free-radical reaction, protecting them from decomposing and thereby slowing the rate of initiation.
2. Water hydrates trace metals catalysts, reducing or completely eliminating their catalytical activity.
3. Water may form insoluble metal hydroxides, eliminating their participation in the reaction.

Similar results demonstrating the protective effect of water on the oxidative losses of carotenoids have been reported by Chou and Breene (1972), Ramakrishnan and Francis (1979), and Kearsley and Rodriguez (1981). In the study by Chou and Breene (1972), decoloration of β-carotene was determined by measuring the reflectance value of β-carotene at 450 nm in microcrystalline cellulose. The rate of β-carotene decoloration, after a brief induction period, similar to that observed with astaxanthine degradation, followed first-order reaction kinetics. The reaction rate, as might be expected, increased with increasing storage temperature. The activation energy was calculated as 9.7 kcal/mole, a value similar in magnitude to the activation energy for lipid oxidation in foods. Ramakrishnan and Francis (1979) included in their study of a model system simulating dehydrated foods not only β-carotene, but also

TABLE 3.2 First Order Rate Constants and Half-Life Values for Astaxanthine in Freeze-Dried Salmon Stored at 37°C

		Monolayer	
Constant	Dry (0.1% RH)	Below (11% RH)	Above (40% RH)
k (hr^{-1})	1.1×10^{-3}	1.0×10^{-3}	0.1×10^{-3}
$T_{1/2}$ (hr)	663	693	6930

Adapted from Martinez and Labuza (1968).

two other widely used food colorants, apo-8′-carotenal and canthaxanthin. The degradation of all three pigments, when exposed to an excess of oxygen, followed first-order reaction kinetics. Table 3.3 gives the decoloration constants and half-life values at 22°C of the three carotenoids in a cellulose model system stored at various relative humidities. The rate of decoloration was highest for β-carotene and lowest for canthaxanthin. The greatest protective effect by water against color losses was at 75% relative humidity where half-life values for β-carotene, apo-8′-carotenal, and canthaxanthin were 17.3, 21.6, and 49.5 days, respectively. The half-life values based on the autooxidation rate constant (Table 3.3) were 12.0, 16.2, and 30.0 days, respectively. The stability of the pigments seem to increase with the polarity of the molecule; however, it is not certain if polarity of the pigment plays a role in the water protective effect. Data reported when a starch system was used rather than a cellulose system suggest that the carotenoids were more stable to autooxidative changes in the starch system. This increased stability may result from the added protective effect of starch, or from the increased moisture content in the system at the same a_w. The monolayer for the cellulose system determined from the adsorption isotherm was observed to occur at a moisture content of 2.8% with a corresponding a_w of 0.18, while for starch the moisture content was 9% and the a_w was 0.27.

More recently Haralampu and Karel (1983), Goldman et al. (1983), and Saguy et al. (1985) developed mathematical kinetic models that describe the effect of a_w on the rate of β-carotene loss in dehydrated foods or systems simulating dehydrated foods.

TABLE 3.3 First-Order Decoloration Constants and Half-Life Values for Three Carotenoids

Rate constant	Relative humidity (%)	β-Carotene k[a]	β-Carotene $T_{1/2}$	Apo-8′-carotenal k	Apo-8′-carotenal $T_{1/2}$	Canthaxanthin k	Canthaxanthin $T_{1/2}$
	dry	9.5	7.3	6.8	10.1	3.3	21.0
	11	8.1	8.6	6.1	11.3	3.0	23.0
Decoloration	23	6.5	10.6	5.1	13.5	2.7	25.6
	52	4.6	15.1	3.9	17.7	1.9	36.4
	75	4.0	17.3	3.2	21.6	1.4	49.5

[a]k = decoloration or autooxidation constant in $day^{-1} \times 10^{-2}$; $T_{1/2}$ = half-life value in days.

Adapted from Ramakrishnan and Francis (1979).

The protective effect of water against the autooxidation of carotenoids, as demonstrated in a number of studies, can be explained by its direct effect on free radicals produced during the pigment's oxidation. The free radical content can be reduced by interaction with water, a reduction which increases with increased moisture content as first suggested by Labuza et al. (1970).

The overall effect of water depends to a great extent on the composition of the food. This was demonstrated in studies involving the loss of carotene in dehydrated red pepper fruits and on oleoresin–cellulose systems (Kanner and Budowski, 1978). The protective effect of water on carotene is demonstrated in both systems (Fig. 3.2), but the time scale and shape of the curves differ. Carotene destruction in the oleoresin system shows an induction period, the length depending on a_w, followed by a rapid onset of oxidation. In powdered

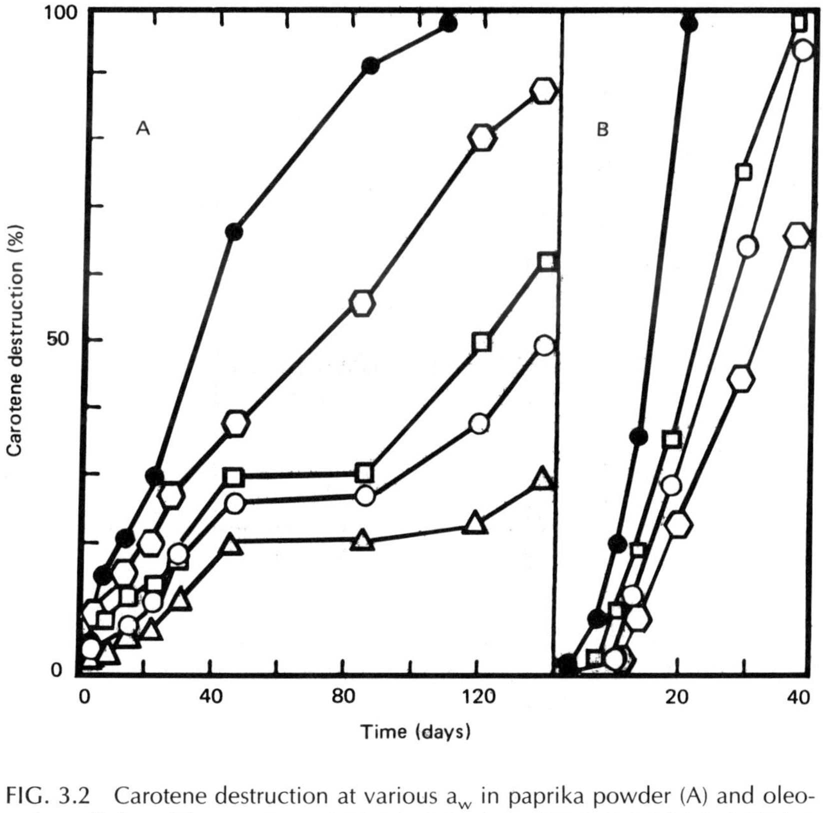

FIG. 3.2 Carotene destruction at various a_w in paprika powder (A) and oleoresin–cellulose (B). a_w values: 0.01 (●); 0.32 (⬡); 0.52 (○); 0.64 (△); 0.75 (□). (*From Kanner et al., 1978.*)

paprika, loss of carotene, or bleaching, occurred in three stages: induction, a stable period, and a period of revived oxidation. The stable period was extended when ascorbic acid and copper, acting as an antioxidant system, were included. Ascorbic acid, depending on concentration and a_w, may act as a prooxidant or antioxidant. Ascorbic acid in a cellulose model system at low a_w (0.01) had neither pro- nor antioxidant activity. At higher a_w values, the activity depended on the ascorbic concentration. At low ascorbic acid concentration (5 μmoles/g cellulose), a prooxidation effect was demonstrated which increases with increasing a_w. At high concentration (100 μmoles/g cellulose) an antioxidant effect was demonstrated, which was more pronounced when copper was present (Fig. 3.3).

In the oleoresin–cellulose system an increase in a_w from 0.01 to 0.75, in the absence of the antioxidant system, increased with induction period from 6 to 12 days. In the presence of additives the induction period increased from 10 to

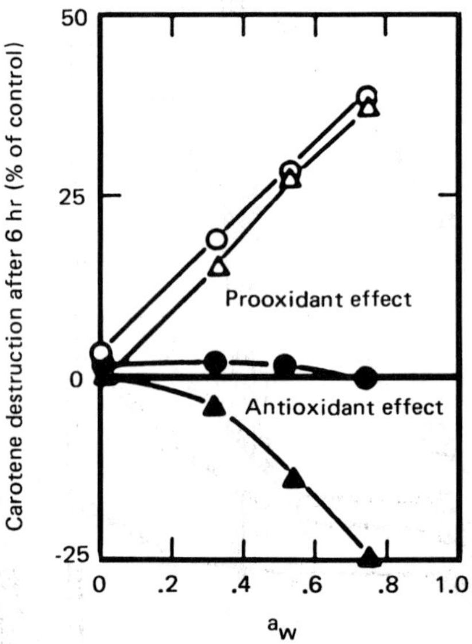

FIG 3.3 Effect of a_w on carotene bleaching in cellulose model in the presence of ascorbic acid and copper ions. Concentrations (mol/g cellulose): ascorbic acid 5 (○); ascorbic acid 5 + Cu^{2+} 0.15 (△); ascorbic acid 100 (●); ascorbic acid 100 + Cu^{2+} 0.15 (▲). (*From Kanner and Budowski, 1978.*)

120 days. It was concluded that ascorbic acid and copper, by being dissolved in high a_w products, play an important part in stabilizing carotenoids and are responsible for the distinct periods of bleaching observed in dry paprika powder. Paprika contains both ascorbic acid and copper within the concentration range of antioxidant activity of the two additives.

CHLOROPHYLLS

A second class of oil-soluble pigments are the chlorophylls. During food processing a number of structural changes can occur which involve, alone or in combination, removal of the central magnesium atom, removal of the phytol and/or carbomethoxy group, isomerization, and oxidation.

Degradation Pathway

The most prevalent degradation pathway of chlorophyll during heat processing is the conversion of chlorophyll into pheophytin, which in turn is partially converted to pyropheophytin (Schwartz et al., 1981; Schwartz and von Elbe, 1983a). This reaction involves water and is acid catalyzed. Elevation of pH in foods to preserve the chlorophylls has been the basis of a number of patents and studies (Blair and Ayres, 1943; Gupte and Francis, 1964; Clydesdale and Francis, 1968). In the study by Gupte and Francis, for example, approximately 80% of the chlorophyll *a* was retained in a spinach puree (pH 8.5) processed at 138°C for 87 sec compared to 43% in puree at a normal pH (6.3) value. Since the reaction is acid catalyzed, the availability of water is essential, and it would suggest that a_w would influence the rate of degradation. This was first pointed out in a study reported by Dutton et al. (1943). These researchers noted that pheophytinization, which occurred in dried spinach containing 18% moisture, did not occur in spinach containing 2.5% moisture, and that the presence or absence of oxygen did not influence the reaction.

Stability of Pigment

Two studies concerned with the degradation of chlorophyll under limited water concentration have been reported by LaJollo et al. (1971) and LaJollo and Marquez (1982). In these studies the degradation of chlorophyll *a* and *b* (C_a; C_b) were observed in blanched, freeze-dried spinach stored at 37 and 55°C in an a_w range of 0.75 to $\simeq$0.01 under either an air or nitrogen atmosphere. Water activity had a definite influence on the rate of chlorophyll degradation. The

TABLE 3.4 Influence of a_w on Rate of Chlorophyll Degradation

	Chlorophyll *a*				Chlorophyll *b*			
	37°C		55°C		37°C		55°C	
a_w	$k \times 10^{-2}$ hr^{-1}	$T_{1/2}$ (hr)	$k \times 10^{-2}$ hr^{-1}	$T_{1/2}$ (hr)	$k \times 10^{-2}$ hr^{-1}	$T_{1/2}$ (hr)	$k \times 10^{-2}$ hr^{-1}	$T_{1/2}$ (hr)
0.75	2.74	25.2	14.7	4.7	0.92	75.3	5.4	12.8
0.62	1.12	61.9	7.2	9.6	0.42	165.0	3.3	21.8
E_a (kcal/g mole)	19		21		20		23	

Adapted from LaJollo et al. (1971).

color of samples stored at both temperatures changed during storage from green to olive-green, indicative of pheophytinization. The formation of pheophytin was confirmed by spectrophotometric and chromatographic techniques. The results further showed that even at an extremely low moisture content enough water was available for the acid-catalyzed reaction to occur. Under all conditions studied only C_a at an a_w of 0.75 showed pseudo–first-order kinetics over the entire range of concentrations. All other reactions, at a_w values below 0.75, deviated from first-order kinetics. There are a number of reasons, as noted by the authors, which may account for the deviation. These may include compartmentalization of chlorophyll and, therefore, at low a_w values, the more firmly bound pigment to be less reactive and more resistant to pheophytinization. Other complex changes, loss of volatile compounds, and formation of organic acids did change the pH of the samples, which may have influenced the kinetics. First-order rate constants and activation energies are given in Table 3.4. These data, as well as previous reports (Schandler et al., 1962; Gupte et al., 1964), indicate a faster reaction time for C_a compared to C_b.

Effect of Water (a_w) on Stability

In samples stored at an a_w less than 0.32, formation of pheophytin was very slow and chlorophyll loss was accompanied by a parallel loss of pheophytin (LaJollo et al., 1971). This is illustrated in Table 3.5. Pure C_a, when subjected to similar conditions, was converted to pheophytin at a faster rate than C_a in the spinach samples, and the reaction was accelerated by oxidative conditions.

TABLE 3.5 Effect of Storage Atmosphere on Retention of Chlorophyll and Pheophytin at 55°C

a_w	Time (days)	Chlorophyll[a]		Total pheophytin[b]	
		Air	N_2	Air	N_2
0.11	0	100	100	100	100
	12	90.8	86.0	86.0	85.0
	40	61.7	60.3	66.4	63.4
	51	56.2	58.8	59.0	66.4
0.01	17	92.4	89.2	91.3	90.0
	40	54.5	54.1	54.0	65.8
	51	—	51.6	—	66.0

[a]Percent of initial.
[b]After acid treatment.
Source: LaJollo et al. (1971).

This would indicate an in vivo protection against degradation, probably because of association of the chlorophyll in a lipoprotein complex.

In the foregoing study the pH value, as indicated, changed during the reaction; e.g., the spinach samples at an a_w value of 0.75 had an initial pH of 6.70, and after storage for 82 hr at 55°C they had a pH of 5.70. It is because of this change in pH that LaJollo and Marquez (1982) studied chlorophyll degradation as a function of pH and a_w. This study confirmed that the rate of chlorophyll degradation increased with an increase in a_w of the system, and that the reaction followed first-order reaction kinetics even in regions of firmly bound water (Fig. 3.4A). Figure 3.4B illustrates the rate of chlorophyll degradation as influenced by pH (a_w 0.75, 38.6°C). The rate increased as the pH value decreased. Temperature at constant a_w had the expected result, e.g., as temperature increased from 38.6 to 56.7°C, the rate increased. This generalization, however, was not true when the temperature was lowered to 32°C from the higher temperature used in this study. In this instance the rate constant increased and at a_w 0.52 was 3.35×10^{-3} hr^{-1} ($T_{1/2} = 207$ hr) and 3.75×10^{-3} ($T_{1/2} = 185$ hr) at 46.0 and 32.0°C, respectively. This difference is explained by the moisture content of the samples needed to maintain the 0.52 a_w value. The moisture contents at 46 and 32°C were 10.9% and 17.7%, respectively. The increase in moisture content at the lower temperature decreased the viscosity of the sample, and probably increased the rate of reaction by allowing for greater mobility of the reactants.

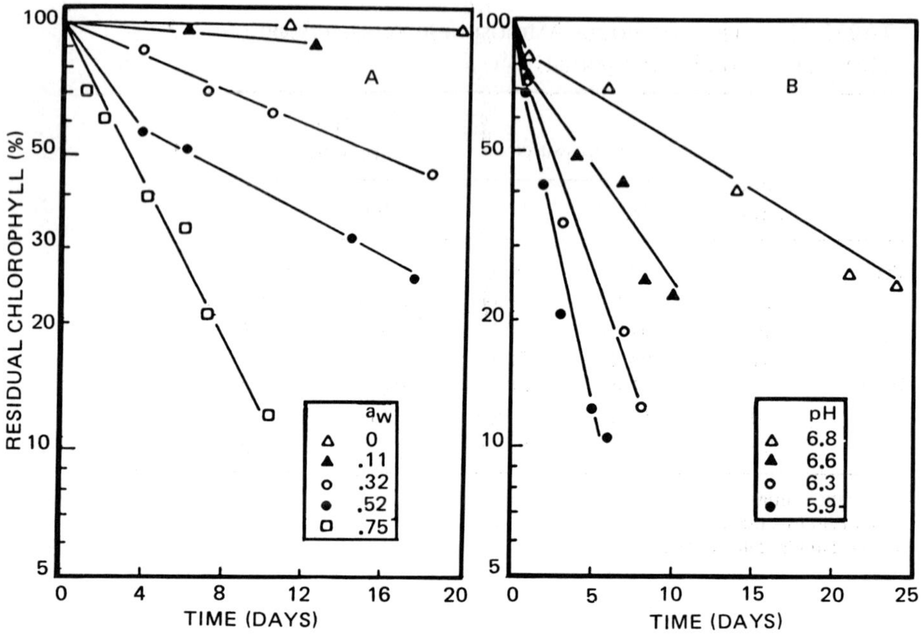

FIG. 3.4 Degradation rate of chlorophyll a in spinach as a function of time: (A) at different a_w (pH = 5.9, 38.6°C); (B) four different pH values (a_w = 0.75, 38.6°C). (*Adapted from LaJollo and Marquez, 1982.*)

LaJollo and Marquez (1982) further noted an increased rate of chlorophyll degradation with the addition of glycerol to the samples. The difference in rates can be attributed to the difference in moisture content of samples at similar a_w values. The first-order rate constant and half-life value for C_a degradation when 15% glycerol was added to a sample at a_w 0.52, pH 5.6 was 3.95×10^{-3} hr^{-1} ($T_{1/2}$ = 175 hr) versus 2.51×10^{-3} hr^{-1} ($T_{1/2}$ = 276 hr) in the absence of glycerol. The moisture contents of the two samples were 10.2 and 12.6%, respectively. In part, an explanation for the increase in rate may be the possibility that glycerol facilitates diffusion of reactants.

BETALAINES AND ANTHOCYANINS

Stability of Pigments

Two classes of water-soluble pigments deserve attention—the betalaines and anthocyanins. The major pigments of the red beet are betanine (red) and vulgaxanthine-I (yellow), collectively called betalaines. In recent years the chemistry of betanine has been studied extensively because of its potential use as a food colorant (von Elbe, 1977). Applications have been limited because the pigment lacks stability when exposed to heat or oxygen (von Elbe et al., 1974; Savolainen and Kuusi, 1978; Saguy, 1979; Sapers and Hornstein, 1979; Wiley et al., 1979; Singer and von Elbe, 1980; Huang and von Elbe, 1985). When an aqueous solution of betanine is heated, the pigment hydrolyzes into betalamic acid (BA) and cyclodopa-5-0-glycoside (CDG) (Kimler, 1972; Schwartz and von Elbe, 1983b). This reaction is partially reversible and involves a Schiff base condensation of the amine of CDG with the aldehyde of BA. It is well known that Schiff base reactions are reversible and dependent on temperature and pH. The activation energies for the forward and reverse reactions were determined to be 17.6 and 0.64 kcal/mole, respectively. Since the degradation reaction does involve water, the greatest stability of betanine has been reported in foods or model systems of low moisture content and/or a_w (Pasch and von Elbe, 1975; Saguy et al., 1980; Cohen and Saguy, 1983; Saguy et al., 1984). Pigment degradation follows first-order kinetics, and stability increases with decreasing a_w.

In the study by Pasch and von Elbe (1975) half-life values of betanine in water–glycerol solutions heated at 75°C were 33 and 124 min at a_w values of 1.0 and 0.37, respectively. A similar trend in stability of betanine in beet powder was reported by Cohen and Saguy (1983). In their study a reduction of a_w from 0.75 to 0.32 resulted in an increase in half-life at 35°C from 8.3 to 133 days and from 10.8 to 126.7 days for betanine and vulgaxanthine-I, respectively. An insignificant difference was reported when comparing kinetic data at a_w 0.32 and 0.41. At a_w 0.12 or lower no deterioration of either pigment was observed. Therefore, an a_w value of 0.12 and a moisture content of 2% (dry weight basis) is recommended for optimal storage stability of the pigment in a beet powder. Comparison of degradation rate data of betanine in beet powder must take into account different moisture contents as related to the adsorption and desorption isotherms. Half-life values for betanine in beet powder at equal a_w values but different moisture levels are limited (Table 3.6). These data illustrate, as was seen with chlorophylls, that a high moisture content (i.e., desorption) causes the degradation rate to increase. Furthermore, specification of a_w alone without the moisture content is not sufficient to predict pigment stability.

Influence of oxygen. The sensitivity of betanine to oxygen has been the subject of several studies. Degradation rate is increased in the presence of oxygen where betanine follows first-order reaction kinetics, but deviation from first-order kinetics occurs in the absence of oxygen (Attoe and von Elbe, 1982). Since the degradation reaction is reversible, to increase the thermostability of betanine the level of oxygen present in the reaction mixture must

TABLE 3.6 Betanine Stability in Terms of Half-Life Values in Beet Powder Stored at 35°C Related to the Adsorption or Desorption Isotherm

	Adsorption		Desorption	
a_w	Moisture (g water/g dry solid)	$T_{1/2}$ (days)	Moisture (g water/g dry solid)	$T_{1/2}$ (days)
0.52	13.8	31.9	21.5	1.3
0.75	25.8	8.3	30.0	0.5

Adapted from Cohen and Saguy (1983).

be minimized and conditions for regeneration of the pigment maximized. The stability of beet pigments in beet powder as affected by oxygen, a_w and moisture content was the subject of a study by Saguy et al. (1984). Data for betanine degradation are presented in Table 3.7. There was an exponential increase in volume of oxygen retained with increasing a_w and a corresponding increase in degradation rate. Degradation was accelerated at 0.5 a_w, when both moisture and oxygen content increased exponentially. The lack of liquid phase mobility governs the reaction below a_w 0.5.

The anthocyanins are responsible for most of the red hues appearing in flowers and fruits. They occur in plant as glycosides. The anthocyanidins have

TABLE 3.7 Effect of Water Activity and Moisture Content on Oxygen Volume and on the Degradation Rate Constant of Betanine Stored at 35°C

a_w	Moisture (% dry basis)	Volume of O_2/100g dry powder	$k \times 10^{-3}$ (day^{-1})
"dry"	0	0.96	N.S.[a]
0.12	2.4	1.22	N.S.
0.23	4.0	1.24	1.16
0.32	5.7	1.89	5.21
0.41	7.3	2.26	4.99
0.52	13.8	2.36	21.73
0.75	25.8	6.06	83.55

[a]N.S. = not significant.
Adapted from Saguy et al. (1984).

TABLE 3.8 Effect of a_w on Color Intensity of Anthocyanins[a] (Absorbance at 540 nm) Held at 43°C for Different Time Periods

Holding time (min)	Water activity						
	1.00	0.95	0.87	0.74	0.63	0.47	0.37
0	0.84	0.85	0.86	0.91	0.92	0.96	1.03
60	0.78	0.82	0.82	0.88	0.88	0.89	0.90
90	0.76	0.81	0.81	0.85	0.86	0.87	0.89
160	0.74	0.76	0.78	0.84	0.85	0.86	0.87
Percent change (0–160 min)	11.9	10.5	9.3	7.6	7.6	10.4	15.5

[a]Concentration 700 mg/100 mL (commercial dried pigment powder).
Adapted from Kearsley and Rodriguez (1981).

been shown to be unstable in water and are much less soluble than the anthocyanins (Timberlake and Bridle, 1966). Glycosylation therefore contributes to their stability and solubility. Though little information relating a_w and anthocyanin stability is available, relative values reported by Kearsley and Rodriguez (1981) show that the color of anthocyanins in a glycerol–water mixture increases in intensity as the a_w is lowered and color losses on holding were least at intermediate a_w (0.63 and 0.79). Losses increased at either lower or higher a_w values (Table 3.8). The color of anthocyanin solutions is highly pH dependent since pH affects the structure of the pigment. According to Brouillard and Delaporte (1977), in an acidic solution at 20°C, four anthocyanin species exist in equilibrium.

(A)		(AH+)		(B)		(C)
quinonoid	⇌	flavylium	⇌	carbinol	⇌	chalcone
blue		red		colorless		colorless

The relative amount of each species present is a function of pH, and heating shifts the equilibrium toward the chalcone. The reversion to the flavylium cation is slow and, depending on the chalcone form (diglycoside vs. monoglycoside), may take 6–12 hr or longer at lower temperatures. The equilibrium between the flavylium and carbinol base requires water and is the reaction responsible for loss of color in slightly acid solutions. From this brief discussion it is clear that water influences the stability, reactivity, and special properties of the various anthocyanin structures.

Hrazdina (1971) studied the thermal degradation of the 3,5-diglycosides of five anthocyanidins in the pH range 3 to 7, and identified a common degradation product, 3,5 di (o-β-D-glucosyl)7-hydroxy-coumarin. The proposed degradation mechanism involves the formation of coumarin from the flavylium cation via several intermediates. Though the exact mechanism of anthocyanin degradation is not fully understood, it is clear that water is an essential component. It should then not be surprising that a_w and water content have a marked influence on the stability of the pigments.

REFERENCES

Attoe, E. L. and von Elbe, J. H. 1982. Degradation kinetics of betanine in solutions as influenced by oxygen. *J. Agric. Food Chem.* 30: 708.

Blair, J. S. and Ayres, T. B. 1943. Protection of natural green pigment in the canning of peas. *Ind. Eng. Chem.* 35: 85.

Brouillard, R. and Delaporte, B. 1977. Chemistry of anthocyanin pigments. 2. Kinetics and thermodynamic study of protein transfer, hydration, and tautomeric reactions of malvidin-3-glucoside. *J. Am. Chem. Soc.* 99: 8461.

Chou, H. E. and Breene, W. M. 1972. Oxidative decoloration of β-carotene in low-moisture model systems. *J. Food Sci.* 37: 66.

Clydesdale, F. M. and Francis, F. J. 1968. Chlorophyll changes in thermally processed spinach as influenced by enzyme conversion and pH adjustment. *Food Technol.* 22: 793.

Cohen, E. and Saguy, I. 1983. Effect of water activity and moisture content on the stability of beet powder pigments. *J. Food Sci.* 48: 703.

Dutton, H. J., Bailey, G., and Kohake, E. 1943. Dehydrated spinach. Changes in color and pigment during processing and storage. *Ind. Eng. Chem.* 35: 1173.

Edwards, C. G. and Lee, C. Y. 1986. Measurement of provitamin A carotenoids in fresh and canned carrots and green peas. *J. Food Sci.* 51: 534.

Eskin, M. N. A. 1979. *Plant Pigments, Flavor and Textures: The Chemistry and Biochemistry of Selected Compounds.* Academic Press, New York.

Goldman, M., Horev, B., and Saguy, I. 1983. Discoloration of β-carotene in model systems simulating dehydrated foods. Mechanism and kinetic principles. *J. Food Sci.* 48: 751.

Gupte, S. M., El-Bisi, H. M., and Francis, F. J. 1964. Kinetics of thermal degradation of chlorophyll in spinach puree. *J. Food Sci.* 29: 379.

Gupte, S. M. and Francis, F. J. 1964. Effect of pH adjustment and high-temperature short-time processing on color and pigment retention in spinach puree. *Food Technol.* 18: 1645.

Haralampu, S. G. and Karel, M. 1983. Kinetic model for moisture dependence of ascorbic acid and β-carotene degradation in dehydrated potato. *J. Food Sci.* 48: 1872.

Hrazdina, G. 1971. Reactions of anthocyanidin-3,5-diglucosides formation of 3,5-di(o-β-D-glucosyl)-7-hydroxy coumarin. *Phytochemistry* 10: 1125.

Huang, A. S. and von Elbe, J. H. 1985. Kinetics of the degradation and regeneration of betanine. *J. Food Sci.* 50: 1115.

Kanner, J. and Budowski, P. 1978. Carotene oxidizing factors in red pepper fruits (*Capsicum annum* L.): Effect of ascorbic acid and copper in a β-carotene-linoleic acid solid model. *J. Food Sci.* 43: 524.

Kanner, J., Mendel, H., and Budowski, P. 1978. Carotene oxidizing factors in red pepper fruits (*Capsicum annum* L.): Oleoresin-cellulose solid model. *J. Food Sci.* 43: 709.

Kearsley, M. W. and Rodriguez, N. 1981. The stability and use of natural colours in foods: anthocyanin, β-carotene and riboflavin. *J. Food Technol.* 16: 421.

Kimler, L. 1972. Betalamic acid and other products of the biotransformation of L-dopa in betalain biogenesis. Ph.D. thesis, Univ. of Texas, Austin.

Labuza, T. P., Tannenbaum, S. R., and Karel, M. 1970. Water content and stability of low and intermediate moisture foods. *Food Technol.* 24: 543.

LaJollo, F. and Marquez, U. M. L. 1982. Chlorophyll degradation in a spinach system at low and intermediate water activities. *J. Food Sci.* 47: 1995.

LaJollo, F., Tannenbaum, S. R., and Labuza, T. P. 1971. Reaction at limited water concentration. 2. Chlorophyll degradation. *J. Food Sci.* 36: 85.

Mackinney, S., Luktoen, A., and Greenbaum, L. 1958. Carotenoid stability in stored dehydrated carrots. *Food Technol.* 12: 164.

Mitsuda, H., Yasumoto, K., and Yamamoto, A. 1967. Inhibition of lipoxygenase by saturated monohydric alcohols through hydrophobic bondings. *Arch. Biochem. Biophys.* 118: 664.

Martinez, F. and Labuza, T. P. 1968. Rate of deterioration of freeze-dried salmon as a function of relative humidity. *J. Food Sci.* 33: 241.

Pasch, J. H. and von Elbe, J. H. 1975. Betanine degradation as influenced by water activity. *J. Food Sci.* 40: 1145.

Paulus, K. and Saguy, I. 1980. Effect of heat treatment on the quality of cooked carrots. *J. Food Sci.* 45: 239.

Penalaks, T. and Murray, T. K. 1970. Effect of processing on the content of carotene isomers in vegetables and peaches. *J. Inst. Can. Technol.* 3(4): 145.

Ramakrishnan, T. V. and Francis, F. J. 1972. Stability of carotenoids in model aqueous systems. *J. Food Quality* 2: 177.

Rhee, K. S. and Watts, B. W. 1966. Effects of antioxidants on lipoxyidase activity in model systems and pea (*Pisum sativum*) slurries. *J. Food Sci.* 31: 669.

Saguy, I. 1979. Thermostability of red beet pigments (betanine and vulgaxanthin-I): Influence of pH and temperature. *J. Food Sci.* 44: 1554.

Saguy, I., Goldman, M., Bord, A., and Cohen, E. 1984. Effect of oxygen retained on beet powder on the stability of betanine and vulgaxanthin-I. *J. Food Sci.* 49: 99.

Saguy, I., Goldman, M., and Karel, M. 1985. Prediction of beta-carotene decolorization in model system under static and dynamic conditions of reduced oxygen environment. *J. Food Sci.* 50: 526.

Saguy, I., Kopelman, J., and Mizrahi, S. 1980. Computer-aided prediction of beet pigment (betanine and vulgaxanthin-I) retention during air-drying. *J. Food Sci.* 45: 230.

Sapers, G. M. and Hornstein, J. S. 1979. Varietal differences in colorants properties and stability of red beet pigments. *J. Food Sci.* 44: 1245.

Savolainen, K. and Kuusi, T. 1978. The stability properties of golden beet and red beet pigments: Influence of pH, temperature, and some stabilizers. *Z. Lebensm. Unters. Forsch.* 166: 19.

Schandler, S. H., Chichester, C. O., and Marsh, B. L. 1962. Degradation of chlorophyll and several derivatives in acid solution. *J. Org. Chem.* 27: 3865.

Schwartz, S. J. and von Elbe, J. H. 1983a. Kinetics of chlorophyll degradation to pyropheophytin in vegetables. *J. Food Sci.* 48: 1303.

Schwartz, S. J. and von Elbe, J. H. 1983b. Identification of betanin degradation products. *Z. Lebensm. Unters. Forsch.* 176: 448.

Schwartz, S. J., Woo, S. L., and von Elbe, J. H. 1981. High performance liquid chromatography of chlorophylls and their derivatives in fresh and processed spinach. *J. Agric. Food Chem.* 29: 533.

Singer, J. W. and von Elbe, J. H. 1980. Degradation rates of vulgaxanthin-I. *J. Food Sci.* 45: 489.

Singleton, V. L., Gortner, W. A., and Young, H. Y. 1961. Carotenoid pigments of pineapple fruit. I. Acid-catalyzed isomerization of pigments. *J. Food Sci.* 26: 49.

Timberlake, C. F. and Bridle, P. 1966. Spectral studies of anthocyanin and anthocyanin equilibrium in aqueous solutions. *Nature* (London) 212: 158.

von Elbe, J. H. 1977. The betalaines. In *Current Aspects of Food Colorants*. Furia, T. E. (Ed.), p. 29. CRC Press Inc., Cleveland, OH.

von Elbe, J. H., Maing, I. Y., and Amundson, C. H. 1974. Color stability of betanine. *J. Food Sci.* 39: 334.

Walter, W. M. and Giesbrecht, F. G. 1982. Effect of lye peeling conditions on phenolic destruction, starch hydrolysis, and carotene loss in sweet potatoes. *J. Food Sci.* 47: 810.

Weckel, K. G., Santos, B., Laferriere, H. E., and Gabelman, W. H. 1962. Carotene components of frozen and processed carrots. *Food Technol.* 16(8): 91.

Weedon, B. C. L. 1971. Occurrence. In *Carotenoids.* Isler, O. (Ed.). Birkhauser, Basel.

Wiley, R. C., Lee, Y. N., Saladini, J. J., Wyss, R. C., and Topalian, H. H. 1979. Efficiency studies of a continuous diffusion apparatus for the recovery of betalaines from the red table beet. *J. Food Sci.* 44: 208.

von Sydow, E., [illegible] and Andersson, C. H. 1974. [illegible] stability [illegible]. J. Food Sci. [illegible]

Walter, W. M., [illegible] 1976. [illegible] sweet potatoes. [illegible]

Wells, [illegible]

[illegible]

[illegible]

4

Effects of Water Activity on Textural Properties of Food

Malcolm C. Bourne

New York State Agricultural Experiment Station
and Institute of Food Science
Cornell University
Geneva, New York

INTRODUCTION

Most foods have an a_w greater than 0.8 at the moment of consumption. There is considerable literature on how moisture content affects textural and viscosity properties of these high a_w foods. However, these effects have been observed and reported in the free water region where binding forces are weak, a large change in percent moisture gives only a small change in a_w, and much of the water may be mechanically trapped in the food. This paper concentrates on the effects of a_w on textural properties of food below a_w about 0.8.

The reason most foods are consumed at a high a_w level is because people like food to be in a moist and tender state for mastication. The adjectives moist, juicy, tender, and chewy describe desirable textural attributes, while hard, dry, tough, and crumbly describe undesirable textural attributes. To obtain desirable textures it is usually necessary to have a high moisture content, which means the a_w is high enough to support microbial growth. When dried to an a_w level that will not support microbial growth, the texture usually becomes too hard, dry, tough, or crumbly.

Rockland (1969) listed food characteristics as a function of their localized moisture sorption isotherms. This list included the following texture terms: (a) local isotherm one (low moisture content)—dry, hard, crisp, shrunken; (b) local isotherm two (intermediate moisture content)—dry, firm, flexible; and (c) local isotherm three (high moisture content)—moist, soft, flacid, swollen, sticky (Fig. 4.1).

A large part of food processing and preparation is directed toward converting a microbiologically stable food with undesirable texture into another form that is texturally desirable but prone to microbial spoilage because of its higher moisture content. Examples include converting wheat grains into bread and cake, or barley into beer. In the home, the cooking of rice, pasta, and beans in water softens and moistens the texture.

A large proportion of our food supply undergoes two sequential storage phases. Foods are first subjected to long-term storage in a dry, stable, inedible form. This phase is followed by short-term storage in a moist, perishable

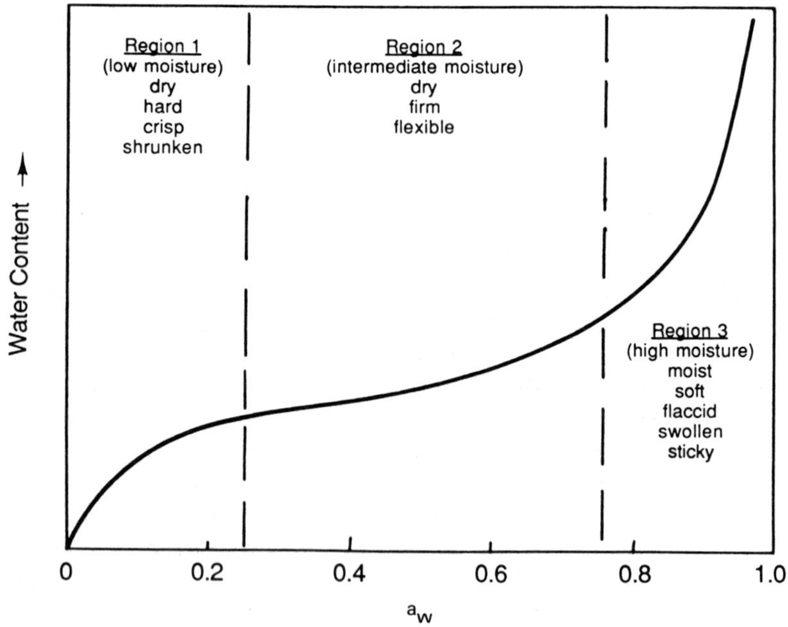

FIG. 4.1 Food textures as a function of localized water sorption isotherms. (*Plotted from data of Rockland, 1969.*)

eating condition. With few exceptions, the problem of producing a food with desirable textural properties at a moisture level that will not support microbial growth has not yet been resolved without resorting to sophisticated and expensive packaging and thermal processing technologies, or by preservatives. Exceptions include dry snack foods, such as potato chips. However, these products must absorb saliva and hydrate rapidly in the mouth to be acceptable. Foods that remain dry during mastication are not relished and are seldom eaten.

A large number of publications exist describing relationships between a_w and microbial growth. It is surprising to find how little literature exists on relationships between a_w and textural properties, even though the achievement of moist and tender textures is the driving force behind many food processing and preparation systems. This may be due in part to a limited understanding of the progress that has been made in the theory and practice of texture measurement in recent years. Useful publications on up-to-date information in the food texture field are listed in the Appendix.

A first consideration is a definition of the word "texture." Although a number of researchers have attempted to define texture, the approach here is not to use this word because it implies a one-dimensional property, and it has been clearly shown that texture is multidimensional in nature (Szczesniak, 1963). The term "textural properties" is preferable because it encompasses and describes a group of properties. Texture cannot be completely specified by a single measurement as can percent moisture or pH.

A useful definition is "the textural properties of a food are that group of physical characteristics that arise from the structural elements of the food, are sensed by the feeling of touch, are related to the deformation, disintegration and flow of the food under a force, and are measured objectively by functions of mass, time, and distance" (Bourne, 1982).

The affect of a_w on food textures is specific to the kind of food under consideration. These relationships will now be discussed, using specific foods as examples.

BEEF

Kapsalis et al. (1970b) cooked beef semimembranosus muscles to 63°C and when cooled, cut them into 12-mm slices. Some slices were frozen and freeze-desiccated in silica gel at a pressure of 0.3 mm Hg, then equilibrated to a series of higher moisture levels after which the maximum cutting-extrusion force was measured using a FTC Texture Press (Kramer Shear Press). Other slices were freeze-dried and isothermally equilibrated to several moisture levels.

Individual 38-mm diameter discs were compressed 25% two times between extensive flat surfaces in an Instron machine at a speed of 2 cm/min and mechanical properties were measured.

FTC Texture Press

The maximum force increased steadily from a_w 0 up to a peak at about a_w 0.8 and then decreased rapidly above a_w 0.8 (Fig. 4.2). Within the range a_w 0–0.85, plots of the maximum extrusion force versus moisture content were rectilinear for each muscle but there was considerable variation from muscle to muscle (Kapsalis et al., 1970b).

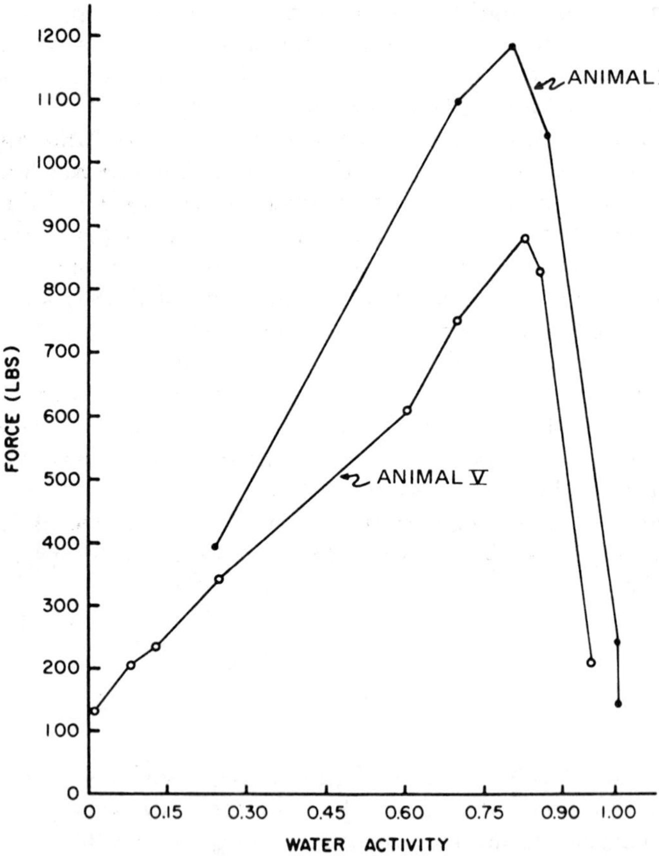

FIG. 4.2 Effect of a_w on maximum cutting-extrusion force of precooked, freeze-desiccated beef. (*From Kapsalis et al., 1970.*)

Instron Compression

The following mechanical properties were measured from the Instron curves. From the first compression:

Secant modulus is the ratio of stress to strain at 3.5% compression.

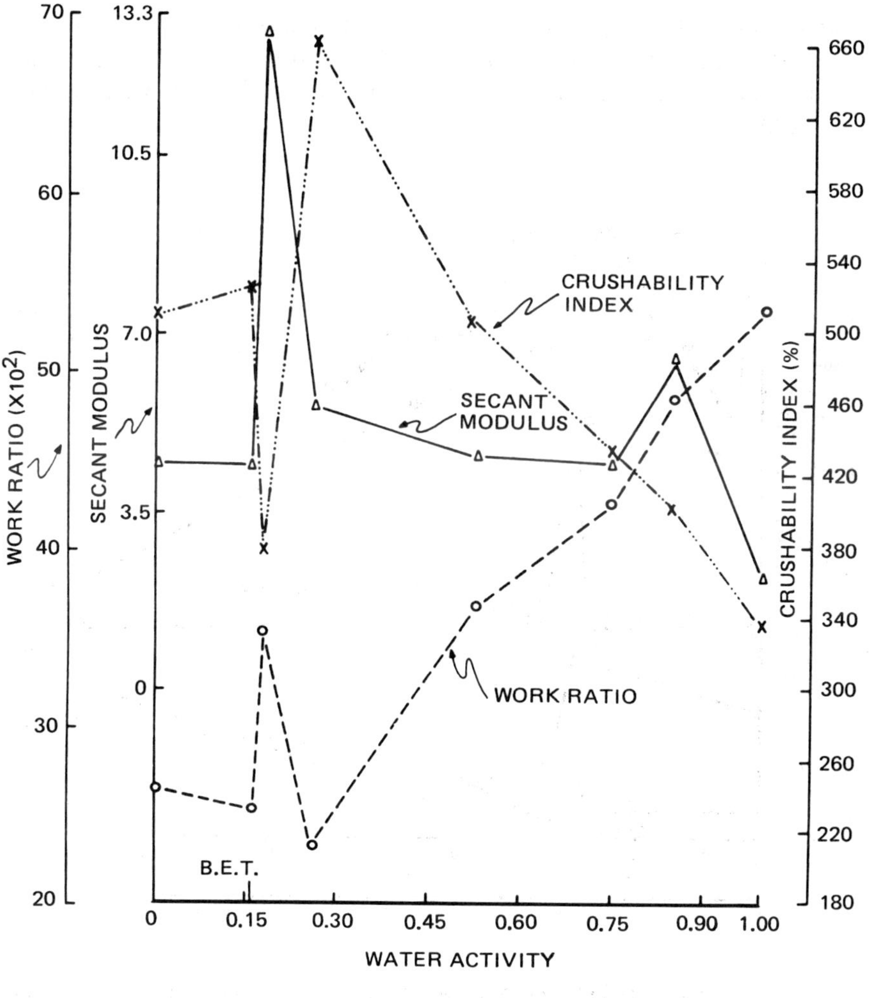

FIG. 4.3 Effect of a_w on mechanical properties of precooked, freeze-dried beef. (*From Kapsalis et al., 1970.*)

Degree of elasticity is the ratio of elastic deformation to the sum of the elastic and plastic deformations derived from the load–unload curve of the first cycle.

Toughness is work per unit volume to compress the sample 25%.

Crushability index is the ratio of nonrecoverable to recoverable work in the first compression.

One more parameter was measured using the two compression curves:

Work ratio is the ratio of the area under the load curve of the second cycle to the area under the first load curve of the first cycle. (This is similar to the cohesiveness parameter of the GF texture profile.) (Kapsalis et al., 1970b)

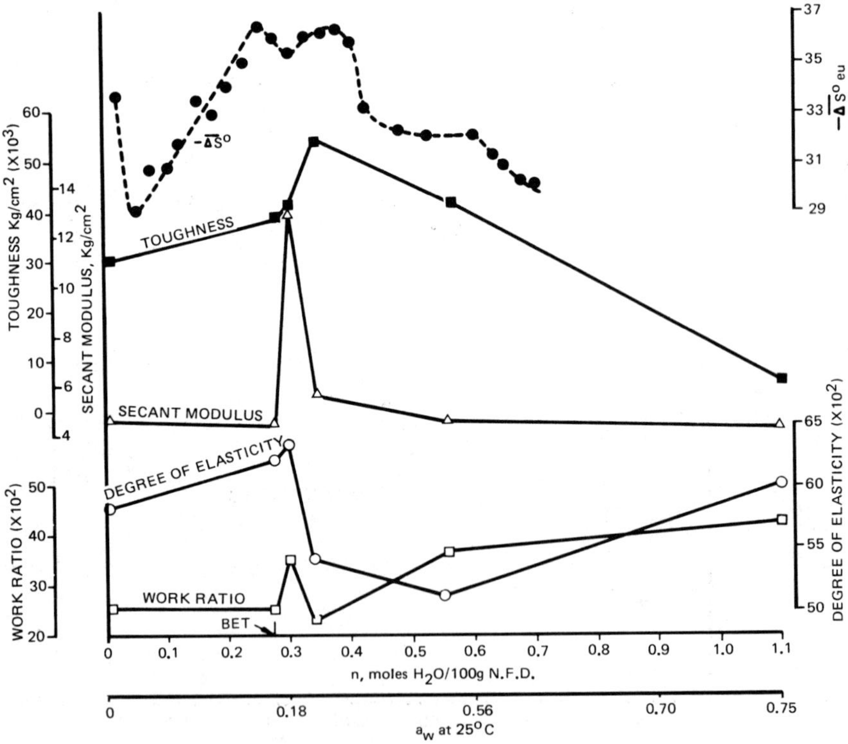

FIG. 4.4 Relationships between textural properties, as revealed by compression testing, and standard differential entropy of water vapor sorption in precooked, freeze-dried beef. NFD = Non Fat Dry solids. (*From Kapsalis et al., 1970.*)

Figure 4.3 shows changes in the work ratio, secant modulus, and crushability index, and Fig. 4.4 in the toughness, degree of elasticity, work ratio, and secant modulus over a wide a_w range. The changes in each of these properties are striking, especially near the Brunauer, Emett, and Teller (BET) monolayer level.

Kapsalis et al. (1970b) pointed out that the sorption of water vapor might cause one of two opposing trends in the standard differential entropy change $-\overline{\Delta S^\circ}$: (a) increased order due to crystallization or similar effects as $-\overline{\Delta S^\circ}$ increases; or (b) decreased order due to increased segmental chain move-

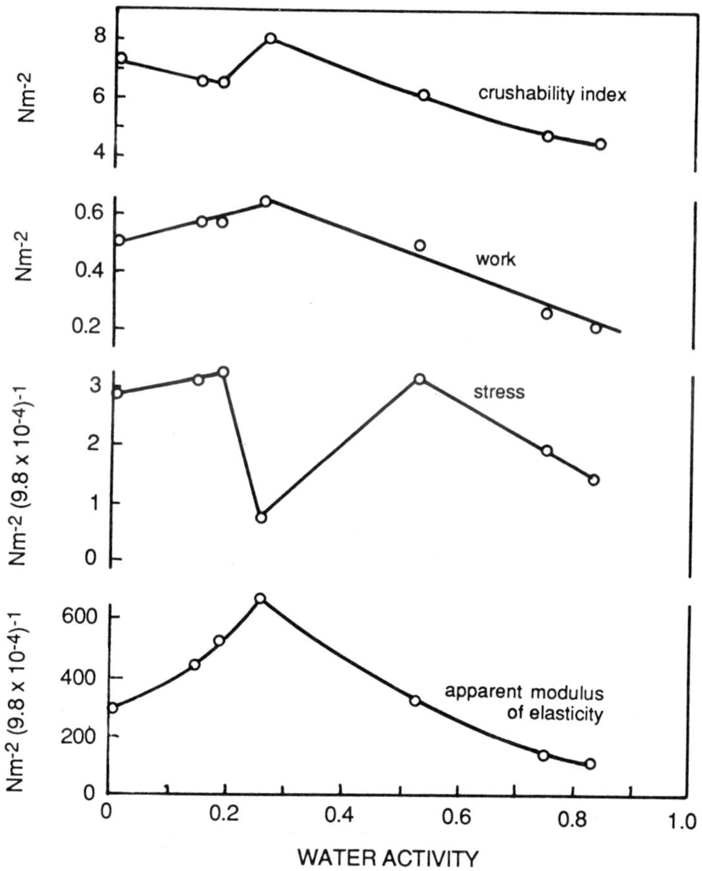

FIG. 4.5 Effect of water activity on mechanical properties of precooked freeze-dried beef. Means of data from four animals. (*Plotted from data of Kapsalis, 1975.*)

ments of the polymer network, unfolding of polypeptide chains, local solubilization or similar effects as $-\overline{\Delta S^\circ}$ decreases. It can be seen from the uppermost curve in Fig. 4.4 that $-\overline{\Delta S^\circ}$ alternately decreases and increases as the a_w increases. This probably accounts for the complex changes in textural properties as the moisture content of the freeze-dried beef increases.

Kapsalis (1975) statistically analyzed data from earlier studies (Kapsalis et al., 1970b) and tabulated values for apparent modulus of elasticity, stress, work, crushability index for each of the four animals used in the experiment. The combined means for the four animals are plotted in Fig. 4.5. Here again, sharp transitions in textural properties are displayed near the BET monolayer.

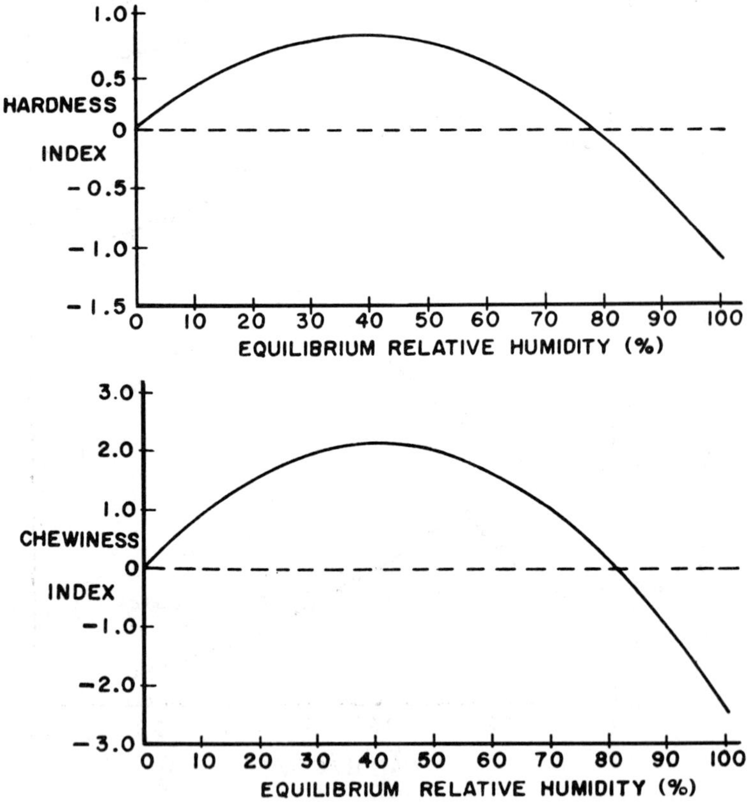

FIG. 4.6 Statistical fit of data from modified texture profile analysis of freeze-dried beef. (*From Reidy and Heldman, 1972.*)

Reidy and Heldman (1972) measured hardness and chewiness of cubes of freeze-dried beef that had been rapidly equilibrated at 0, 20, 40, 60, 80, and 100% relative humidity. They used a modified texture profile analysis test, compressing the beef cubes 35–40% in an Instron at a speed of 2 cm/min. They observed some variation resulting from the plate temperature in the freeze drier and the conditioning temperature. Nevertheless, a statistical fit of all the data showed that both hardness and chewiness increased from a_w 0 up to 0.4 and decreased from a_w 0.4 to a_w 1.0 (Fig. 4.6).

In a subsequent study on freeze-dried beef, Heldman et al. (1973) reported that after six months' storage both hardness and chewiness were directly related to the a_w of the beef (Fig. 4.7).

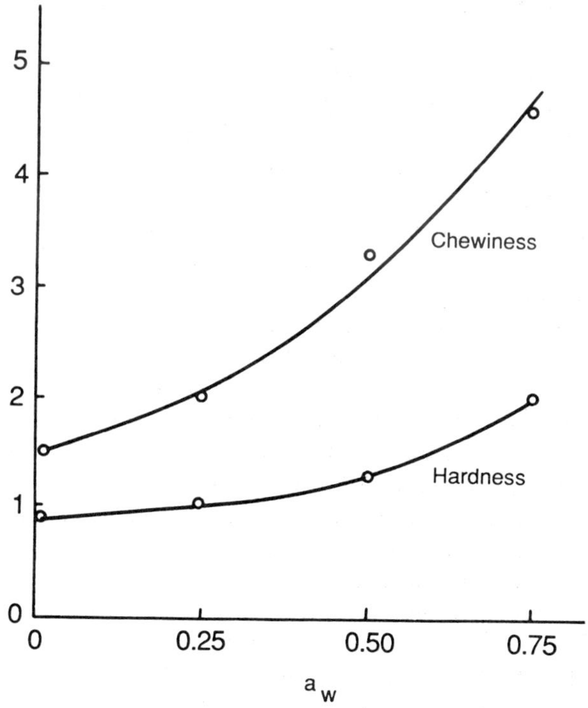

FIG. 4.7 Effect of a_w on hardness and chewiness of freeze-dried beef after 6 months storage at 38°C. (*Plotted from data of Heldman et al., 1973.*)

APPLES

Bourne (1986) measured Texture Profile Analysis (TPA) parameters of fresh apple, and apple equilibrated to a_w levels from 0.85 to 0.01 using a procedure developed by Friedman et al. (1963) and Szczesniak et al. (1963) and adapted to the Instron by Bourne (1968). Cubes 10 mm on each side were cut from apple chunks after equilibration and subjected to 90% compression two times in an Instron testing machine at a speed of 5 cm/min.

Traces of typical force-time curves obtained during two compressions of the apple cubes are shown in Fig. 4.8. (The first and second compressions have been moved closer together on the time axis to conserve space.) The fresh apple with a_w approximately 0.99 is fairly rigid as shown by the initial steep slope of the force curve, and it shows a definite fracturability peak. The area under the second compression curve is much less than that under the first compression curve, which indicates low cohesiveness. In contrast, the apple at a_w 0.85 shows no fracturability but has high deformability as shown by the very low initial slope of the force curve.

The TPA curves for apple at a_w 0.75, 0.65, 0.55, 0.44, 0.33, and 0.23 are qualitatively similar, all showing high deformability, no fracturability, high cohesiveness, and high springiness. At a_w 0.12, the texture profile is similar to those at the higher a_w levels except that cohesiveness is lower and hardness is extraordinarily high. There is an abrupt change in texture profile between a_w 0.12 and a_w 0.01. At a_w 0.01 the fracturability peak reappears and the TPA curve qualitatively resembles that of fresh apple at a_w 0.99 but with higher fracturability and much higher hardness.

Figure 4.9 shows changes in texture parameters of note as a function of a_w. The texture curve can be divided into three stages: In the first stage [a_w 0.99 (fresh apple) to a_w 0.85], moisture reduction causes a complete loss of turgor pressure characteristic of fresh apple, resulting in total loss of fracturability, a great increase in deformability, and a moderate increase in cohesiveness. In the second stage (a_w 0.85 to a_w 0.23), apple undergoes a progressive increase in hardness, springiness, gumminess, and chewiness, and a decrease in deformation. These changes are probably caused by decreases in the number of multimolecular layers of water associated with the apple solids. The third stage (a_w 0.23 to a_w 0.01) is characterized by sharp increases in hardness and gumminess from a_w 0.23 to 0.12 followed by similarly sharp decreases between a_w 0.12 and 0.01. Chewiness falls sharply between a_w 0.12 and 0.01. Springiness and cohesiveness decrease sharply between a_w 0.23 and 0.12. Fracturability is 0 at a_w 0.23 and 0.12, and becomes large at a_w 0.01.

The BET monolayer value was calculated to be a_w 0.14. Stage 3 covers the a_w range from a little above to below the BET monolayer. Figure 4.9 illustrates

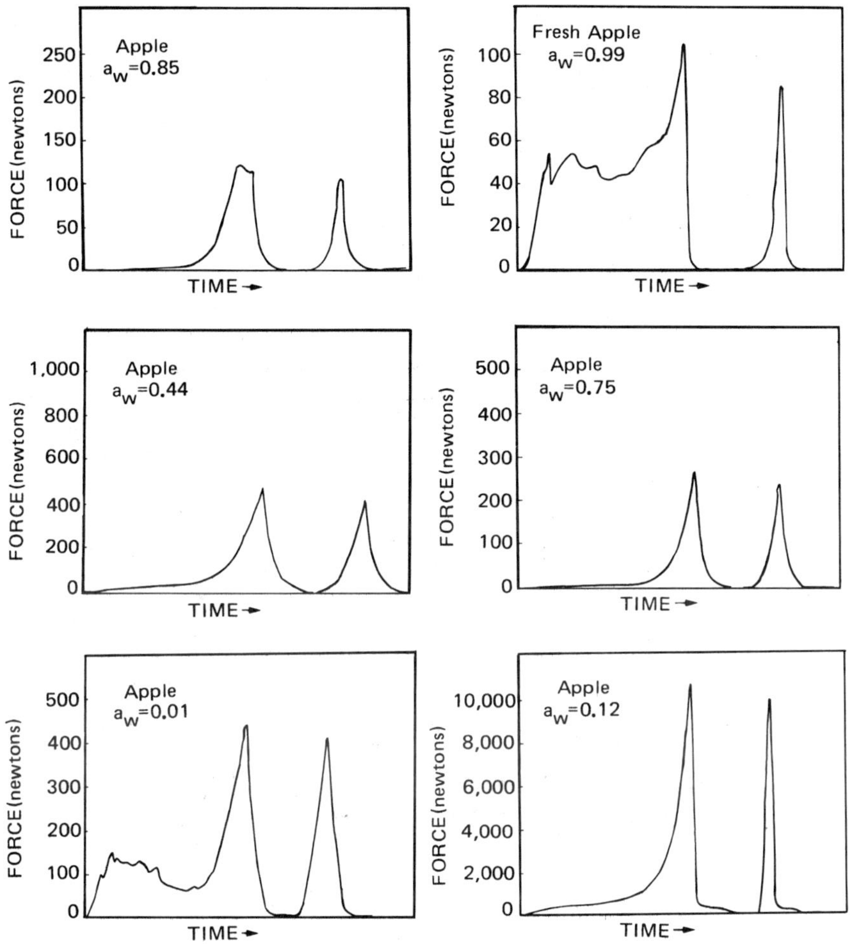

FIG. 4.8 Typical force time curves traced from Instron charts of texture profile analyses (two 90% compressions) on apple tissue dried to various water activity levels. Note the differences in the force scales used at different a_w. (*From Bourne, 1986.*)

that major textural changes occur in apple flesh as it undergoes the transition from multilayer to monolayer to less than a monolayer of adsorbed water.

At the lowest a_w (0.01) the textural parameters resemble that of fresh apple but with much higher hardness and fracturability. At all intermediate levels of a_w the texture is tough, leathery, deformable, and chewy with no fracturability and crispness.

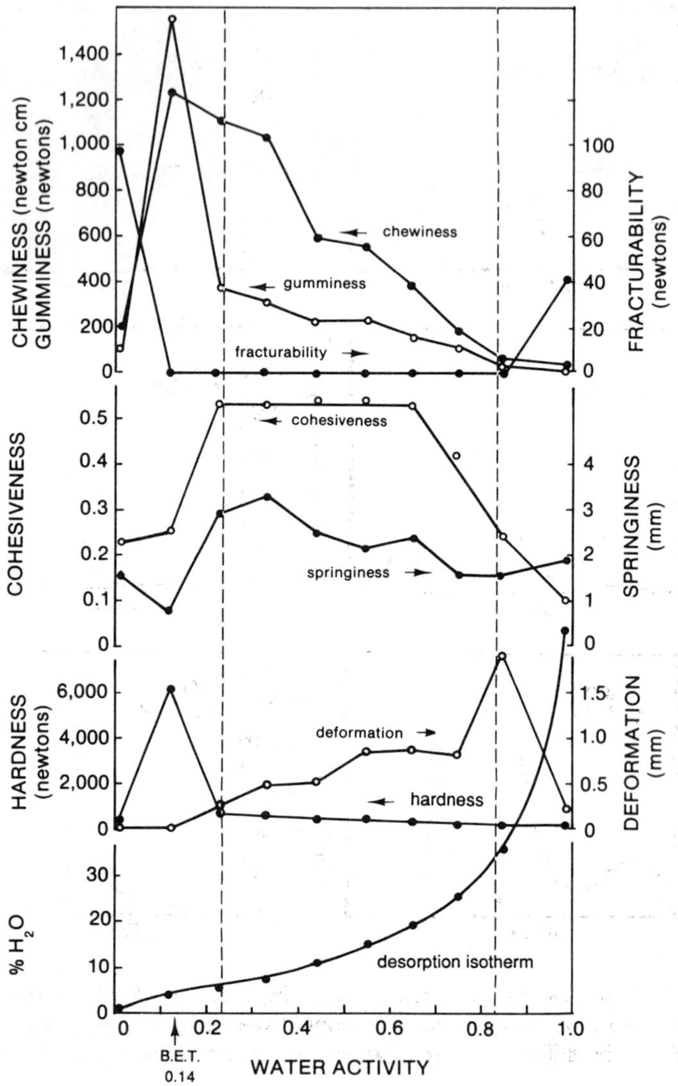

FIG. 4.9 Moisture desorption isotherm and texture profile parameters of apple flesh equilibrated to various a_w levels. The BET monolayer is at a_w 0.14. (*Plotted from data of Bourne, 1986.*)

CEREAL GRAINS

Chattopadhyay et al. (1979) fashioned single rice grains of the cultivar IR8 into cylinders, equilibrated them to various moisture levels, and compressed them to failure in an Instron machine. Three compression speeds (0.254, 0.0508, and 0.0127 cm/min) and five test temperatures (25, 36, 47, 58, and 69°C) were used. The data obtained at 36°C and 0.254 cm/min compression speed are plotted in Fig. 4.10. These researchers expressed the moisture content on a percent dry basis. For illustrative purposes, values for percent moisture were converted to a_w using the water sorption isotherm published by Hogan and Karon (1955). There is a fivefold decrease in failure stress as the a_w

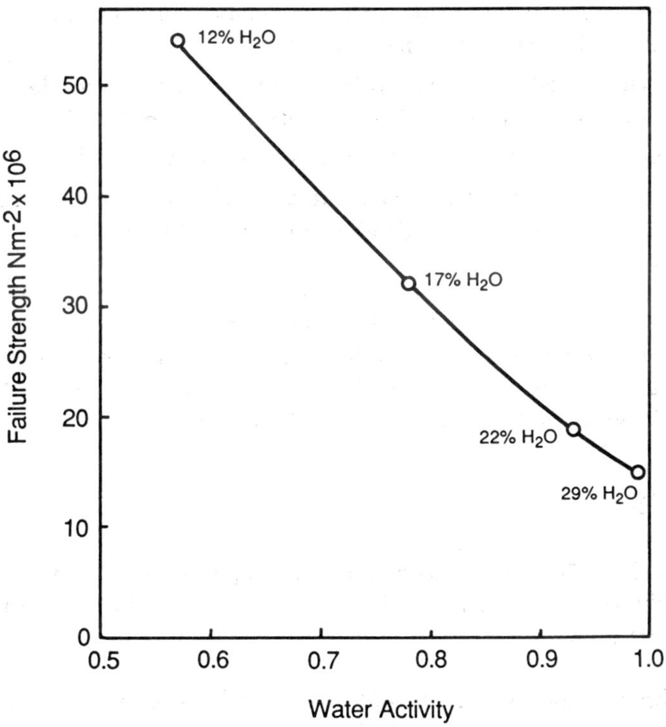

FIG. 4.10 Failure strength of dehulled rice grains compressed at 0.254 cm/min in an Instron at 36°C. (*Plotted from data of Chattopadhyay et al., 1979.*) [Conversion from percent moisture (db) to a_w is based on data of Hogan and Karon (1955).]

increases from about 0.57 to about 0.93. Similar trends were observed at the other temperatures and compression speeds.

Obuchowski and Bushuk (1980) used six different techniques to measure hardness of wheat grains at 9.5, 11.0, 12.5, 14.0, and 15.5% moisture. Three kinds of wheat (durum, hard red spring, and soft white) were studied. According to the water sorption isotherm published by Hubbard et al. (1957) this range of moisture content is equivalent to an a_w range 0.44–0.75. The parameters studied to characterize grain kernel hardness were:

A. Wheat hardness index, which is the maximum torque produced by grinding in a two-step Brabender Hardness Tester divided by the percent yield of flour.
B. Wheat hardness index measured in a one-step Brabender Hardness Tester.
C. Time in seconds to grind a standard sample of wheat into flour in a two-step Brabender Hardness Tester.
D. Pearling resistance index, which is the weight in grams after pearling 20g of wheat grains for 20 sec in a Strong Scott Barley Pearler.
E. Torque needed to grind grain in a two-step Brabender Hardness Tester.
F. Particle size index, which is the percent of flour passing through a 177 micrometer sieve after grinding in a two-step Brabender Hardness Tester.

The results are shown in Fig. 4.11. Three of the indices (A, B, and E) decrease with increasing a_w; two indices (C and D) increase. One index (F, particle size) is relatively unchanged, indicating that this a_w range has little effect on the locus of the failure planes in the wheat kernels.

WALNUTS

Veeraju et al. (1978) studied some physical properties of shelled walnut kernels equilibrated over a range of 10–85% relative humidity. Three test procedures were used:

1. Vibration test: Lots (250g) of kernels were placed in 1-L tinplate lever-lid cans, which were attached to a table vibrating at 7 Hz and 5 mm amplitude for 15 min. The kernels were then emptied onto a tray and the chaff was carefully brushed off, collected, and weighed. The percent chaff, which is composed of pellicle and abraded kernel meat, was calculated.

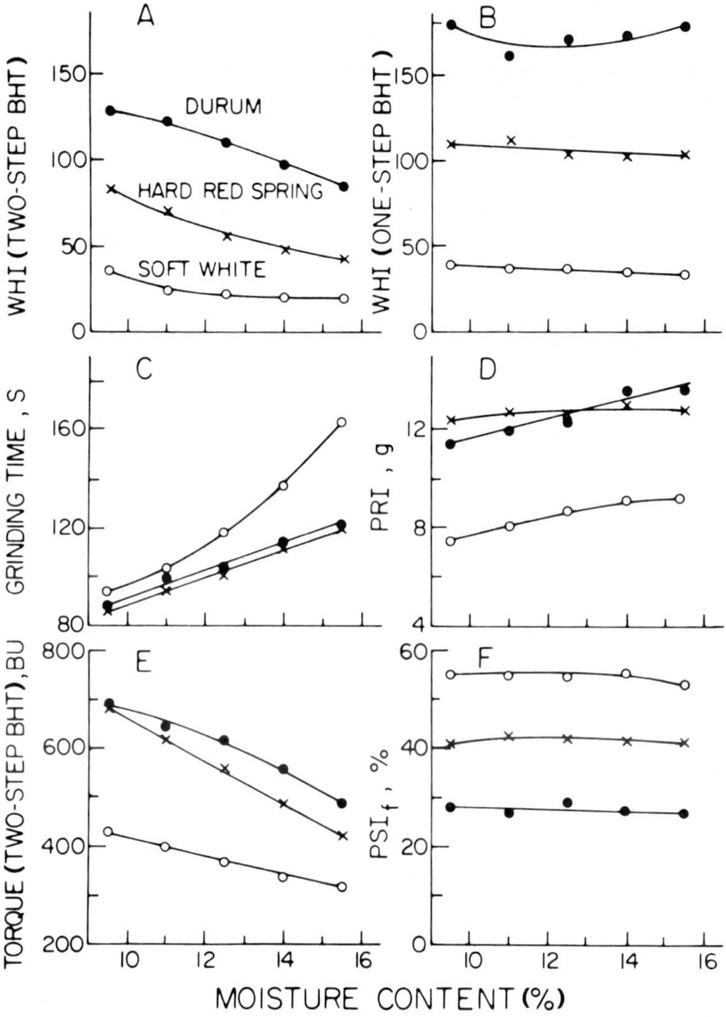

FIG. 4.11 Hardness indices for three classes of wheat as a function of moisture content equivalent to a_w 0.44–0.75. Details of the six tests are given in the text. (*From Obuchowski and Bushuk, 1980.*)

2. The pellicle content in the chaff was obtained by extracting lipid from the chaff with petroleum ether and weighing the dry residue. The amount of abraded meat (calculated on the basis of the extracted lipid) was subtracted from the amount of chaff to give the amount of pellicle rubbed off.
3. Drop test: Cans of unbroken kernels were dropped a distance of 915 mm two times onto a concrete floor. The broken kernels were separated and weighed to give a percent of total walnut meats.

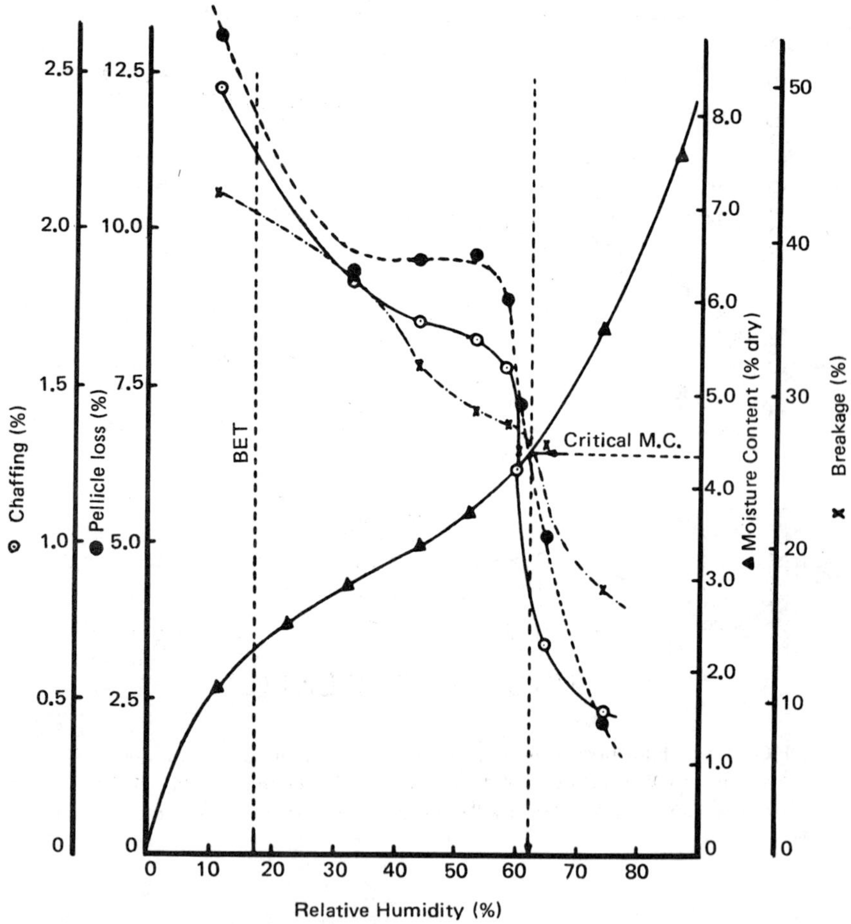

FIG. 4.12 Effect of moisture content of walnut kernels on pellicle loss, chaffing, and kernel breakage. (*From Veerraju et al., 1978.*)

The results of these tests are shown in Fig. 4.12. The percent breakage decreases continuously from 10 to 75% relative humidity. Chaffing and pellicle loss decreased sharply from 10 to 35% relative humidity, slowly from 35 to 60% relative humidity, and very sharply above 60% relative humidity.

SUGARS

When molten sugars are rapidly cooled or solutions of sugars are spray-dried or freeze-dried they usually solidify as a glassy amorphous solid, which is in a metastable state. Under suitable conditions, the amorphous sugars change into a thermodynamically more stable crystalline state. Fine powders of glassy sugars are less sticky and less prone to cake than are crystalline powders of equivalent size.

Makower and Dye (1956) exposed glassy powders of sucrose and glucose to atmospheres of various relative humidities at 25°C and measured their moisture contents over a period of 2–3 years. For sucrose, they observed that at relative humidities of 4.6, 8.6, and 11.8%, the equilibrium moisture content was directly proportional to the relative humidity of the atmosphere, and after equilibrium was reached (about 20 days), there was no further change in moisture content during more than 800 days storage. These powders showed practically no evidence of crystallinity and maintained their free-flowing property during the entire storage period.

At higher relative humidities during storage, Makower and Dye (1956) observed that glassy sugar powders gained moisture up to a maximum, and then lost moisture at a rate that increased with increasing relative humidity, eventually reaching an almost anhydrous condition with a moisture content of about 0.1% (Fig. 4.13). The decrease in moisture content occurs as the amorphous sucrose is transformed into the crystalline form. The crystalline form caked badly even though it has a much lower moisture content than the amorphous form.

Makower and Dye (1956) also showed similar changes of moisture content with storage time for amorphous glucose powder which crystallized, dried, and became sticky when stored at a relative humidity of about 8.6% but remained free flowing and in the glassy state when stored at 4.6% relative humidity.

The work of Makower and Dye was confirmed by Iglesias et al. (1975). In a subsequent report, Iglesias and Chirife (1978) showed that the addition of gums and other hydrophilic polymers to sucrose delayed crystallization and loss of moisture from the sugar, but had a negligible effect on the moisture content in the sugar after crystallization was complete.

The above evidence shows that, in the case of sugars, changes in a_w may either inhibit or promote a physical change in the nature of the sugar that, in

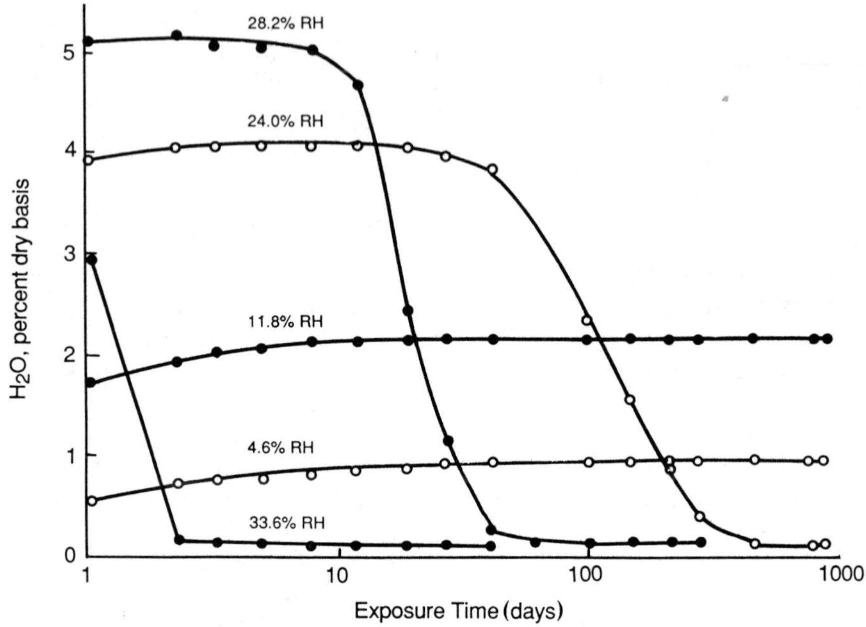

FIG. 4.13 Changes in moisture content of amorphous sucrose powder stored in atmospheres of various relative humidities (RH) at 25°C. (*Plotted from data of Makower and Dye, 1956.*)

turn, affects the texture. A low a_w maintains the sugar in the form of a glassy, amorphous, free-flowing powder at a moisture level of several percent, while a medium a_w level promotes crystallization into a sticky caking powder that has a much lower moisture content than the amorphous sugar.

Warburton and Pixton (1978) reported that lactose in the form of an amorphous glass was stable and gave a conventional water sorption isotherm when stored at relative humidities below 42%. Between 42 and 52% relative humidity, this amorphous glass crystallized and released water, giving a sharp break in the water sorption isotherm.

POWDERS

Moreyra and Peleg (1980) studied the compression and decompression behavior of fine food powders. They observed that the cohesiveness of finely powdered foods increased with increasing moisture content. Food powders with a higher moisture content had a lower bulk density, higher compressi-

TABLE 4.1 Some Physical Characteristics of Food Powders

Material[a]	Moisture (%)	Bulk density (g/cc)	Compressibility	Relaxation constant
baby formula	dry	0.49	6.4	2.6
baby formula	1.0	0.46	8.1	1.5
bran	2.8	0.55	5.0	2.0
bran	5.4	0.42	6.7	1.4
onion powder	0.7	0.66	8.6	2.0
onion powder	1.2	0.55	9.5	1.4
onion powder	2.3	0.50	8.1	1.0
sucrose	dry	0.77	4.5	4.0
sucrose	0.17	0.71	6.0	2.5

[a]All material had a particle size −100 + 200 mesh.
Source: Moreyra and Peleg (1980).

bility, and lower relaxation constant than lower moisture content powders (Table 4.1).

When a cohesive powder is poured into a container it forms an open structure supported by interparticle forces and contains many voids. In contrast, a noncohesive powder contains smaller voids created by random orientation of the particles. The open bed structure of cohesive powders has a low bulk density, is mechanically weak, and collapses under small loads, resulting in high compressibility. After compression, high-moisture powder has a higher tensile strength than low-moisture powder, reflecting greater interparticle attractive forces and a greater number and area of contact points between the higher-moisture particles (Peleg, 1985).

Peleg and Mannheim (1969) reported that the rate of caking of onion powder with a moisture content of 4–5% increased rapidly with increasing temperature (Table 4.2). This moisture level is equivalent to a_w 0.3–0.35 at

TABLE 4.2 Caking of Onion Powder[a]

Storage temperature (°C)	Caking tendency
15	No change in 6 months
25	Agglomerates in 7–10 days
30	Caking in 7–10 days
35	Caking in 3 days

[a]Moisture content 4–5%.
Source: Peleg and Mannheim (1969).

20°C. In view of the well known fact that a_w increases with rising temperature at constant moisture content, it seems probable that the degree of caking is directly related to a_w. These researchers noted that no anticaking agents were effective in powder at 7% moisture, while at 3% moisture the powder remained free flowing without anticaking agents, even after 30 days' storage at 35°C.

OTHER FOODS

Kapsalis et al. (1970a) reported on three textural parameters of space foods that were given a single 80% compression in a modified texture profile analysis test. The space foods used were bite-sized beef sandwich, chicken sandwich, cheese sandwich, and chicken bites. Hardness and cohesiveness increased and the crushability index decreased as the a_w increased (Fig. 4.14). At low a_w, space foods were excessively crumbly, as indicated by a high crushability index and low cohesiveness. It was concluded that textural properties of carbohydrate materials seem to be affected to a greater extent than protein materials by changes in a_w.

Katz and Labuza (1981) studied the crispness of several starch-based snack foods by sensory and instrumental tests. They reported that sensory crispness decreased with increasing a_w (Fig. 4.15) and most products became unacceptable in the a_w 0.35–0.50 range.

Zabik et al. (1979) equilibrated freshly baked sugar-snap cookies made from two different flours for five days in atmospheres with relative humidities ranging from 11 to 93%, and then measured their texture. The effect of a_w on the breaking strength (measured by a single blade mounted in the Allo-Kramer Shear Press) is shown in Fig. 4.16. The breaking strength of cookies made from low quality flour was always less than for cookies made from high quality flour. Both types of cookies showed a sharp decrease in breaking strength with increasing relative humidity in the atmosphere. The correlation coefficient between breaking strength and relative humidity was $r = -0.94$ for high quality cookie flour and $r = -0.89$ for low quality cookie flour.

The tenderness of these cookies, measured with a standard Allo-Kramer ten-blade cell, decreased in a similar manner to the breaking strength, i.e., a sharp decrease between 11 and 33% relative humidity followed by a slower rate of decrease between 33 and 93% relative humidity (Zabik et al., 1979).

Additional tests were performed on a second lot of sugar-snap cookies stored in relative humidity atmospheres ranging from 52 to 79% using an Instron to measure breaking strength and compressibility. All texture parameters showed a decreasing trend with increasing relative humidity of storage (Zabik et al., 1979).

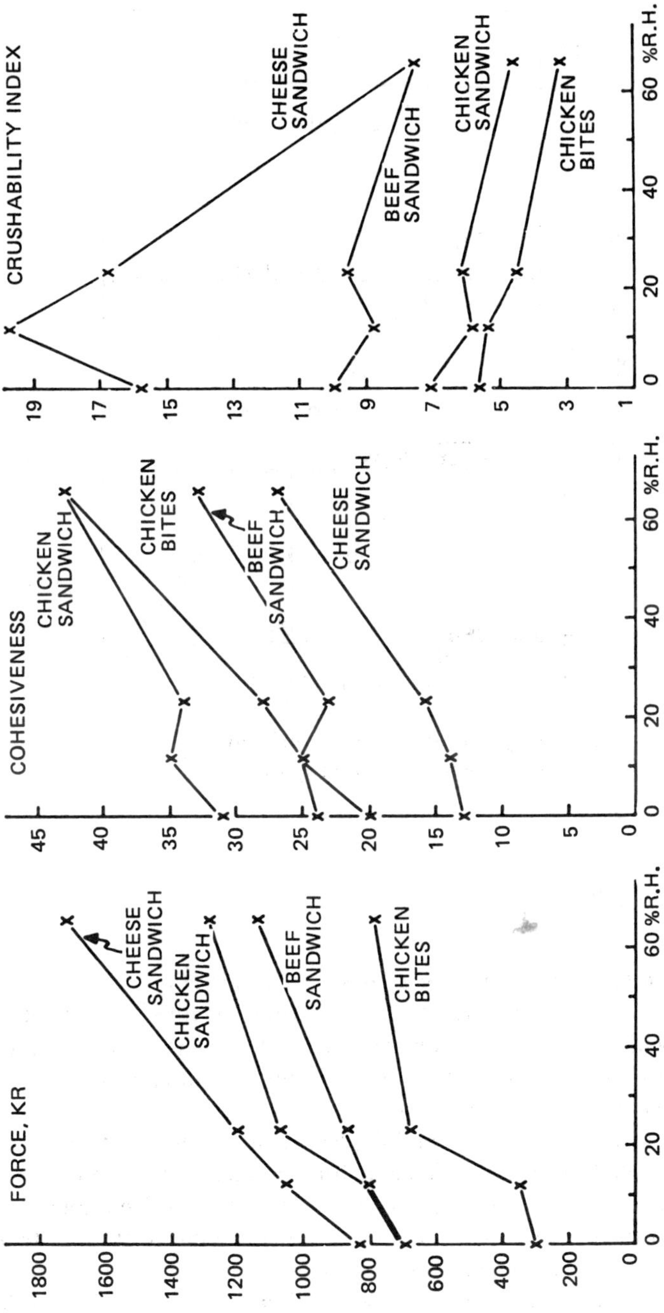

FIG. 4.14 Hardness, cohesiveness, and crushability index vs. relative humidity for dehydrated space foods. (*From Kapsalis et al., 1970a.*)

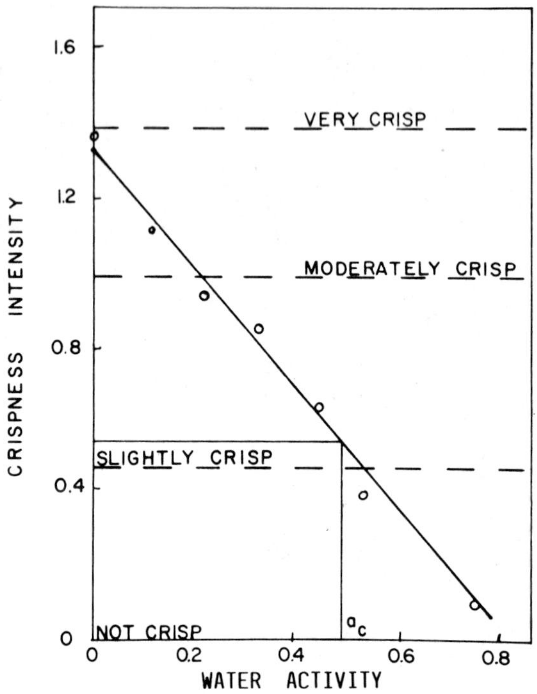

FIG. 4.15 Sensory crispness intensity of popcorn vs. a_w. *(From Katz and Labuza, 1981.)*

CONCLUSIONS

Evidence shows conclusively that a_w has a major effect on textural properties of foods. However, the amount of data describing texture is relatively sparse compared with, for example, data relating a_w to water content or microbial growth. Much more research is needed to build a comprehensive data base on water activity–texture relationships in foods.

In some instances (e.g., beef and apples) rapid changes in textural properties near the BET monolayer demand further study with more data points clustered around the a_w level corresponding to the BET monolayer. Sharp changes near the BET monolayer offer a tantalizing glimpse of the potential for the use of textural parameters to develop sensitive methods to study the effects of water at the molecular level, including chemical binding, energy factors, and other physicochemical aspects of the nature of water in foods.

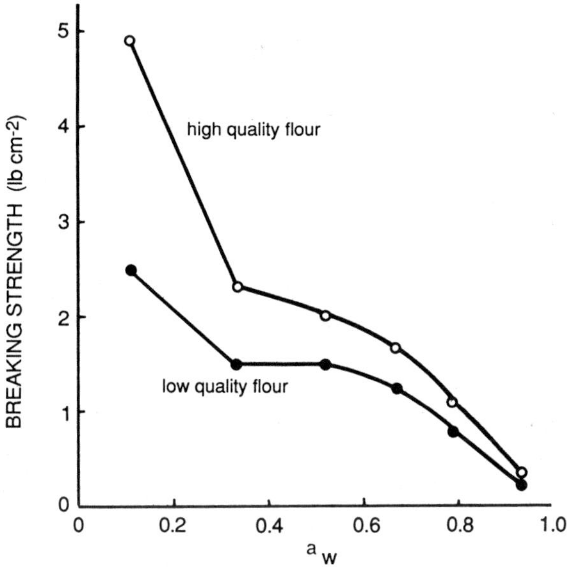

FIG. 4.16 Breaking strength of sugar-snap cookies made from a high quality cookie flour (soft red winter wheat) and a low quality cookie flour (hard red winter wheat). (*Plotted from data of Zabik et al., 1979.*)

Presently there are insufficient data to predict what the textural properties of a given type of food will be at a given a_w. Various physicochemical changes can be invoked to account for the changes after the data are collected, but there are no sound theories to predict in advance the textural properties of a food at a given a_w. Such theories do not exist simply because there is not enough known about relationships between a_w and texture. Although the data base shows interesting trends, it is inadequate to construct testable theories.

REFERENCES

Bourne, M. C. 1968. Texture profile of ripening pears. *J. Food Sci.* 33: 223.

Bourne, M. C. 1982. *Food Texture and Viscosity. Concept and Measurement.* Academic Press, New York.

Bourne, M. C. 1986. Effect of water activity on texture profile parameters of apple flesh. *J. Texture Studies* 17:331.

Chattopadhyay, P. K., Hammerle, J. R., and Hamann, D. D. 1979. Time, temperature, and moisture effects on the failure strength of rice. *Cereal Foods World* 24: 514.

Connell, J. J. 1957. Some aspects of the texture of dehydrated fish. *J. Sci. Food Agric.* xx: 520.

Friedman, H. H., Whitney, J. E., and Szczesniak, A. S. 1963. The Texturometer—a new instrument for objective texture measurement. *J. Food Sci.* 28: 390.

Heldman, D. R., Reidy, G. A., and Palnitkar, M. P. 1973. Texture stability during storage of freeze-dried beef at low and intermediate moisture contents. *J. Food Sci.* 38: 282.

Hogan, J. T. and Karon, M. L. 1955. Hygroscopic equilibria of rough rice at elevated temperatures. *J. Agric. Food Chem.* 3: 385 (cited by Iglesias and Chirife, 1982).

Hubbard, J. E., Earle, F. R., and Senti, F. R. 1957. Moisture relations in wheat and corn. *Cereal Chem.* 34: 422 (cited by Iglesias and Chirife, 1982).

Iglesias, H. A. and Chirife, J. 1978. Delayed crystallization of amorphous sucrose in humidified-freeze dried model systems. *J. Food Technol.* 13: 137.

Iglesias, H. A. and Chirife, J. 1982. *Handbook of Food Isotherms: Water Sorption Parameters for Food and Food Components.* Academic Press, New York.

Iglesias, H. A., Chirife, J., and Lombardi, J. L. 1975. Comparison of water vapour sorption by sugar beet root components. *J. Food Technol.* 10: 385.

Kapsalis, J. G. 1975. The influence of water on textural parameters in foods at intermediate moisture levels. In *Water Relations in Foods.* Duckworth, R. B. (Ed.), p. 627. Academic Press, New York.

Kapsalis, J. G., Drake, B., and Johansson, B. 1970a. Textural properties of dehydrated foods. Relationships with the thermodynamics of water vapor sorption. *J. Texture Studies* 1: 285.

Kapsalis, J. G., Walker, J. E. Jr., and Wolf, M. 1970b. A physicochemical study of the mechanical properties of low and intermediate moisture foods. *J. Texture Studies* 1: 464.

Katz, E. E. and Labuza, T. P. 1981. Effect of water activity on the sensory crispness and mechanical deformation of snack food products. *J. Food Sci.* 46: 403.

Makower, B. and Dye, W. B. 1956. Equilibrium moisture content and crystallization of amorphous sucrose and glucose. *J. Agric. Food Chem.* 4: 72.

Moreyra, R. and Peleg, M. 1980. Compressive deformation patterns of selected food powders. *J. Food Sci.* 45: 864.

Obuchowski, W. and Bushuk, N. 1980. Wheat hardness: Comparison of methods of its evaluation. *Cereal Chem.* 57: 421.

Peleg, M. 1985. The role of water in the rheology of hygroscopic food powders. In *Properties of Water in Foods*. Simatos, D. and Multon, J. L. (Ed.). Martinus Nijhoff Publishers, Dordrecht.

Reidy, G. A. and Heldman, D. R. 1972. Measurement of texture parameters of freeze dried beef. *J. Texture Studies* 3: 213.

Rockland, L. B. 1969. Water activity and storage stability. *Food Technol.* 23(10): 11.

Szczesniak, A. S. 1963. Classification of textural characteristics. *J. Food Sci.* 28: 385.

Szczesniak, A. S., Brandt, M. A., and Friedman, H. H. 1963. Development of standard rating scales for mechanical parameters of texture and correlation between the objective and sensory methods of texture evaluation. *J. Food Sci.* 28: 397.

Veerraju, P., Hemuvathy, J., and Prabbhakar, J. V. 1978. Influence of water activity on pellicle chaffing, color and breakage of walnut (*Juglans regia*) kernels. *J. Food Proc. Preserv.* 2: 21.

Warburton, S. and Pixton, S. W. 1978. The moisture relations of spray dried skimmed milk. *J. Stored Product Res.* 14: 143.

Zabik, M. E., Fierke, S. G., and Bristol, D. K. 1979. Humidity effects on textural characteristics of sugar-snap cookies. *Cereal Chem.* 56: 29.

APPENDIX

Useful Reference Literature on Food Texture

Bourne, M. C. 1982. *Food Texture and Viscosity. Concept and Measurement.* Academic Press, New York.

deMan, J. M., Voisey, P. W., Rasper, V. F., and Stanley, D. W. 1976. *Rheology and Texture in Food Quality*. Avi Publishing Co., Westport, CT.

Food Microstructure published twice yearly by Scanning Electron Microscopy Inc., P.O. Box 66507, AMF O'Hare, IL 60666.

Holcomb, D. N. and Kalab, M. 1981. *Studies of Food Microstructure.* Scanning Electron Microscopy Inc., P.O. Box 66507, AMF O'Hare, IL 60666.

Journal of Texture Studies published quarterly by Food & Nutrition Press, P.O. Box 71, Westport, CT 06881.

Kramer, A. and Szczesniak, A. S. (Ed.). 1973. *Texture Measurements of Foods.* Reidel Pub. Co., 306 Dartmouth St., Boston, MA 02116.

Rha, C. H. (Ed.). 1974. *Theory, Determination and Control of Physical Properties of Food Materials.* Reidel Publishing Co., Dordrecht, Holland.

Sherman, P. (Ed.). 1979. *Food Texture and Rheology*. Academic Press, New York.

5

Adaptation and Growth of Microorganisms in Environments with Reduced Water Activity

John A. Troller

Procter & Gamble Company
Winton Hill Technical Center
Cincinnati, Ohio

INTRODUCTION

Perhaps the most widely sought parameter with regard to the effects of a_w on microorganisms is that relating to the minimal level for growth. Theoretically, this point defines the limit below which a microorganism or group of microorganisms can no longer reproduce. If the minimal a_w level is obtained in a food, the product is then refractory to growth by the particular organism involved, although others may subsist with little difficulty under these circumstances and cause spoilage or compromise the safety of the product. The specificity and uniqueness of minimal levels can be illustrated by the example of *Staphylococcus aureus*. This foodborne pathogen has a published minimal a_w for growth of from 0.83 to 0.86. In fact, the "casual" investigator would be hard pressed to obtain growth of most strains much below 0.89 a_w and, if determined in many foods, the minimal a_w for growth of *S. aureus* might be as high as 0.90–0.91. Other examples exist, but the point to be made is that in most foods it is more important to know the factors (including a_w) which impinge on microbial growth rather than placing too much confidence in a specific minimal number. Organisms in foods rarely encounter environments in which the a_w is poised at or below the minimal level for growth and more frequently are faced with the need to grow and, even, prosper at levels intermediate between minimal and optimal. Growth and the mechanisms evolved to achieve this growth at low a_w will be the subject of this text.

GROWTH

Microorganisms elicit a wide range of responses to reductions in the a_w of specific environments in which they exist (Troller and Christian, 1978). These responses are modulated by the physical and chemical nature of the environment, the species of organism involved and a host of other factors that determine the extent of growth that occurs under a given set of circumstances. Space limitations do not permit delving into these factors in any detail; however, it should be acknowledged that discussions of growth refer to systems in which all other growth factors, such as oxidation-reduction potential, pH, temperature, etc., have been optimized, when, in fact, it is extremely difficult to isolate responses to a_w alone—especially in foods. The combination of a_w and other inhibitory factors will be discussed in the following text.

Some generalizations about growth of microorganisms can be made, namely, that bacteria usually are capable of subsisting at higher a_w ranges than are yeasts and molds. The point here, and as stated above, is not the absolute a_w at which growth occurs, but rather the fact that the minima cover a rather wide range of a_w values.

Under optimal a_w levels for microbial growth, a bacterial growth curve might appear generally as shown (control plot) in Fig. 5.1. When, however, a humectant is included in the culture medium, changes are observed that may vary in magnitude from phase to phase and humectant to humectant, but in general appear to follow an inhibitory pattern. The initial portion of this growth curve is composed of a lag phase. At one time this part of the curve or plot was termed the resting phase, but in fact there is very little resting being done by cells newly introduced into an osmotically hostile environment. On the contrary, it is at this time that the cell is preparing itself for later growth and to this end the physiological machinery is being created for this event. If a subinhibitory level of humectant is introduced, the duration of the lag phase is extended. Table 5.1 illustrates this effect on *S. aureus* using either NaCl or a mixture of NaCl, Na_2SO_4, and KCl. Obviously, there is a strong negative correlation between the a_w of these systems and the lag time with either solute system.

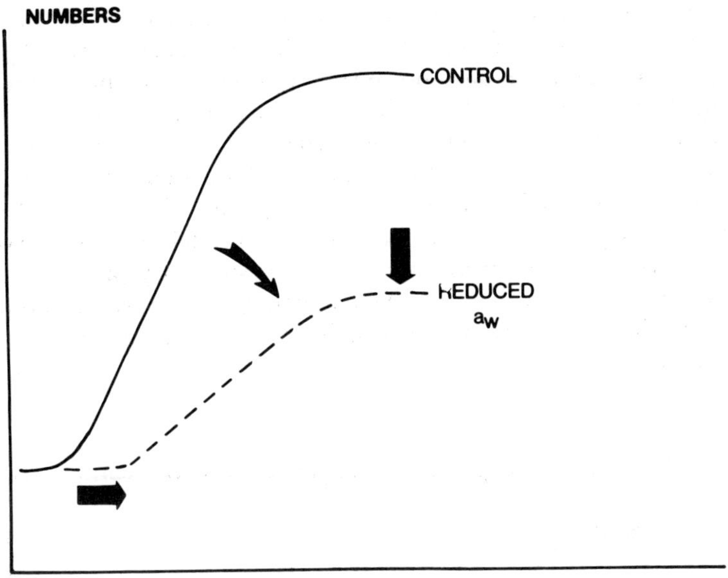

FIG. 5.1 Hypothetical curves showing effects (arrows) of a_w reduction on growth of bacteria.

TABLE 5.1 Effect of a_w on Lag Phase[a]—*Staphylococcus aureus*

	Lag time (hr)	
a_w	NaCl	Salt mixture (NaCl, Na_2SO_4, KCl)
0.99	1	1.5
0.97	2	—
0.96	—	2.5
0.94	3.5	4
0.92	8	4
0.90	13	9

[a]pH 6.8, incubated at 37°C.

The effect of reduced a_w on the next and most important growth phase, the logarithmic phase, is also shown in Fig. 5.1. By decreasing the rate of growth, the logarithmic curve is rotated downward, thus extending the time required for individual cells to reproduce. This is reflected in fewer cell divisions per hour, as shown in Fig. 5.2, or in a longer generation time. An actual plot of the growth of *S. aureus* is shown in Fig. 5.3, in which the slope of the logarithmic phase decreases with a decrease in a_w. At some point in the growth of bacteria, the rate of division begins to abate, the rate of cell death begins to increase, and a steady-state condition termed the maximum stationary phase is reached. This phase is a determinant of the total population of organisms that is present and, like the lag and logarithmic phases, the maximum stationary phase also is influenced by a_w.

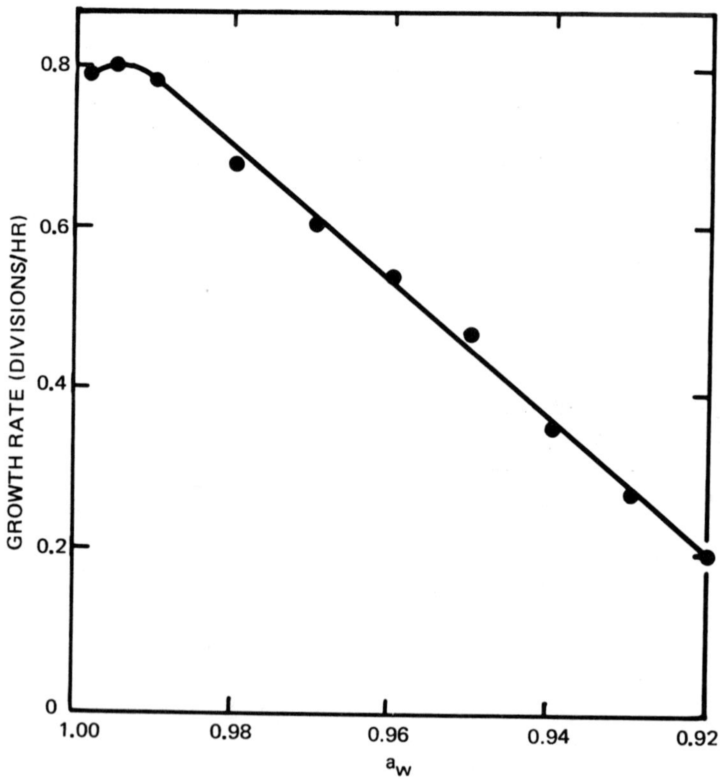

FIG. 5.2 Effect of a_w on the growth rate of *Staphylococcus aureus*. (*From Troller and Christian, 1978.*)

Obviously these data can be used to predict the effects of environments of low a_w on food stability. For example, an extension of the lag phase to some hypothetical point after consumption of a food at risk means that the food, for all intents and purposes, is refractory to microbial spoilage. What it may *not* do is provide assurance of safety. If, for example, a product with low a_w contains viable cells of an infectious foodborne pathogen such as a *Salmonella,* this organism could survive conditions of low a_w and, when ingested, grow in the alimentary tract and cause symptoms of the disease. In fact, organisms poised at low a_w levels may maintain viability for relatively long periods of time. Similarly, many species are much more resistant to thermal inactivation at low a_w levels (Corry, 1974), although there seems to be significant variation among strains (Goepfert et al., 1970).

At a specific a_w one humectant may tend to be more inhibitory than another. Said another way, organisms will grow at a lower a_w when the a_w is

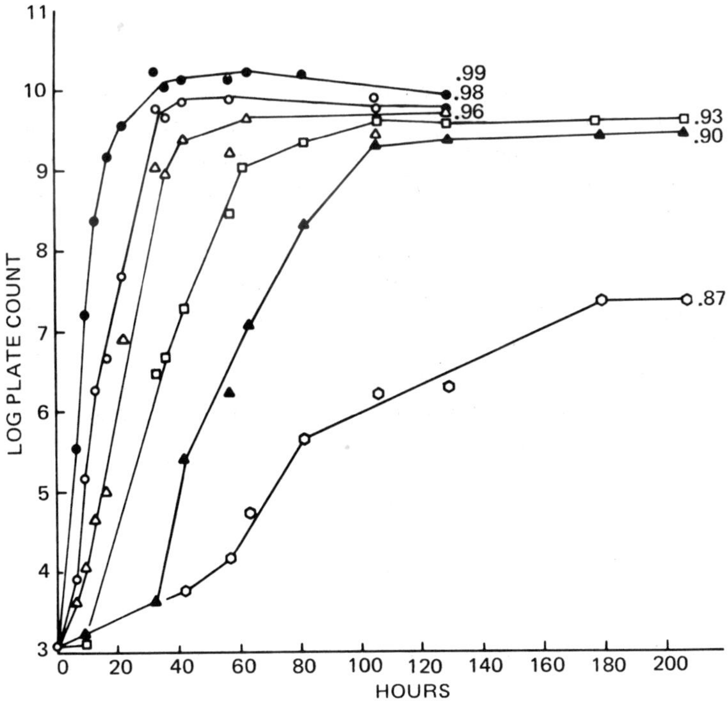

FIG. 5.3 Growth curves of *Staphylococcus aureus* at various a_w levels.

TABLE 5.2 Growth Limits of Bacteria at Low Water Activities

Organism	Limiting a_w in media adjusted with	
	Sodium chloride	Glycerol
Pseudomonas fluorescens	0.97	0.95
Salmonella oranienburg	0.95	0.935
Escherichia coli	0.95	0.935
Clostridium botulinum	0.945	0.93
Bacillus megaterium	0.945	0.925
Micrococcus lysodeikticus	0.93	0.93
Bacillus cereus	0.92	0.92
Bacillus subtilis	0.90	0.92
Staphylococcus aureus	0.85	0.89

Source: Gould and Measures (1977).

reduced with one humectant as compared to another. This phenomenon is referred to as a "solute effect." For example, *Clostridium perfringens* appears to be much less inhibited when glycerol is used as a solute than when sucrose or NaCl is used (Kang et al., 1969). The opposite situation, in which glycerol is more inhibitory than either NaCl or sucrose, is much more rare, although *S. aureus* exhibits this response. A summary table of these effects previously reported by Gould and Measures (1977) is shown in Table 5.2. At one time these differences were thought to be a function of the solute type, its dissociation characteristics, specific gravity, or solubility. However, as a molecular basis for growth inhibition at low a_w has developed, it appears likely that solute effects reflect, to at least some extent, the result of the ability or inability of the cell membrane to exclude certain humectants or to cope with humectants that are not excludable on an intracellular basis.

COMBINATIONS OF a_w AND OTHER FACTORS

The exploitation of combinations of environmental factors together with lowered a_w levels are common in the food industry (Leistner et al., 1981). Generally, as the minimal a_w for growth of a microorganism is approached, alteration of other environmental factors will have a greater impact on growth rates (Troller, 1980). Although only binary combinations with a_w will be discussed here, ternary and quaterary combinations are common in many food products.

pH

The use of combinations of pH and a_w to preserve foods is most typically expressed in products such as semi-dry sausages and cheeses. In addition to pH, the type of acid is also a factor which influences the extent of inhibition. Generally citric and acetic acids tend to be more inhibitory in combination with a_w reduction than do hydrochloric or phosphoric acids (Troller, 1985). Dissociation constants and pH levels are critical in such studies because organic acids tend to be effective primarily in the undissociated form. The general effect of a_w and pH on growth of bacteria is shown in Fig. 5.4.

Oxygen

The removal of oxygen and/or reduction in oxidation-reduction potential tends to sensitize most aerobic bacteria to reductions in a_w level. For example, Scott (1957) in an early review of the water relations of food spoilage microor-

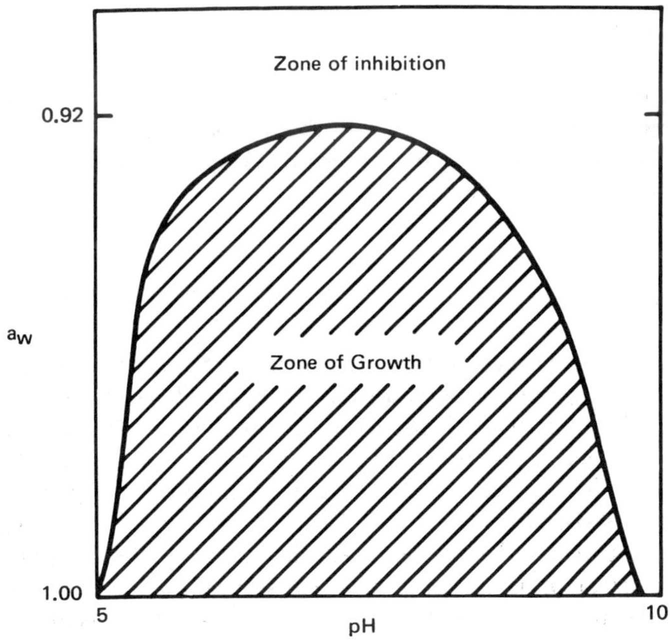

FIG. 5.4 Interacting effects of pH and a_w on growth of bacteria.

ganisms stated that the minimal a_w for growth of *S. aureus* is 0.86 aerobically and 0.90 anaerobically. Lund and Wyatt (1984) studied the combined effect of oxidation-reduction potential and NaCl concentration on the probability of growth and toxin formation by *Clostridium botulinum*. They observed that at high a_w (control) there was little inhibition of growth within a range of +60 to −400 mV. However, the addition of 3.25 and 4% NaCl at −400 mV significantly decreased the probability of outgrowth of spores.

Preservatives

Many commonly used chemical preservatives, e.g., sorbic acid, propionic acid, nitrates, and nitrites, interact synergistically with reductions in a_w to effect growth of microorganisms. In some cases, e.g., nitrite addition to meats at low a_w, the effect is one of synergism to prevent growth of *C. botulinum*. Alternatively, the stability of many types of intermediate moisture foods depends on a combination of factors in which bacterial growth is inhibited by a_w reduction and mold growth is prevented by preservatives such as sorbic acid.

Temperature

Reduction in a_w usually increases the resistance of microorganisms to thermal destruction. Corry (1974) reported that various sugar solutions and polyols were particularly effective in this regard. This topic is discussed in greater detail in the following chapter.

Radiation

Although little has been done to investigate the combined effects of radiation and a_w on growth of microorganisms, it would not be surprising if increased research activity in this area evolves from recent FDA approvals to use gamma irradiation to sanitize certain foods. In earlier studies, Roberts et al. (1965) had indicated that 3 and 6% NaCl solutions did not materially affect the radiation resistance of spores of *Clostridium sporogenes*. However, increasing doses of gamma radiation increased the sensitivity of spores to inhibition by NaCl after radiation. Doses of radiation sufficient to carry out a desirable 12D process resulted in deleterious flavor changes. Treatments such as reduced a_w might permit a reduction in doses of radiation without a concomitant reduction in safety margins.

OSMOREGULATORY MECHANISMS

Whenever a bacterial cell is exposed to an environment of low a_w it loses its internal turgor pressure rapidly and becomes flaccid, a process called plasmolysis. In this condition cells cannot reproduce and will either die or remain dormant (Sperber, 1983). To regain its potential for growth, the cell must decrease its internal a_w and regain its turgor pressure (deplasmolysis). If this is accomplished successfully, the water content of the cytoplasm will be increased and metabolism will not be impaired (Gould, 1985). Of course, depending on the organism and type of humectant, reductions in a_w can be obtained which overwhelm the cell's ability to cope, physiologically, with an environment of low a_w, resulting in a slowing or cessation of growth.

The osmoregulatory process that controls cell reaction to environments of low a_w can be described as a sequence of four events: an initial sensing step, a second, translational step in which a response is triggered or initiated, a third, physiological accommodation step and, finally, growth.

It should be noted that these steps may not occur in all bacteria and fungi. Similarly, K^+ accumulation, for example, may be the primary mechanism by which some organisms (halophiles, for example) accommodate to osmotic changes, whereas in others it may act only as a trigger or as a co-reactant to neutralize and/or drive reactions to equilibrium.

Sensing Osmotic Changes

It is the initial surge of intracellular K^+ that may serve as the primary sensing step and, in some species, as the modulator of metabolic response to low a_w. Other functions attributed to K^+ are shown in Table 5.3. Sensing of low a_w environments has been somewhat more difficult to determine with certainty. LeRudelier et al. (1984) indicates that the task of sensing osmotic perturbations in the cell's environment is a function of porins in the outer cell membrane which form passive diffusion pores. These pores permit hydrophilic compounds to cross the outer membrane. Csonka (1982), on the other

TABLE 5.3 Potassium Functions During Osmotic Stress

1. Primary maintenance of turgor pressure
2. Co-factor in reactions synthesizing intracellular amino acid pools
3. Maintain cell contents at neutral charge
4. Possible sensing of environments of low a_w

hand, believes that osmotically unbalanced environments stimulate a proline permease in the cell membrane, the advantages of which will be discussed very shortly.

An even more important role of the membrane may be to exclude Na^+, which, if permitted to enter the cell, would quickly inactivate any number of vital enzymatic systems. Kanemasa et al. (1976) attributed the barrier properties of *S. aureus* membranes to Na^+ alterations in the types and amount of phospholipids within the membrane. Beyond these possibilities, little has appeared in the literature that has pertained directly to the mechanisms involved in the events which sense and initiate a cellular response to osmotic changes in the cell's environment.

Translation

While K^+ may or may not be the trigger which commences the process of osmoregulation, its transport into the cell is the primary modulatory event (Harold, 1982). Helmer et al. (1982), in a series of elegant experiments, demonstrated that at least two and probably four K^+ transport systems exist in *Escherichia coli* (Fig. 5.5). Perturbations in the osmotic conditions which confront the bacterial cell appear to be modulated or translated through changes in K^+ uptake. This is accomplished, in part, by an operon or series of three high-affinity genes labeled kdp A, B, and C plus a fourth kdp gene, kdp D, a membrane-bound protein which acts as a promoter. The kdp A, B, and C genes code for inner-membrane proteins of various molecular weights which act as the gatekeepers and span the membrane between the periplasmic space and the cytoplasm. Under conditions of low a_w, the synthesis of kdp proteins is turned on and the kdp D protein specifically alters its conformation in a manner which permits it to act with the kdp operon to increase and intensify transcription to maintain cytoplasmic K^+ content (Epstein and Laimins, 1980).

In addition to this high affinity K^+ transport system, there exists a low-affinity, high-rate system consisting of three components, the so-called Trk system. This system is constitutive and requires ATP and a proton motive force to supply energy for net K^+ uptake. Helmer et al. (1982) identified a proton motive force as supplying the primary energy to drive this reaction, whereas ATP supplies the energy to turn off K^+ transport. If K^+ requirements are met by this primary or Trk system, the kdp system described above is not expressed and only the Trk system functions.

There are, in addition, two systems that have received only minimal investigation. One of these, the K^+ export model, has only been postulated and is of some interest because of the potential existence of export-blocking proteins that might be synthesized by the cell in response to osmotic challenge. In a

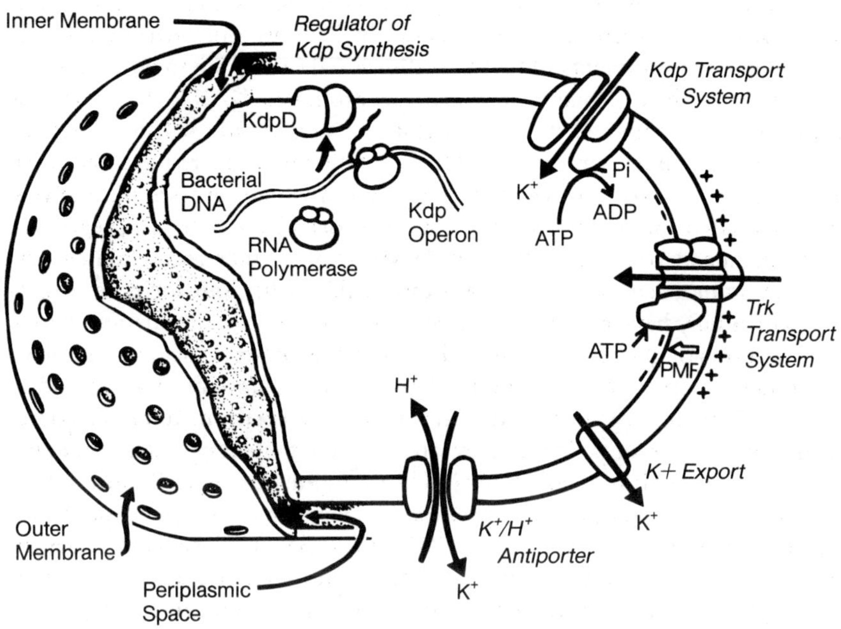

FIG. 5.5 Possible locations of K^+ uptake and control in bacteria. (*From Helmer et al., 1982.*)

system such as this, K^+ would not be pumped out of the cell and would be retained to trigger a metabolic response or to provide primary isoosmotic conditions across the membrane. The so-called antiporter system appears to function primarily as a regulator of the cell's internal pH.

Accommodation

With little doubt, K^+ fills a vital osmoregulatory function and probably is the dominant substance involved in sensing as well as translating what is sensed into a response. There is, however, some disagreement concerning the role of K^+ during accommodation. The essence of this argument seems to revolve around whether K^+ is the primary intracellular solute in nonhalophilic bacteria as it most assuredly is in halophiles (Christian and Waltho, 1961) or whether it participates in accommodation in another way. Measures (1975) and Gould and Measures (1977) showed that K^+ was required to maintain electrical neutrality or the balance of charges within cells exposed to environments of low a_w in which various pool amino acids, such as α-ketoglutarate and glutamic acid, accumulate intracellularly. The physiological response of this model to environments of low a_w is illustrated in Fig. 5.6 in which a series of reactions is shown, many of which are promoted by K^+. The principal reaction involves conversion of α-ketoglutarate to glutamic acid by glutamate dehydrogenase, an enzyme activated by K^+. Glutamic acid reduces the intracellular a_w which, in turn, results in rehydration which reverses plasmolysis, reducing relative amounts of K^+ and glutamate dehydrogenase. This leaves the cell at a balanced, osmotic "null-point" by virtue of the increased glutamic acid pool. For some bacteria the process stops at this point, but for other organisms, which may have relatively large amino acid pools, accumulation of high concentrations of glutamic acid would require concomittant acquisition of K^+ to maintain the system at neutrality. This excessive amount of K^+ could be detrimental to the organism and at the very least costly in terms of energy expenditure. These organisms, therefore, convert glutamic acid to γ-aminobutyric acid or proline, neither of which are highly charged. Proline can be synthesized as shown in Fig. 5.6 or it can be exogenously supplied and incorporated through the cell membrane (Christian, 1955). In either case, it has the effect of reversing osmotic inhibition. Both γ-aminobutyric acid and proline are remarkably efficient at reducing intracellular a_w without interfering in the cell's metabolism and, for this reason, have been termed compatible solutes by Brown and Simpson (1972). Compatible protoplasmic solutes in bacteria include glycylbetaine, proline, glutamic acid, γ-aminobutyric acid, and glycerol. This list probably will be expanded as investigations continue in the future.

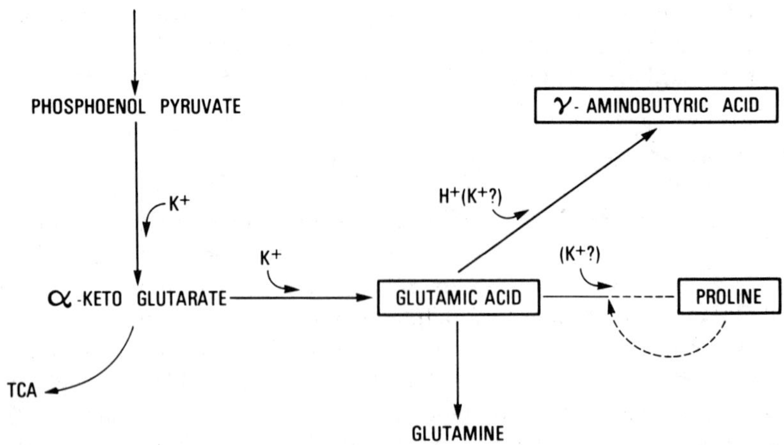

FIG. 5.6 Interrelationships of primary compatible solutes in bacteria. (*From Gould and Measures, 1977.*)

Compatible solutes attract water and thereby restore or partially restore isoosmotic conditions across the cell membrane so that essential metabolic reactions can be maintained. Such solutes are termed compatible because even at very high relative concentrations they do not appreciably interfere with the metabolic and reproductive functions of the cell.

While amino acids appear to be the most common compatible solutes in bacteria, polyols of various types (Table 5.4) are the predominant protoplasmic solutes in many fungi (Brown, 1978), although Luard (1982) has indicated that K^+ and/or hexoses predominate in some mold species. Like amino acids in bacteria, these compounds are synthesized by the organism or they may be obtained from exogenous sources. Unlike bacteria, fungi probably do not rely on K^+ for accumulation and control of protoplasmic solutes, although much remains to be learned about osmoregulation in eucaryotes. Solutes do appear to function both as osmoregulators and food reserves not only in fungi, but in higher species as well.

Of the compatible solutes, proline has perhaps been the most thoroughly studied. Interestingly, this amino acid functions as an osmoregulant in many types of osmotolerant higher plants as well as in many bacteria. The role of proline was first suggested by the work of Christian (1955) and Christian and Waltho (1964), who observed that growth of *Salmonella oranienburg* at low a_w

TABLE 5.4 Compatible Protoplasmic Solutes in Fungi

Solute	Genus
Mannitol	*Geotrichum*
	Platymonas
	Aspergillus
	Dendryphiella
	Penicillium
Cyclohexanetetrol	*Monochrysis*
Arabitol	*Dendryphiella*
	Saccharomyces
Sorbitol	*Stichococcus*
d-Galactosyl-(1,1)-glycerol	*Ochromonas*
Glycerol	*Chlamydomonas*
	Aspergillus
	Dunaliella
	Saccharomyces
	Debaromyces
Erythritol	*Aspergillus*
	Peniccilium

could be stimulated by the addition of this amino acid. Perhaps the most dramatic effect observed when exogenous proline is supplied to bacteria growing at low a_w is a rapid reversal of plasmolysis. Although uptake from media may be one method of accumulating proline in response to water stress, most organisms appear to be able to synthesize this amino acid. In fact, synthesis probably is the most common mechanism for accumulating proline in osmotically inhibited bacteria. From a mechanistic point of view, the accumulation of compatible solutes as opposed to relying on K^+ buildup to equalize the osmotic pressure across the membrane, is far more efficient. In the latter case, as exemplified by the evolutionarily primitive halophilic bacteria, the cell must adapt to high intracellular K^+ levels by synthesizing enzymes capable of functioning under these circumstances. This is a far less efficient way of dealing with this problem than synthesizing solutes via existing pathways that are compatible with existing metabolic machinery. In this way, nothing (other than minute amounts of RNA and related protein) has to be changed. Only additional amounts of one or two substances under tight feedback control are required.

Exactly how these solutes are able to avoid interference is not fully understood; however, Gould (1985) suggests that specific binding between solutes and intracellular enzymes is *not* the mechanism. Wyn Jones and Pollard (1985), in a discussion indicating that solutes are effective by virtue of the fact that they are excluded from the hydration sphere of proteins, suggest that "benign solutes" might more accurately describe the nonparticipatory nature of these materials.

To conclude this discussion on osmoregulatory mechanisms, it can be stated that:

1. A reduction in turgor pressure with consequent plasmolysis initiates K^+ accumulation within the cell.
2. K^+ transport occurs via specific sites in the inner cell membrane.
3. K^+ is the primary osmoprotectant in halophiles and may fulfill a similar function in *E. coli* and perhaps other bacteria.
4. Osmoregulation is accomplished in the majority of nonhalophilic prokaryotes by amino acids.

GENETIC ADAPTATION

The genetic components controlling osmoregulation in microorganisms have been investigated for several years. *Escherichia coli* appears to have evolved a particularly advanced scheme for protection against osmotic stress through a proline-overproducing mutation which confers osmotolerance. These mutations have been mapped (LeRudelier et al., 1984) and submitted to molecular cloning techniques. Csonka (1981) has described a proline permease which functions only under conditions of osmotic stress and provides a reasonable rationale for the earlier reports by Christian (1955), in which he described the stimulation of growth of osmotically inhibited bacteria by exogenous proline. LeRudelier and Valentine (1982) reported a transfer of the plasmid responsible for expression of the proline permease to a nitrogen-fixing strain of *Klebsiella pneumoniae* resulting in an increase in intracellular free proline when this organism is exposed to high levels of NaCl. This, in turn confers an enhanced level of osmoresistance on the organism which, among other things, results in its ability to fix nitrogen while under osmotic stress. Work also is currently underway similarly to transform nitrogen-fixing *Rhizobium* spp. and eventually perhaps even to produce plant varieties with enhanced tolerance to the absence of moisture.

The existence of genetically engineered nitrogen-fixing bacteria would have a significant impact on the growth of legumes in salt-containing soils that now lie fallow. An even more significant impact would result from an increased tolerance of nonleguminous plants to saline or very dry environments, which could then prosper in soils that formerly might have been inhospitable to their growth, a prospect that has been explored in a fascinating review by LeRudelier et al. (1984). Scientists currently are working in related areas of research throughout the world. The potential benefits to mankind, of course, are enormous.

REFERENCES

Brown, A. D. and Simpson, J. R. 1972. The water relations of sugar-tolerant yeasts: The role of intracellular polyols. *J. Gen. Microbiol.* 2: 589.

Brown, A. D. 1978. Compatible solutes and extreme water stress in eukaryotic microorganisms. *Adv. Microb. Physiol.* 17: 181.

Christian, J. H. B. 1955. The water relations of growth and respiration of *Salmonella oranienburg* at 30°. *Aust. J. Biol. Sci.* 8: 490.

Christian, J. H. B. and Waltho, J. A. 1964. The sodium and potassium content of nonhalophilic bacteria in relation to salt tolerance. *J. Gen. Microbiol.* 25: 97.

Corry, J. E. L. 1974. The effect of sugars and polyols on the heat resistance of salmonellae. *J. Appl. Bacteriol.* 37: 31.

Csonka, L. N. 1981. Proline over-production results in enhanced osmotolerance in *Salmonella typhimurium*. *Mol. Gen. Genet.* 182: 82.

Epstein, W. and Laimins, L. 1980. Potassium transport in *Escherichia coli*: Diverse systems with common control by osmotic forces. *Current Trends Biochem.* 5: 21.

Goepfert, J. M., Iskander, J. M., and Amundson, C. H. 1970. Relation of the heat resistance of salmonellae to the water activity of the environment. *Appl. Microbiol.* 19: 429.

Gould, G. W. and Measures, J. C. 1977. Water relations in single cells. *Phil. Trans. R. Soc. Lond.* B. 278: 151.

Gould, G. W. 1985. Present state of knowledge of a_w effects on microorganisms. In *Properties of Water in Foods.* Simatos, D. and Multon, J. L. (Ed.). Martinus Nijhoff Pub., Dordrecht, The Netherlands.

Harold, F. M. 1982. Pumps and currents: A biological perspective. *Curr. Topics Membrane Transp.* 16: 485.

Helmer, G. L., Laimins, L. A., and Epstein, W. 1982. Mechanisms of potassium transport in bacteria. In *Membranes and Transport,* Vol. 2, Martonosi, A. N. (Ed.). Plenum Press, New York.

Kanemasa, Y., Katayama, T., Hayashi, H., Takatsu, T., Tomochika, K., and Okabe, A. 1976. The barrier role of cytoplasmic membrane in salt tolerance mechanisms in *Staphylococcus aureus. Zentralbl. Bakteriol. Parasitenk. Infekt. Hyg. Abt. Suppl.* 5: 189.

Kang, C. K., Woodburn, M., Pagenkopf, A., and Chency, R. 1969. Growth, sporulation and germination of *Clostridium perfringens* in media of controlled water activity. *Appl. Microbiol.* 18: 798.

Leistner, L., Rodel, W., and Krispien, K. 1981. Microbiology of meat and meat products in high and intermediate moisture ranges. In *Water Activity: Influences on Food Quality.* Rockland, L. B. and Stewart, G. F. (Ed.). Academic Press, New York.

LeRudelier, D. and Valentine, R. C. 1982. Genetic engineering in agriculture: Osmoregulation. *Trends Biochem. Sci.* 7: 431.

LeRudelier, D., Strom, A. R., Dandekar, A. M., Smith, L. T., and Valentine, R. C. 1984. Molecular biology of osmoregulation. *Science* 224: 1064.

Luard, E. J. 1982. Accumulation of intracellular solutes by two filamintous fungi in response to growth at low steady state osmotic potential. *J. Gen. Microbiol.* 128: 2563.

Lund, B. M. and Wyatt, G. M. 1984. The effect of redox potential and its interaction with sodium chloride concentration on the probability of growth of *Clostridium botulinum* type E from spore inocula. *Food Microbiol.* 1: 49.

Measures, J. C. 1975. Role of amino acids in osmoregulation of nonhalophilic bacteria. *Nature* 257: 398.

Roberts, T. A., Ditchett, P. J., and Ingram, M. 1965. The effect of sodium chloride on radiation resistance and recovery of irradiated anaerobic spores. *J. Appl. Bacteriol.* 28: 336.

Scott, W. J. 1957. Water relations of food spoilage microorganisms. *Adv. Food Res.* 7: 83.

Sperber, W. H. 1983. Influence of water activity on food-borne bacteria—A review. *J. Food Prot.* 46: 142.

Troller, J. A. and Christian, J. H. B. 1978. *Water Activity and Food.* Academic Press, New York.

Troller, J. A. 1980. Influence of water activity on microorganisms in foods. *Food Technol.* 34: 76.

Troller, J. A. 1985. Effects of a_w and pH on growth and survival of *Staphylococcus aureus.* In *Properties of Water in Foods.* Simatos, D. and Moulton, J. L. (Ed.). Martinus Nijhoff Pub., Dordrecht.

Wyn Jones, R. G. and Pollard, A. 1985. Towards a physical chemical characterization of compatible solutes. In *Biophysics of Water.* Franks, F. and Mathias, S. F. (Ed.). John Wiley & Sons, Chichester.

6

Survival and Death of Microorganisms as Influenced by Water Activity

Lawrence M. Lenovich

Hershey Foods Corporation
Hershey, Pennsylvania

INTRODUCTION

The survival of microorganisms in foods is of prime concern from both safety and economic perspectives. Techniques for processing and the storage stability of foods are influenced by the microflora present. Microbiologists have studied how microorganisms behave under various conditions of temperature, pH, and a_w. Of all the factors affecting the survival and death of microorganisms in food systems, the influence of a_w has been one of those most extensively studied by food microbiologists in recent years.

Factors influencing the growth of microorganisms may also influence death and survival. The basis for survival and death of microorganisms as influenced by a_w is complex. Multiple factors, both intrinsic and extrinsic, influence this relationship but differ within food types and processes and among types of flora involved.

ANALYTICAL CONSIDERATIONS FOR SURVIVAL STUDIES

The study of microbial survival requires an understanding of methods used to collect and interpret data. A few basic considerations are noteworthy. Microbial death tends to follow first-order chemical inactivation. Graphic illustrations of this phenomenon can be obtained by plotting the logarithm of survivors against time. Plots are generally linear, but several exceptions, including "shoulder" and "tailing" effects, may result. Troller and Christian (1978) and Mossel (1975) described the basis for survivor curves and have explained these anomalies in general terms. Cell clumps or chains, cell injury, physiological age, and heterogeneity of cell populations may influence the nonlinearity of survivor curves. Survivor curves are used to describe the effects of processing, freezing, heating, and dehydration on viable microbial populations. These treatments, in combination with a_w, affect the survival of microorganisms. Cells subjected to physical or chemical "insults" may become injured or sublethally stressed. Recovery of sublethally stressed cells requires specific

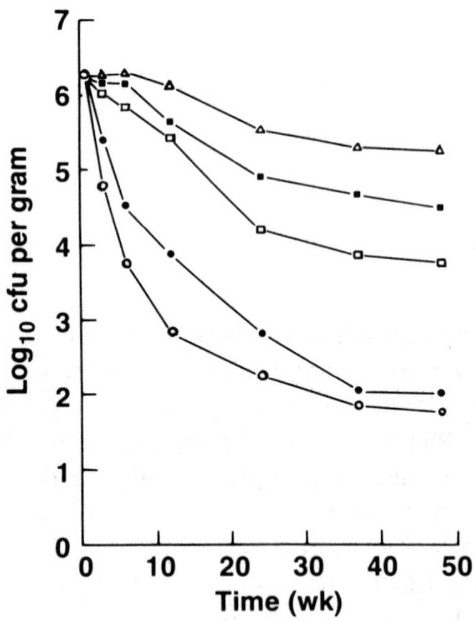

FIG. 6.1 Survival of *Aspergillus flavus* in lemon-flavored gelatin at a_w 0.78 (○), 0.66 (●), 0.51 (□), 0.42 (■), and 0.32 (△), stored at 21°C for 48 wk. (*From Beuchat, 1979.*)

methodology and has serious consequences in the assessment of survivor curves. The assessment of survivors in foods with a reduced a_w requires additional consideration. The methods associated with these determinations are discussed elsewhere. An example of survivor curves as influenced by a_w at sublethal temperatures is illustrated in Fig. 6.1.

INTRACELLULAR FACTORS INFLUENCING a_w EFFECTS

When a microorganism is transferred to a new environment, there are two possible outcomes: survival or death. A generalized growth response of a microorganism subjected to a reduced a_w is depicted in Fig. 6.2. The ability of an organism to adapt to an environment is one of the primary factors contributing to its subsequent survival or death. Survival is accomplished by adaptation that occurs in a two-stage process (Brown, 1976). In stage 1, changes that occur are associated with the cell's thermodynamic adjustment. If change in a_w occurs, this involves a change in osmotic stress. Stage 2 of adaptation occurs when the cell begins to grow and involves an alteration of metabolic activities, such as enzyme synthesis and enzyme activity. Changes during stage 2 occur more slowly than during stage 1.

Brown (1976) suggested that when the cell attains a physiological steady state (i.e., similar intracellular and extracellular a_w levels), there are two theoretical mechanisms for tolerance to a reduced a_w. In mechanism I, he suggests that intracellular proteins (enzymes) of a tolerant organism are different

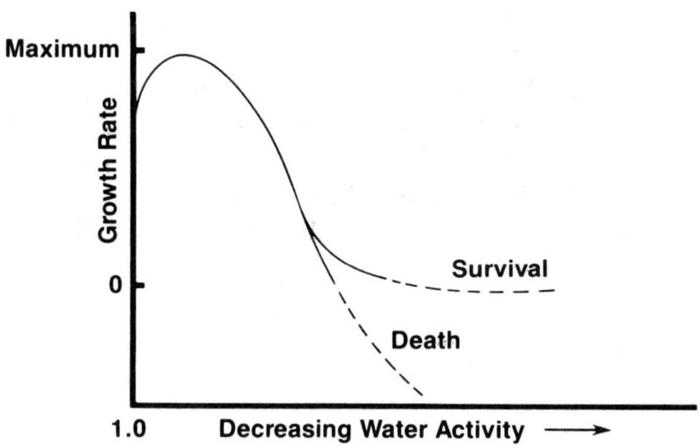

FIG. 6.2 Influence of a_w on bacterial growth rate. Generalized response. (*From Sperber, 1983.*)

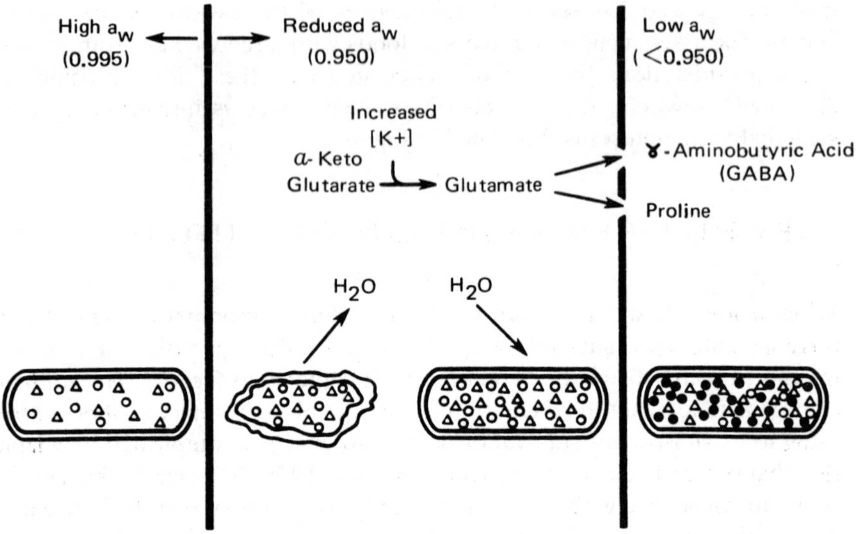

FIG. 6.3 Intracellular adaptation to reduced a_w: potassium ions (△), glutamate (○), GABA or proline (●). (*From Sperber, 1983.*)

from those in a nontolerant organism and are intrinsically more capable of functioning under extreme environmental conditions. In mechanism II, he further suggests that the proteins of nontolerant and tolerant organisms are basically the same, but in tolerant species, intracellular conditions are modified so that enzyme function is not impaired. Although some species may fall into each of these classes, tolerance to reduced a_w is clearly not explainable by two distinct mechanisms.

Adjustment of the intracellular environment in response to low a_w occurs in xerotolerant microorganisms. The production of "compatible solutes" by halophilic bacteria (Kushner, 1968), yeasts (Brown, 1977), and osmotolerant bacteria (Measures, 1975; Gould and Measures, 1977) is a general mechanism used by cells to adapt to low a_w levels. Fig. 6.3 illustrates the mechanism used by cells to resist low a_w.

INTRINSIC FACTORS INFLUENCING a_w EFFECTS

Whether a microorganism survives or dies in a low a_w environment is influenced by intrinsic factors that are also responsible for its growth at a higher a_w. These factors include water-binding properties, nutritive potential, pH, E_h, and the presence of antimicrobial compounds. The influences exerted by these factors interact with a_w both singularly and in combination.

Water-Binding Properties

The type of solute present in a food influences the water-binding properties of that system. Figure 6.4 lists some solute types that affect the availability of water for microbial growth. Microbial growth and survival are not entirely ascribed to reduced a_w but also to the nature of the solute. Anand and Brown (1968) reported differences in growth rates of yeasts using various solutes to reduce the a_w of media to similar levels. They observed generally that polyethylene glycol (PEG) was more inhibitory to yeast growth than were glucose and sucrose at a similar a_w. Marshall et al. (1971) evaluated the inhibitory effects of NaCl and glycerol at the same a_w on 16 species of bacteria. They found that glycerol was more inhibitory than NaCl to relatively salt-tolerant bacteria and less inhibitory than NaCl to salt-sensitive species. While studying the interactive effect of solute type and preservative in *Saccharomyces rouxii*, Lenovich et al. (1986) showed that the type of solute influences resistance to sorbate (Fig. 6.5).

The exact nature of the role that water plays in the mechanism of cell survival is not clearly understood. Experimental results discussed above cannot be adequately explained by any current physical or biological theory. Lang and Steinberg (1983) differentiated between states of bound water—polymer and solute water. Water adsorbed by macromolecules was called "polymer" water and water adsorbed by solutes was referred to as "solute" water. They

Ionic compounds (Na and K salts of HCl and citric acids)

↓

Low molecular weight nonionic compounds (sugars)

↓

High molecular weight compounds with ionic groups (protein, peptides)

↓

High molecular weight compounds with polar groups (dextrin, starch, pectins)

FIG. 6.4 Decreasing order of water-retaining capacity of compounds at the same molecular concentration. (*Adapted from Mossel, 1975.*)

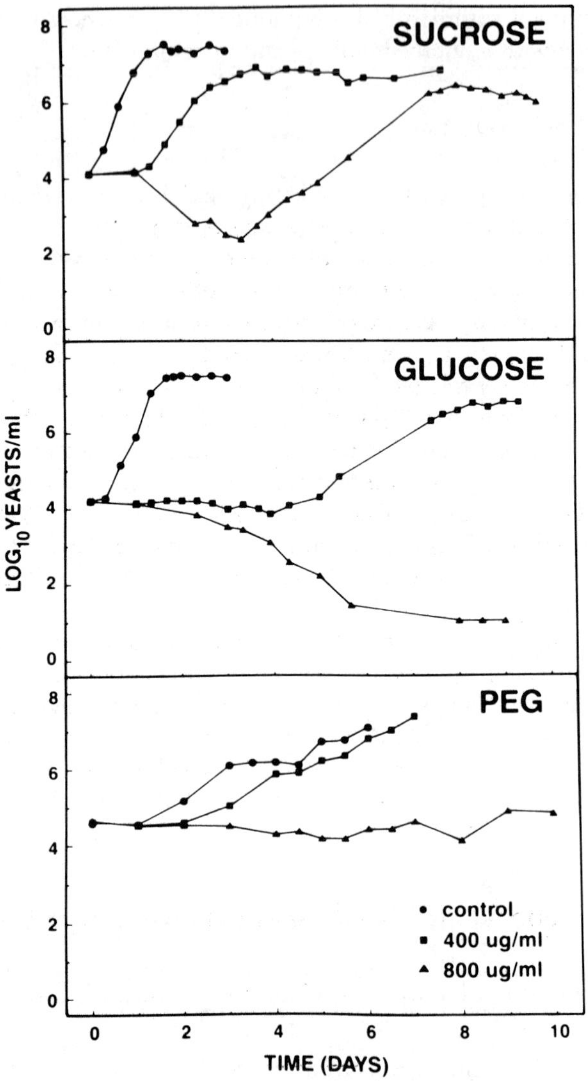

FIG. 6.5 Growth of *Saccharomyces rouxii* at 30°C in defined media at 0.92 a_w at pH 5.0 with sorbate in three different solutes—sucrose, glucose, and polyethylene glycol; sorbate: Control (●), 400 (■), 800 (▲) μg sorbate per mL. (*From Lenovich, 1986.*)

further hypothesized that a food simultaneously contains water in these different states and that it is polymer or solute water that may be important to microbial stability in a particular food.

Mugnier and Jung (1985) studied the survival of bacteria and fungi in relation to a_w and nutritive solutes using bipolymer gels. They concluded that a_w alone was not sufficient to explain the death of test flora. They suggested previously proposed concepts advanced by Duckworth and Kelly (1973) and Seow (1975) to explain the effects of interactions of water and bipolymer gels on the survival of the test organisms. Concepts proposed by Duckworth and Kelly (1973) and Seow (1975) have been called the "discontinuity of properties of water" and the "point of mobilization of solutes," respectively (Fig. 6.6). Basically, Mugnier and Jung (1985) described two fractions of water having distinct differences in their behavior. One fraction of water is not a solvent for solute material. Above a certain amount of hydration, referred to as the mobilization point, there exists a second fraction of water in the polymer system which can serve as a "true" solvent for the solute under study. These concepts were used to explain the survival of microorganisms in polymer systems at various a_w levels.

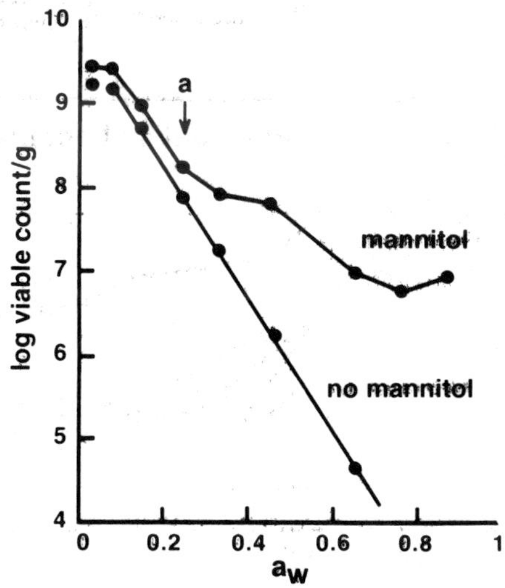

FIG. 6.6 Effect of a_w value on survival of *Rhizobium japonicum* USDA 138 entrapped in xanthan-carob inocula with or without mannitol after 30 days of storage at 28°C (a = theoretical mobilization point of mannitol). (*From Mugnier and Jung, 1985.*)

Nutritive Potential

Corry (1973) reported that the survival of vegetative bacteria is influenced by nutrients in the food matrix. These influences show no consistent inhibitory pattern and are greatly affected by the nature of the suspending matrix. Mugnier and Jung (1985), however, reported that for a given a_w and organism, there are large differences in survival rate as a function of the nutritive solutes used to culture the microorganisms. While comparing a Gram-positive bac-

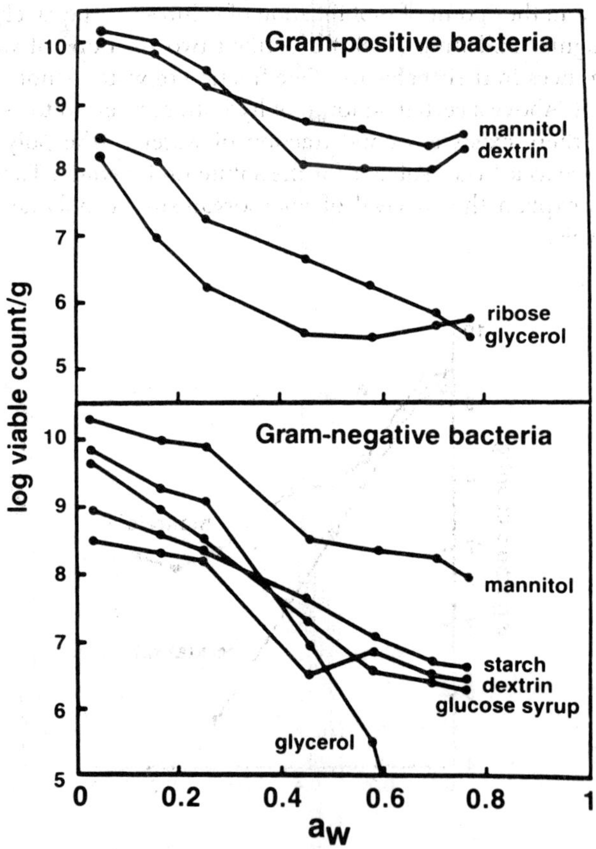

FIG. 6.7 Effect of a_w value and type of solute on survival of a Gram-negative bacterium (*Rhizobium japonicum*) USDA 138 and a Gram-positive bacterium (*Arthrobacter* sp.) entrapped in xanthan-carob inocula after 30 days of storage at 28°C. (*From Mugnier and Jung, 1985.*)

terium (*Rhizobium* sp.) and a Gram-negative bacterium (*Arthrobacter* sp.), they concluded that low molecular weight compounds (C_3 to C_5) had a deleterious effect on survival compared to higher molecular weight compounds (C_6 to C_{12}), which had a protecting effect (Fig. 6.7). Death of *Rhizobium* sp. was more pronounced in glycerol than in ribose and dextrin. The survival of *Anthrobacter* sp. as a function of a_w showed a response similar to that of *Rhizobium* sp. There were progressively decreasing degrees of deterioration with glycerol, ribose, dextrin, and mannitol.

pH

The combined inhibitory effects of pH and a_w on survival of microorganisms are clearly additive. The effects of reduction of a_w on the survival of pathogenic and spoilage organisms are enhanced by a reduction in the pH (Christian and Stewart, 1973; Beuchat, 1979; Hauschild and Hilsheimer, 1979). That is, at any given a_w, survival decreases with a reduction in pH. This effect is enhanced as the a_w is reduced.

Antimicrobial Compounds

The effects of antimicrobial compounds in foods as influenced by a_w on survival and death of microorganisms are of significance. Compounds that exert these effects can be present as naturally occurring components, compounds formed from processing or prolonged storage of food, or additives. There is a lack of information concerning the interaction of antimicrobial compounds and a_w. Restaino et al. (1983) studied the growth and survival of *Saccharomyces rouxii* at a_w levels ranging from 0.82 to >0.995 in the presence of 0–0.15% sorbate. Generally, they found that lowering the a_w enhanced the resistance of *S. rouxii* to increased concentrations of sorbate. Lenovich (1986) studied the interactive effects of sorbate and reduced a_w on the growth and survival of *S. rouxii*. Adaptation to sorbate increased the resistance of *S. rouxii* to the preservative more at reduced a_w than high a_w.

EXTRINSIC FACTORS INFLUENCING a_w EFFECTS

There are several key extrinsic factors relative to a_w that influence microbial deterioration in foods. Temperature, oxygen, chemical treatments, and irradiation are all important in terms of food processing operations and storage of raw and finished products.

Temperature

The effects of temperature on survival of microorganisms have been widely documented. The heat resistance of vegetative cells and spores as influenced by a_w is probably the most extensively studied area in terms of inactivation of microorganisms. The influence of a_w on cell survival has been studied extensively in the context of heat processing and freezing of foods. The effect of a_w on survival at various temperatures has also been reviewed by Troller and Christian (1978).

Elevated temperatures. In general, vegetative cells (Gibson, 1973) and spores of fungi (Beuchat, 1983) are more heat resistant as the a_w of the heating menstruum is reduced (Fig. 6.8). For bacterial spores, heat resistance is greatest at about 0.2–0.4 a_w and decreases at a_w values less than 0.2 (Murrell and Scott, 1966). The physicochemical composition of the heating menstruum profoundly affects rates of heat inactivation and resistance of microorganisms. The type of solute used to adjust a_w to the same level may result in significant differences in heat resistance of a given organism. Small molecular weight compounds such as NaCl and glycerol were shown to decrease the resistance of heat resistant strains of salmonella whereas the heat resistance of heat-sensitive strains was increased (Baird-Parker et al., 1970). Larger molecular weight solutes, such as sucrose, were shown to exert a more protective effect against heat inactivation.

Juven et al. (1978) concluded that an increase in sugar content in orange concentrate cannot be solely responsible for the increased heat resistance of spoilage yeasts. Other juice components, such as citric acid, were thought to be involved in the increased heat resistance. Differences in cellular permeability are suggested as a possible basis for the resulting variability in heat resistance (Beuchat, 1983). Other differences related to type of solute and degree of heat resistance were reported by Härnulv and Snygg (1972). Spores of *Bacillus subtilis* showed maximal heat resistance at low a_w in water vapor and glycerol solutions, whereas only small variations in heat resistance were noted with increasing concentrations of NaCl, glucose, and LiCl.

A high fat content in foods has been shown to have a significant influence on heat resistance. Milk chocolate, which is high in fat and low in moisture, was shown to protect *Salmonella senftenberg* 775W and *Salmonella typhimurium* during extensive heat treatments (Goepfert and Biggie, 1968). $D_{70°C}$ values (Table 6.1) ranged from 6 to 8 hr for *S. senftenberg* and 12–18 hr for *S. typhimurium*. The addition of 2% moisture to milk chocolate reduced the $D_{72°C}$ value of *Salmonella anatum* from 20 hr to 4 hr (Barille and Cone, 1970). The protective effect of fat in these foods was reported to be due to an increased solubility of water in fat during heating, thus resulting in a decrease in a_w of the fat. This, in turn, resulted in an increase in heat resistance of organisms present in the fat (Senhaji, 1977; Senhaji and Loncin, 1977).

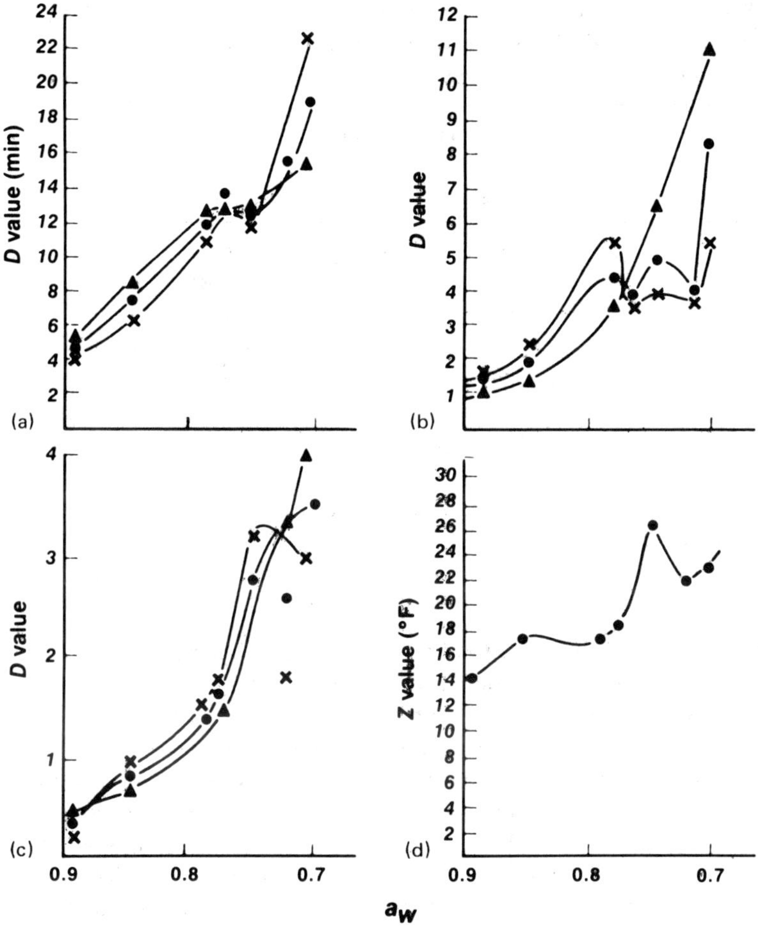

FIG. 6.8 Computed D values plotted against a_w for *Salmonella senftenberg* at (a) 65.5°C, (b) 70°C, and (c) 75°C. ●, average D values; ▲, D values for Exp. a; x, D values for Exp. b. (d) computed z values against a_w for *S. senftenberg*: ●, z values for average D values. (*From Gibson, 1973.*)

Freezing temperatures. Food processing by freezing is an alternate preservation technique that affects the survival of microbial cells differently than exposing them to elevated temperatures. Generally, Gram-positive bacteria are more resistant to inactivation during frozen storage than are Gram-negative bacteria and vegetative cells of yeasts and molds. Fungal spores have good resistance to inactivation at freezing temperatures, whereas bacterial spores are least affected (Troller and Christian, 1978).

TABLE 6.1 D Values of *S. senftenberg* 775W and *S. typhimurium* in Milk Chocolate (results of three trials at each temperature)

Temperature (°C)	D values (min)	
	S. senftenberg 775W	*S. typhimurium*
70	360	720
	480	678
	480	1050
80	144	222
	96	222
	108	222
90	36	78
	30	72
	42	

Source: Goepfert and Biggie (1968).

During freezing, the a_w of a food is substantially reduced. Water separates in the form of ice leaving an increasingly concentrated solution. As ice and solution are in equilibrium at a given temperature, the vapor pressure of ice is equal to that of the solution, and the resulting a_w is a function of the temperature (Christian, 1963). The a_w of a frozen food can only be adjusted by altering the temperature. For example, at −5°C and −10°C, the corresponding a_w values of a given food are 0.9526 and 0.9074, respectively (Christian, 1963). Although the composition of the food does not entirely control the a_w, the components do have an influence on the survival of microorganisms during frozen storage. Murdock and Hatcher (1978) compared yeast survival in 45° and 65° Brix orange concentrate at −17.8, −9.4, −1.1, and 4.4°C (Fig. 6.9). Yeasts died more rapidly in 45° Brix concentrate at −17.8°C than at −9.4 or −1.1°C. In 65° Brix concentrate, survival was enhanced at the lower temperatures.

The rate of microbial inactivation during frozen storage is reduced as the temperature is both reduced and does not fluctuate (Leistner and Rodel, 1979). These researchers suggested that optimal microbiological quality of frozen foods could be attained by initial storage of foods at −10°C (a_w = 0.90) to reduce the number of undesirable organisms followed by freezing at very low temperatures (i.e., −30°C).

Although significant death of microorganisms is associated with frozen storage, freezing under the proper conditions has been used as a preservative

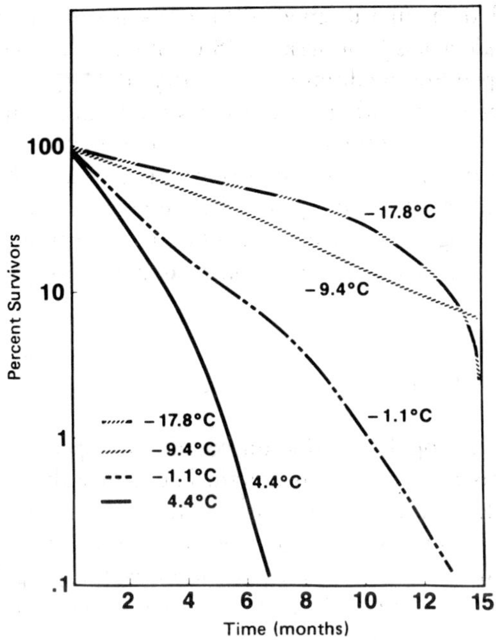

FIG. 6.9 Survival of a yeast in 65° Brix orange concentrate during bulk storage at −17.8, −9.4, −1.1, and 4.4°C. (*From Murdock and Hatcher, 1978.*)

technique for microbial cultures (Marshall et al., 1973). Freeze-drying of microorganisms with a nonelectrolyte such as glycerol or sugar reduces mortality during dehydration, storage, and rehydration (Brown, 1976). It appears that in this situation, the nonelectrolyte functions directly by substituting as a "solvating" molecule for water removed by dehydration. This sequence of events ultimately prevents protein inactivation. Cell viability is enhanced upon rehydration of the culture.

Gases

The survival of *Salmonella newport* and *Staphylococcus aureus* inoculated in cake mix and dehydrated onion soup at 25°C was enhanced when products were stored under vacuum rather than air in the range 0.00–0.22 a_w (Christian and Stewart, 1973). Data indicated that the deleterious effect of air was similar to the effect of 1.5% oxygen and 98.5% nitrogen. Marshall et al. (1973) provided additional data on the effect of various gases on the survival of *S. newport* and *Pseudomonas fluorescens* at 25°C. Storage in air caused rapid death whether the suspending medium was very dry (0.00 a_w) or moistened (0.40 a_w). Survival was greater when cells were stored under carbon dioxide and argon and reduced when storage was in nitrogen or in vacuo. The composition of the storage medium also affected survival. The observed small differences between gases and variations in a_w were less important. Except for *Penicillium* sp., tolerance of fungi to atmosphere containing oxygen and carbon dioxide was reduced by decreasing a_w and storage temperature (Magan and Lacey, 1984).

Chemical Treatments

The lethality of propylene oxide on conidia of *Aspergillus niger* and *Penicillium thomii* was studied by Tawaratani and Shibasaki (1972). Fungicidal activity of propylene oxide was increased with increasing relative humidity in the atmosphere. The survival of *Escherichia coli* on pecans was investigated by Beuchat (1973) under various storage conditions and propylene oxide treatments. He showed that survival increased as pecan moisture decreased.

Irradiation

There is little information available on the effects of irradiation on the survival and death of microorganisms as affected by a_w. Generally, microorganisms are more resistant to inactivation when dry than in the presence of water. Although this may be an oversimplified view, several reports substantiate this claim. Radiosensitivity of *S. aureus* and *Streptococcus faecium* was decreased as the moisture of the suspending medium was decreased (Webb, 1964; Christensen and Sehested, 1964). Härnulv and Snygg (1973) reported that the radiation resistance of *Bacillus subtilis* and *Bacillus stearothermophilus* spores increased slightly with decreasing a_w. In glycerol solutions, a phase of rapid increase in radiation resistance with decreasing a_w was followed by a slower increase as a_w was reduced.

IMPLICATIONS AND FUTURE RECOMMENDATIONS

It is clear from numerous investigations that a_w plays a significant role in the survival and death of microorganisms in foods. The impact on the food industry can be seen directly in the quality of raw materials during storage, the quality of foods produced under various treatments, and the shelf life of foods stored under various conditions. Consideration must be given relative to microbial stability during the development of new food products. Composition of a new product, storage conditions after manufacturing, and distribution practices are of extreme importance in predicting survival and death of potential spoilage microorganisms.

Of equal importance to the survival and death of microorganisms are the implictions of a_w and the mechanisms directly responsible. More research into this area is warranted. Interesting possibilities exist regarding the properties and interactions of bound water and their specific effects on the survival of microorganisms. Little information is available regarding chemical and structural changes in cell composition in response to reduced a_w and subsequent survival mechanisms. Additional investigations into the physiological mechanisms that may be responsible for cell survival are warranted.

REFERENCES

Anand, J. C. and Brown, A. D. 1968. Growth rate patterns of the so-called osmophilic and nonosmophilic yeasts in solutions of polyethylene glycol. *J. Gen. Microbiol.* 52: 205.

Baird-Parker, A. C., Boothroyd, M., and Jones, E. 1970. The effect of water activity on the heat resistance of heat sensitive and heat resistant strains of salmonellae. *J. Appl. Bacteriol.* 33: 515.

Barrile, J. C. and Cone, J. F. 1970. Effect of added moisture on the heat resistance of *Salmonella anatum* in milk chocolate. *Appl. Microbiol.* 19: 177.

Beuchat, L. R. 1973. *Escherichia coli* on pecans: Survival under various storage conditions and disinfection with propylene oxide. *J. Food Sci.* 38: 1063.

Beuchat, L. R. 1979. Survival of conidia of *Aspergillus flavus* in dried foods. *J. Stored Prod. Res.* 15: 25.

Beuchat, L. R. 1983. Influence of water activity on growth, metabolic activities and survival of yeasts and molds. *J. Food Prot.* 46: 135.

Brown, A. D. 1976. Microbial water stress. *Bacteriol. Rev.* 40: 803.

Brown, A. D. 1977. Compatible solutes and extreme water stress in eukaryotic microorganisms. *Adv. Microbiol. Physiol.* 17: 181.

Christensen, E. A. and Sehested, K. 1964. Radiation resistance of *Streptococcus faecium* and spores of *Bacillus subtilis* dried in various media. *Acta Path. Microbiol. Scand.* 62: 448.

Christian, J. H. B. 1963. Water activity and the growth of microorganisms. In *Biochemistry and Biophysics in Food Research.* Butterworth, London.

Christian, J. H. B. and Stewart, B. J. 1973. Survival of *Staphylococcus aureus* and *Salmonella newport* in dried foods, as influenced by water activity and oxygen. In *The Microbiological Safety of Foods.* Hobbs, B. C. and Christian, J. H. B. (Ed.). Academic Press, New York.

Corry, J. E. L. 1973. The water relations and heat resistance of microorganisms. *Prog. Ind. Microbiol.* 12: 73.

Duckworth, R. B. and Kelly, C. E. 1973. Studies of solution processes in hydrated starch and agar at low moisture levels using wide-line nuclear magnetic resonance. *J. Food Technol.* 8: 105.

Gibson, B. 1973. The effect of high sugar concentrations on the heat resistance of vegetative microorganisms. *J. Appl. Bacteriol.* 36: 365.

Goepfert, J. M. and Biggie, R. A. 1968. Heat resistance of *Salmonella typhimurium* and *Salmonella senftenberg* 775W in milk chocolate. *Appl. Microbiol.* 16: 1939.

Gould, G. W. and Measures, J. C. 1977. Water relations in single cells. *Phil. Trans. Royal Soc. London* 278: 151.

Härnulv, B. G. and Snygg, B. G. 1972. Heat resistance of *Bacillus subtilis* spores at various water activities. *J. Appl. Bacteriol.* 35: 615.

Härnulv, B. G. and Snygg, B. G. 1973. Radiation resistance of spores of *Bacillus subtilis* and *B. stearothermophilus* at various water activities. *J. Appl. Bacteriol.* 36: 677.

Hauschild, A. H. W. and Hilsheimer, R. 1979. Effect of salt and pH on toxigenesis by *Clostridium botulinum* in caviar. *J. Food Prot.* 42: 245.

Juven, B. J., Kanner, J., and Weisslowicz, H. 1978. Influence of orange juice composition on the thermal resistance of spoilage yeasts. *J. Food Sci.* 43: 1074.

Kushner, D. J. 1968. Halophilic bacteria. *Adv. Microbiol.* 10: 73.

Lang, K. W. and Steinberg, M. P. 1983. Characterization of polymer and solute bound water by pulsed NMR. *J. Food Sci.* 48: 517.

Leistner, L. and Rodel, W. 1979. Microbiology of intermediate moisture foods. In *Proceedings of the International Meeting on Food Microbiology and Technology.* Jarvis, B., Christian, J. H. B., and Michener, H. D. (Ed.). Medicina Viva Servizio Congress: S.r.l.-Parma, Italy.

Lenovich, L. M. 1986. Mechanism of sorbate resistance in *Saccharomyces rouxii* at reduced water activity. Ph.D. thesis, Drexel Univ., Philadelphia, PA.

Lenovich, L. M., Buchanan, R. L., and Worley, N. J. 1986. Effect of solute type on sorbate resistance in *Saccharomyces rouxii* at reduced water activity.

Paper #134, 46th Annual Meeting of Inst. of Food Technologists, Dallas, TX, June 15–18.

Lerici, C. R. and Guerzoni, M. E. 1979. Survival of *Saccharomyces cerevisiae* in intermediate moisture systems: Influence of physicochemical factors. In *Proceedings of the International Meeting on Food Microbiology and Technology*. Jarvis, B., Christian, J. H. B., and Michener, H. D. (Ed.), p. 73. Medicina Viva Servizio Congress: S.r.l.-Parma, Italy.

Magan, N. and Lacey, J. 1984. Effects of gas composition and water activity on growth of field and storage fungi and their interactions. *Trans. Br. Mycol. Soc.* 82: 305.

Marshall, B. J., Ohye, D. F., and Christian, J. H. B. 1971. Tolerance of bacteria to high concentrations of NaCl and glycerol in the growth medium. *Appl. Microbiol.* 21: 363.

Marshall, B. J., Coote, G. G., and Scott, W. J. 1973. Effects of various gases on the survival of dried bacteria during storage. *Appl. Microbiol.* 26: 206.

Measures, J. C. 1975. Role of amino acids in osmoregulation of nonhalophilic bacteria. *Nature* 257: 398.

Mossel, D. D. A. 1975. Water and microorganisms in foods—a synthesis. In *Water Relations of Foods*. Duckworth, R. B. (Ed.), p. 347. Academic Press, London.

Mugnier, J. and Jung, G. 1985. Survival of bacteria and fungi in relation to water activity and solvent properties of water in biopolymer gels. *Appl. Environ. Microbiol.* 50: 108.

Murdock, D. I. and Hatcher, W. S. Jr. 1978. Effect of temperature on survival of yeast in 45° and 65° Brix orange concentrate. *J. Food Prot.* 41: 689.

Murrell, W. and Scott, W. J. 1966. The heat resistance of bacterial spores at various water activities. *J. Gen. Microbiol.* 43: 411.

Restaino, L., Bills, S., Tscherneff, K., and Lenovich, L. M. 1983. Growth characteristics of *Saccharomyces rouxii* isolated from chocolate syrup. *Appl. Environ. Microbiol.* 45: 1614.

Senhaji, A. F. 1977. The protective effect of fat on the heat resistance of bacteria (II). *J. Food Technol.* 12: 217.

Senhaji, A. F. and Loncin, M. J. 1977. The protective effect of fat on the heat resistance of bacteria. *Food Technol.* 12: 203.

Seow, C. C. 1975. Reactant mobility in relation to chemical reactivity in low and intermediate moisture systems. *J. Sci. Food Agric.* 26: 535.

Sperber, W. H. 1983. Influence of water activity on foodborne bacteria—a review. *J. Food Prot.* 46: 142.

Tawaratani, T. and Shibasaki, I. 1972. Effect of moisture content on the microbicidal activity of propylene oxide and the residue in foodstuffs. *J. Ferment. Technol.* 50: 349.

Troller, J. A. and Christian, J. H. B. 1978. Microbial survival. Ch. 7. In *Water Activity and Food.* Academic Press, New York.

Webb, R. B. 1964. In *Physical Processes in Radiation Biology*. Augenstein, L., Mason, R., and Rosenberg, R. (Ed.), p. 267. Academic Press, New York.

7

Influence of Water Activity on Sporulation, Germination, Outgrowth, and Toxin Production

Larry R. Beuchat

University of Georgia
Agricultural Experiment Station
Experiment, Georgia

INTRODUCTION

Production of various types of spores by some genera of bacteria and fungi is a normal part of metabolic processes. Induction of sporulation occurs in response to many factors, including nutrient availability, adverse pH, suboptimal temperature, accumulation of inhibitory metabolites, and osmotic stress, i.e., changes in a_w in the surrounding environment. Bacterial endospores and some types of fungal spores are characterized by their need for special requirements to initiate germination and outgrowth. Among these needs are optimum a_w values.

Secondary metabolites are produced by some microorganisms that are highly toxic and/or carcinogenic to humans. Production of these metabolites is influenced by many of the same factors that effect sporulation, including a_w. The following text summarizes the effects of a_w on spore formation and germination as well as toxin production by microorganisms most commonly associated with foods and food spoilage. Minimal a_w values for growth and toxin production by microorganisms of public health significance are listed in Table 7.1.

TABLE 7.1 Minimal a_w for Growth and Toxin Production by Microorganisms of Public Health Concern

Microorganism	Minimal a_w for Growth	Minimal a_w for Toxin production	Toxin	Reference
Bacteria				
Clostridium botulinum	0.93		type A	Baird-Parker and Freame (1967)
	0.95	0.95	type A	Ohye and Christian (1967)
		0.94	type A	Kautter et al. (1979)
	0.93		type B	Baird-Parker and Freame (1967)
	0.94	0.94	type B	Ohye and Christian (1967)
		0.94	type B	Kautter et al. (1979)
	0.95		type E	Baird-Parker and Freame (1967)
	0.97	0.97	type E	Ohye and Christian (1967)
	0.95		type E	Ohye et al. (1967)
	0.972		type E	Emodi and Lechowich (1969)
	0.965	0.965	type G	Briozzo et al. (1986)
Clostridium perfringens	0.93–0.95			Kim (1965)
	0.95			Kang et al. (1969)
Bacillus cereus	0.95			Scott (1957)
	0.93			Jakobsen et al. (1972)
	0.95			Raevuori and Genigeorgis (1975)
Staphylococcus aureus	0.86			Scott (1957)
	0.86			Marshall et al. (1971)
		<0.90	Enterotoxin A	Troller (1972)
		0.87	Enterotoxin A	Lotter and Leistner (1978)
		0.97	Enterotoxin B	Troller (1971)
	0.87			Notermans and Heuvelman (1983)
Molds				
Aspergillus flavus	0.78			Ayerst (1969)
		0.84	aflatoxin	Diener and Davis (1970)
	0.80	0.83–0.87	aflatoxin	Northolt et al. (1977)
A. parasiticus	0.82	0.87	aflatoxin	Northolt et al. (1978)
	0.82			Lotzsch and Trapper (1978)

A. ochraceus		0.85	ochratoxin	Bacon et al. (1973)
	0.83	0.83–0.87	ochratoxin	Northolt et al. (1979a)
	0.77		ochratoxin	Pitt and Christian (1968)
Penicillium cyclopium	0.81	0.87–0.90	ochratoxin	Northolt et al. (1979a)
	0.82			Ayerst (1969)
	0.83			Snow (1949)
	0.85			Pelhate (1968)
P. viridicatum	0.83	0.83–0.86	ochratoxin	Northolt et al. (1979a)
P. ochraceus	0.81	0.88	penicillic acid	Northolt et al. (1979b)
		0.80	penicillic acid	Bacon et al. (1973)
	0.76	0.81	penicillic acid	Troller (1980)
P. cyclopium	0.87	0.97	penicillic acid	Northolt et al. (1979b)
	0.82			Ayerst (1969)
P. martensii	0.83	0.99	penicillic acid	Northolt et al. (1979b)
	0.79			Ayerst (1969)
P. islandicum	0.83			Ayerst (1969)
P. urticae	0.83–0.85	0.95	patulin	Northolt et al. (1978)
	0.81			Orth (1976)
	0.83			Mislevic and Tuite (1970)
		0.85	patulin	Troller (1980)
P. expansum	0.83–0.85	0.99		Northolt et al. (1978)
	0.83			Ayerst (1969)
	0.83			Mislevic and Tuite (1970)
A. clavatus	0.85	0.99	patulin	Northolt et al. (1978)
Byssochlamys nivea	0.84			Orth (1976)
Alternaria alternata		<0.90	altenuene, alternariol, alternariol monomethyl ether	Magan et al. (1984)
Stachybotrys atra	0.94	0.94	stachybotrym	Jarvis (1971)
Trichothecium roseum	0.90		trichothecine	Pelhate (1968)

Adapted from Beuchat (1983).

INFLUENCE OF a_w ON SPORULATION AND GERMINATION

Bacteria

Bacterial spores and some types of fungal spores are characterized by their extreme dormancy and requirement for heat shock or other severe treatment to initiate germination and outgrowth. Among the factors influencing bacterial spore dormancy is the dehydrated state of the spore protoplast. The effect of a_w within spores and the role of solutes used to control a_w in the medium in which spores are suspended have received comparatively minor attention with regard to events leading to spore activation. However, reports have been made describing germination of *Bacillus* and *Clostridium* spores as influenced by a_w.

Jakobsen et al. (1972) studied minimal a_w permitting germination and outgrowth of *Bacillus cereus* spores. Dimethylsulfoxide, glycerol, erythritol, sorbitol, glucose, fructose, KCl, and NaCl were used as solutes. The type of solute greatly influenced the minimal a_w at which heat-shocked spores germinated and grew. The effects of NaCl and KCl were similar and most pronounced at high a_w; germination practically ceased at $a_w \leq 0.95$. Partial germination in fructose-supplemented media was observed at a_w 0.94–0.90. Sorbitol and glucose were less inhibitory than NaCl, KCl, and fructose, while dimethylsulfoxide, glycerol, and erythritol were much less inhibitory than the other test solutes, especially with regard to germination. For example, more than 10% of the spores germinated in glycerol media at a_w 0.85. Regardless of the minimal a_w for germination, outgrowth (vegetative cell division) was not observed in media containing test solutes at $a_w < 0.93$. Thus the mechanisms of germination and outgrowth are influenced by both physical and chemical effects of solutes. Furthermore, these solutes may exert their effects independently on processes involved in either germination or outgrowth of *B. cereus* spores.

Germination of *Bacillus stearothermophilus* spores in nutrient broth in relation to a_w was studied by Anagnostopoulos and Sidhu (1981). The a_w-controlling solutes were glycerol, sucrose, NaCl, and KCl. Sucrose was most inhibitory to germination of heat-shocked spores, while glycerol was most favorable, and NaCl and KCl were intermediate.

Gomez et al. (1980) investigated the effects of glycerol and sucrose in heating media on recovery of activated *Clostridium perfringens* spores. Heating in sucrose solution increased recovery, presumably due to an increase of spore germination following solute permeation during heating. The same effect was also observed with glycerol but only when lysozyme was included in the recovery medium.

Spores injured by physical or chemical stresses may suffer damage to permeability barriers and thus react differently than healthy cells to a_w in recovery media or in foods. Germination of injured spores usually occurs within a more narrow range of a_w values than does germination of healthy spores. This is true for both bacterial and fungal spores. For example, heat-stressed conidia of *Aspergillus flavus* have increased sensitivity to NaCl and sucrose (Adams and Ordal, 1976; Beuchat, 1981).

Fungi

Minimal a_w values for sporulation of several genera of molds have been determined, and data suggest that higher a_w is required for spore formation than for germination of spores. Pitt and Christian (1968) reported that *Monascus bisporus* (= *Xeromyces bisporus*) produced aleurospores at a_w 0.66 and *Chrysosporium* spp. produced aleurospores at a_w 0.70. Some species of *Eurotium* produced phialospores at a_w 0.70.

The type of solute used to adjust a_w can influence growth and sporulation of fungi. Kushner et al. (1979) observed that optimum a_w for growth and sporulation of *Eurotium rubrum* followed the same patterns. Optimum a_w in media containing glucose, fructose, and arabinose were 0.962, 0.962, and 0.954, respectively.

Water requirements for teleomorph production often differ from those for anamorph production. For example, *Phytophthora* species can form oospores at lower a_w than that required for sporangia or zoospores (Sneh and McIntosh, 1974; Reeves, 1975). The minimum a_w permitting asexual sporulation of *Chrysosporium fastidium*, a mold not uncommon to dried fruit, is usually higher than that required for germination (Pitt and Christian, 1968). Sexual sporulation often requires even higher a_w than does asexual spore production. Water activity also may influence the size and morphology of spores.

Germination of fungal spores as influenced by a_w has been of interest to plant pathologists for many years. More recently, food mycologists have investigated germination kinetics of fungal spores associated with spoilage of foods. Observations reported by Tomkins as early as 1929 have since been confirmed and expanded by other researchers (Bonner, 1948; Clayton, 1942; Mislivec et al., 1975; Snow, 1949), and can be generalized to a wide range of fungal spores. Pitt (1981) presented an extensive summary of minimal a_w at which spores of several genera of molds will germinate.

At any given temperature, a reduction in a_w causes a decrease in the rate of germination. The presence of appropriate nutrients tends to broaden the range of a_w and temperature at which spore germination and growth will occur, and the a_w range permitting germination is greatest at an optimum temperature. For example, *Alternaria citri* spores will germinate at a_w 0.838 at 30°C, 0.876 at 18 and 37°C, 0.908 at 10°C, and 0.942 at 5°C (Tomkins, 1929).

Pitt and Hocking (1977) studied spore germination times and growth rates of six xerophilic fungi as affected by type of solute and pH. Most strains grew most vigorously in media adjusted to reduced a_w with a glucose/fructose mixture but were completely or partially inhibited by NaCl. Less influence on growth was exerted by pH 4.0 and 6.5 than by solute. The authors suggested that a universal isolation medium for xerophilic fungi could be based on glycerol or a mixture of glucose and fructose but not on NaCl as a a_w-limiting solute.

Alternating the pH from 6.5 to 4.0 increased the minimum a_w for germination of selected field fungi by about 0.02 a_w at optimum temperatures and 0.05 a_w at marginal temperatures (Magan and Lacey, 1984). The change in pH also increased the time required before germination occurred.

Loss of germination potential by *Neurospora crassa* conidia due to release of a dialysable, thermostable substance was first described by Charlang and Horowitz (1971). The release of this substance was enhanced in media of low a_w. This germination-essential component was later characterized as ninhydrin-positive (Charlang and Horowitz, 1974). It was suggested that membrane damage occurs in media at low a_w and that an increase in permeability is responsible for the release of cellular components. Conidia damaged at low a_w were able to recover when transferred to nutrient solution at elevated a_w.

INFLUENCE OF a_w ON TOXIN PRODUCTION

Bacteria

Clostridium botulinum. Spore-forming bacteria that are also capable of causing foodborne intoxication are of special interest with regard to minimal a_w requirements. The effects of reduced a_w on germination and outgrowth of spores and on toxin production by *C. botulinum* have been studied extensively (Sperber, 1982). The minimal a_w for germination and outgrowth is similar to that required for toxin production. This value is influenced by temperature, pH, and nutrient availability. The type of solute also influences the minimal a_w at which spores will germinate and grow. Emodi and Lechowich (1969) reported that ionic solutes such as NaCl and KCl prevented growth of *C. botulinum* type E in laboratory media at $a_w \leq 0.975$ (5.0%, w/v) and 0.974 (6.0%), respectively. No growth was detected in the same broth containing sucrose and glucose at concentrations yielding $a_w \leq 0.976$ (38.5%) and 0.970 (22.5%), respectively.

Glycerol, a "compatible solute," permits germination of *C. botulinum* spores (a_w 0.89) and growth of types A, B (a_w 0.93), and E (a_w 0.94) at lower values than NaCl (a_w 0.93, 0.96, and 0.97, respectively) (Baird-Parker and Freame, 1967).

Spores of *C. botulinum* types A and B are capable of germinating and growing at lower a_w than are spores of type E. Minimum a_w for growth of types A and B has been observed at a_w 0.94–0.96 (Baird-Parker and Freame, 1967; Denny et al., 1969; Marshall et al., 1971; Scott, 1957; Tanaka et al., 1979; Ohye and Christian, 1967). For example, pasteurized cheese spreads at pH > 5.7 have been demonstrated to support growth and toxin production by *C. botulinum* types A and B (Kautter et al., 1979). The a_w of these spreads was 0.936–0.953.

Imitation cheeses (a_w 0.942–0.973) also have been studied for their ability to support growth of *C. botulinum* types A and B at 26°C (Kautter et al., 1981). Results indicated that these cheeses posed no hazard from botulism, even under conditions of abuse.

The combined effect of a_w and pH on growth and toxin production by *C. botulinum* type G was investigated by Briozzo et al. (1986). The minimum a_w at which growth and toxin formation at 32°C occurred was 0.965 in media at pH 6.5–6.9 in which the a_w was adjusted with NaCl or sucrose. Growth and toxin production was delayed in media containing sucrose. The optimum a_w and pH values for toxin production were 0.99 and 6.9, respectively.

Staphylococcus aureus. Among the bacteria capable of causing foodborne poisoning, *S. aureus* is exceptionally tolerant to NaCl. Thus, its growth and enterotoxin production in substrates adjusted to low a_w through the addition of NaCl has been studied in several laboratories. Enterotoxin A-producing strains are capable of producing enterotoxin at a lower a_w value than that required for enterotoxin B production (Troller, 1971; 1972). Water activities of 0.95–0.96 would easily prevent production of enterotoxin B but not the production of enterotoxins A or C in laboratory media. Enterotoxin A was produced at a_w 0.90.

The minimum a_w for growth and enterotoxin A production is influenced by temperature, oxygen, and pH. Lotter and Leistner (1978) reported that *S. aureus* grew and formed enterotoxin A minimally between a_w 0.864 and 0.867 at 30°C; these values were increased to 0.870 and 0.887 at 25°C. A mixture of NaCl, KCl, and Na_2SO_4 was used to adjust the a_w. *Staphylococcus aureus* stored aerobically at 37°C grows at $a_w \geq 0.84$ but requires a minimum of 0.88 at 20°C (Lee et al., 1981). Under anaerobic storage at 37°C, the minimum a_w requirement for growth increased to 0.90.

The limiting a_w for growth of *S. aureus* in sealed cans of precooked bacon at an oxygen concentration of 5.5% was 0.87 at 37°C and 0.91 at 20°C (Silverman et al., 1983), values intermediate to those for aerobic and anaerobic storage reported by Lee et al. (1981). Enterotoxin A production was detected for populations exceeding 10^6 CFU/g of bacon.

Notermans and Heuvelman (1983) studied the combined effect of a_w (0.99–

0.87), pH (4.0–7.0), and suboptimal temperature (8–30°C) on growth and enterotoxin production by *S. aureus*. Growth was not observed at a_w 0.85, at pH 4.3, or at 8°C. Production of enterotoxin B appears to be dictated by a_w; at a_w 0.96 enterotoxin was produced at all temperatures allowing growth, while at a_w 0.93 enterotoxin was hardly produced. Production of enterotoxin A occurred at nearly all conditions allowing growth. Production of enterotoxins C and F was rarely observed at a_w 0.93. The authors concluded that the fact that enterotoxin A can be produced under more adverse conditions may explain the observation that about 80% of staphylococcal intoxications are caused by enterotoxin A-producing strains.

Fungi

Growth of molds on foods has taken on new significance since the discovery of aflatoxins in the early 1960s. Since that time, researchers have isolated and characterized hundreds of other mold metabolites that can cause toxic effects in humans. In addition to aflatoxins, which have been studied most extensively, the effect of a_w on production of ochratoxins, sterigmatocystin, patulin, penicillic acid, citrinin, zearalenone, penitrim A, the ergot alkaloids, and altenuene have also been investigated. For a review of toxic fungal metabolites in foods, see Watson (1985).

The presence of high populations of mycotoxin-producing molds in foods does not necessarily indicate that mycotoxins are present in those foods. Conversely, the absence of molds capable of producing mycotoxins does not necessarily mean that mycotoxins are not present. If the history of a given commodity is known, i.e., if information is available on its a_w and/or moisture content as well as the length of time it was stored at known temperatures, then the potential as a human health hazard can be more adequately assessed.

Aflatoxin. In general, molds are capable of growing over a wider a_w range than are bacteria. The range depends on many factors, including temperature, nutrient availability, pH, and a_w. The optimum and limiting conditions for production of aflatoxin have been determined for several strains of *Aspergillus flavus* and *A. parasiticus*. Northolt et al. (1977) reported optimum temperatures for aflatoxin B_1 production by *A. flavus* at a_w 0.90–0.99 to be 13–16, 16–24, and 31°C, depending upon the strain being tested. Strains with a low temperature optimum for aflatoxin B_1 production grew rapidly at 37°C without aflatoxin production. At $a_w \leq 0.95$ together with a moderate or low temperature, toxin production was inhibited more than growth. The lowest a_w at which aflatoxin was produced was between 0.83 and 0.87.

In another study (Northolt et al., 1976), various solutes were evaluated for their ability to inhibit aflatoxin production by *A. parasiticus* at 10–32°C. The optimum a_w for growth in submerged cultures was lower when a_w was adjusted with glycerol and glucose than with NaCl, KCl, and Na_2SO_4. When surface-cultured on media containing sucrose or glycerol, the optimum a_w for growth was 0.99. In no instance was aflatoxin B_1 formed at a_w 0.83 or at 10°C.

In other studies, the lower limit for aflatoxin production has been reported to be 0.70–0.75 in rice (Boller and Schroeder, 1974), 0.84 in corn (Hunter, 1969), and 0.85 in peanuts (Diener and Davis, 1967). Storage of salami (a_w 0.94) at 13°C was considered sufficient to prohibit aflatoxin and sterigmatocystin production (Incze and Frank, 1976). It is difficult to compare results from various laboratories because of the use of widely different substrates, incubation temperatures, and analytical procedures.

Ochratoxin. The effects of a_w and temperature on growth of and ochratoxin A production by *Aspergillus ochraceus, Penicillium cyclopium,* and *P. viridicutum* have been studied (Northolt et al., 1979a). The minimum a_w values were 0.83–0.87, 0.87–0.90, and 0.83–0.86, respectively, for the three molds when cultured on an agar medium containing sucrose or glycerol. Optimum temperatures for ochratoxin A production by *A. ochraceus* were 31 or 37°C, depending upon the a_w, whereas *P. cyclopium* produced maximum quantities at 24°C. *P. viridicutum* produced more at 24°C than at 16°C when cultured on Czapek maize extract agar supplemented with sucrose, but no difference was noticed on the same base medium supplemented with glucose. Ochratoxin A production by *P. cyclopium* on Edam cheese at a_w 0.95 was observed at 20–24°C. On barley at a_w 0.95, *P. viridicatum* grew at 0–31°C but produced ochratoxin A only at 12–24°C.

Thus the a_w range for ochratoxin A synthesis by *A. ochraceus* and Penicillia may be similar, although the optimum temperature for production is higher for *A. ochraceus.* Since Penicillia, among them *P. cyclopium,* are the predominant molds on hard cheeses in warehouses, shops, and households, there is cause for concern about the presence of ochratoxin when spoilage of these cheeses occurs.

Penicillic acid. The range of temperatures and a_w at which penicillic acid is produced varies, depending upon the mold strain. On laboratory agar media in which a_w had been adjusted by addition of sucrose or glycerol, the minimum a_w for production by *P. cyclopium* and *A. ochraceus* was 0.97 and that of *Penicillium martensii* was 0.99 (Northolt et al., 1979b). On Gouda and Tilsiter cheeses (a_w 0.96–0.98), the temperature range for growth of *P. cyclopium* was 0–24°C and 4–16°C, respectively. On poultry feed, *A. ochraceus* produced penicillic acid at a_w as low as 0.88, whereas the minimum a_w for penicillic acid production by *P. cylopium* was 0.97.

While Penicillia capable of producing penicillic acid are sometimes used to "ripen" dry sausages, the absence of the toxin in these products may be explained by the improbability of its production at a_w less than 0.95. The chemical composition of the substrate greatly influences the minimal a_w at which various strains of molds will produce penicillic acid, however, and caution should be taken when assessing the probability of a given product or commodity becoming contaminated during storage.

Patulin. Patulin is a metabolite of *Penicillium, Aspergillus,* and *Byssochlamys* species. Penicillia capable of producing the toxin are commonly found on apples and other fresh fruits as well as on molded bread and cheese. Minimum a_w values for patulin production by *Penicillium expansum, P. urticae* (*patulum*), and *Aspergillus clavatus* are reported to be 0.99, 0.95, and 0.99, respectively (Northolt et al., 1978). The optimum temperatures for patulin production at high a_w by *P. urticae* varied with the strain and were 8 and 31°C. Since patulin is produced only at relatively high a_w, its presence in breads and hard cheeses is detected only occasionally even though molds capable of producing the toxin may proliferate.

***Alternaria* toxins.** *Alternaria* can grow at refrigeration temperatures and may be associated with extensive spoilage of fruits and vegetables during transport and storage. It also causes discoloration of wheat kernels. *Alternaria alternata* produces at least four secondary metabolites (altenuene, alternariol, alternariol monomethyl ether, and tenuazonic acid) that are toxic to mammalian systems. Magan et al. (1984) studied the effects of a_w and temperature on production of *A. alternata* mycotoxins on wheat grain. Greatest production of altenuene, alternariol, and alternariol monomethyl ether occurred at a_w 0.98 and 25°C. Little toxin was produced at a_w 0.90. All three metabolites were produced at 5°C and a_w 0.98–0.95 and at 30°C and a_w 0.98–0.90.

The production of tenuazonic acid on cottonseed by *Alternaria tenuissima* as influenced by a_w was studied by Young et al. (1980). Maximum production occurred at 20°C and 37% moisture ($a_w \geq 0.99$). Production was halved when the a_w was reduced to 0.95. It is likely that biosynthesis of toxic metabolites by *Alternaria* species is favored by different temperatures and a_w.

Stability of mycotoxins. Most studies concerning stability of mycotoxins in foods have been directed toward determining the effects of temperature, pH, and chemicals on rates of inactivation. The rates of disappearance of patulin and citritin from grains at a_w 0.70 and 0.90 were investigated by Harwig et al. (1977). At a_w 0.70, estimates of half-lives for patulin in barley, corn, and wheat were 12.7, 4.4, and 4.4 days, respectively. At 0.90 a_w, half-lives were 6.8, 2.4, and 1.9 days. For citrinin, these values were 7.8, 15.5, and 11.9 days at a_w 0.70, and 1.8, 10.4, and 3.0 days at a_w 0.90. Although differences in disappearance rates were attributed to differences in grain composition, changes were not due solely to pH, since there was no apparent correlation between half-lives and pH of grains.

SUMMARY

Every microorganism has limiting a_w values below which it will not grow, form spores, or produce toxic metabolites. These values may be different for each metabolic process. Sporulation may occur at or slightly below the minimum a_w for growth. The minimum a_w for ascospore formation by fungi is generally higher than that required for formation of asexual spores. In contrast, germination of spores of some microorganisms may occur at a_w values below that required for growth. Without exception, the minimum a_w for growth of microorganisms is less than or equal to the minimum a_w for toxin production. In many instances, toxin production occurs only in a range of a_w values considerably higher than that required for growth. Optimal conditions of temperature, pH, oxygen tension, and nutrient availability are necessary to permit sporulation, germination, and toxin production at reduced a_w.

REFERENCES

Adams, G. H. and Ordal, Z. J. 1976. Effects of thermal stress and reduced water activity on conidia of *Aspergillus parasiticus*. *J. Food Sci.* 41: 547.

Anagnostopoulos, G. D. and Sidhu, H. S. 1981. The effect of water activity and the a_w controlling solute on spore germination of *Bacillus stearothermophilus*. *J. Appl. Bacteriol.* 50: 335.

Ayerst, G. 1969. The effects of moisture and temperature on growth and spore germination in some fungi. *J. Stored Prod. Res.* 5: 127.

Bacon, C. W., Sweeney, V. G., Robbins, J. D., and Burdick, D. 1973. Production of penicillic acid and ochratoxin A on poultry feed by *Aspergillus ochraceus*: temperature and moisture requirements. *Appl. Microbiol.* 26: 155.

Baird-Parker, A. C. and Freame, B. 1967. Combined effect of water activity, pH and temperature on the growth of *Clostridium botulinum* from spore and vegetative cell inoculum. 30: 420.

Beuchat, L. R. 1981. Combined effects of solutes and food preservatives on rates of inactivation of and colony formation by heated spores and vegetative cells of molds. *Appl. Environ. Microbiol.* 41: 472.

Beuchat, L. R. 1983. Influence of water activity on growth, metabolic activities and survival of yeasts and molds. *J. Food Prot.* 46: 135.

Boller, R. A. and Schroeder, H. W. 1974. Influence of relative humidity on production of aflatoxin in rice by *Aspergillus parasiticus*. *Phytopathology* 64: 17.

Bonner, J. T. 1948. A study of the temperature requirements of *Aspergillus niger*. *Mycologia* 40: 728.

Briozzo, J., Amato de Lagarde, E., Chirife, J., and Parada, J. L. 1986. Effect of water activity and pH on growth and toxin production by *Clostridium botulinum* type G. *Appl. Microbiol.* 51: 844.

Charlang, G. and Horowitz, N. H. 1971. Germination and growth of *Neurospora* at low water activities. *Proc. Natl. Acad. Sci.* 68: 260.

Charlang, G. and Horowitz, N. H. 1974. Membrane permeability and the loss of germination factor from *Neurospora crassa* at low water activities. *J. Bacteriol.* 117: 261.

Clayton, C. N. 1942. The germination of fungus spores in relation to controlled humidity. *Phytopathology* 32: 921.

Denny, C. B., Goeke, D. J., and Sternberg, R. 1969. Inoculation tests of *Clostridium botulinum* in canned breads with specific references to water activity. Res. Rpt. No. 4-69. Natl. Canners Assoc., Washington, DC.

Diener, U. L. and Davis, N. D. 1967. Limiting temperature and relative humidity for growth and production of aflatoxin and free fatty acids by *Aspergillus flavus* in sterile peanuts. *J. Am. Oil Chem. Soc.* 44: 259.

Diener, U. L. and Davis, N. D. 1970. Limiting temperature and relative humidity for aflatoxin production by *Aspergillus flavus* in stored peanuts. *J. Am. Oil Chem. Soc.* 47: 347.

Emodi, A. and Lechowich, R. V. 1969. Low temperature growth of type E *Clostridium botulinum* spores. 2. Effects of solutes and incubation temperature. *J. Food Sci.* 34: 82.

Gomez, R. F., Gombas, D. E., and Herrero, A. 1980. Reversal of radiation-dependent heat sensitization of *Clostridium perfringens* spores. *Appl. Environ. Microbiol.* 39: 525.

Harwig, J., Blanchfield, B. J., and Jarvis, G. 1977. Effect of water activity on disappearance of patulin and citrinin from grains. *J. Food Sci.* 42: 1225.

Hunter, J. H. 1969. Growth and aflatoxin production in shelled corn by the *Aspergillus flavus* group as related to relative humidity and temperature. Ph.D. Diss., Purdue Univ., West Lafayette, IN.

Incze, K. and Frank, H. K. 1976. Is there a danger of mycotoxins in Hungarian salami? I. The influence of substrate, a_w value and temperature on toxin production in mixed cultures. *Die Fleischwirtschaft* 56: 219.

Jakobsen, M., Filtenborg, O., and Bramsnaes, F. 1972. Germination and outgrowth of the bacterial spore in the presence of different solutes. *Lebensm. Wiss. Technol.* 5: 159.

Jarvis, B. 1971. Factors affecting the production of mycotoxins. *J. Appl. Bacteriol.* 34: 199.

Kang, C. K., Woodburn, M., Pagenkopf, A., and Cheney, R. 1969. Growth, sporulation and germination of *Clostridium perfringens* in media of controlled water activity. *Appl. Microbiol.* 18: 798.

Kautter, D. A., Lilly, T., Lynt, R. K., and Solomon, H. M. 1979. Toxin production by *Clostridium botulinum* in shelf-stable pasteurized process cheese spreads. *J. Food Prot.* 42: 784.

Kautter, D. A., Lynt, R. K., Lilly, T., and Solomon, H. M. 1981. Evaluation of the botulism hazard from imitation cheeses. *J. Food Sci.* 46: 749.

Kim, C. H. 1965. Substrate factors for growth and sporulation of *Clostridium perfringens* in selected foods and simple systems. Ph.D. Diss., Purdue Univ., West Lafayette, IN.

Kushner, L., Rosenzweig, W. D., and Stotzky, G. 1979. Effects of salts, sugars, and salt-sugar combinations on growth and sporulation of an isolate of *Eurotium rubrum* from pancake syrup. *J. Food Prot.* 41: 706.

Lee, R. Y., Silverman, G. J., and Munsey, D. T. 1981. Growth and enterotoxin A production by *Staphylococcus aureus* in precooked bacon in the intermediate moisture range. *J. Food Sci.* 46: 1687.

Lotter, L. P. and Leistner, L. 1978. Minimal water activity for enterotoxin A production and growth of *Staphylococcus aureus*. *Appl. Environ. Microbiol.* 36: 377.

Lotzsch, R. and Trapper, D. 1978. Aflatoxin and patulin production as a function of water activity (a_w-values). *Die Fleischwertschaft* 58: 2001.

Magan, N., Cayley, G. R., and Lacey, J. 1984. Effect of water activity and temperature on mycotoxin production by *Alternaria alternata* in culture and on wheat grain. *Appl. Environ. Microbiol.* 47: 1113.

Magan, N. and Lacey, J. 1984. Effect of temperature and pH on water relations of field and storage fungi. *Trans. Brit. Mycol. Soc.* 82: 71.

Marshall, B. J., Ohye, D. F., and Christian, J. H. B. 1971. Tolerance of bacteria to high concentrations of NaCl and glycerol in the growth medium. *Appl. Microbiol.* 21: 363.

Mislivec, P. B., Dieter, C. T., and Bruce, V. R. 1975. Effect of temperature and relative humidity on spore germination of mycotoxic species of *Aspergillus* and *Penicillium*. *Mycologia* 67: 1187.

Mislivec, P. B. and Tuite, J. 1970. Temperature and relative humidity requirements of species of *Penicillium* isolated from yellow dent corn kernels. *Mycologia* 62: 75.

Northolt, M. D., van Edmond, H. P., and Paulsch, W. E. 1977. Differences between *Aspergillus flavus* strains in growth and aflatoxin B_1 production in relation to water activity and temperature. *J. Food Prot.* 40: 778.

Northolt, M. D., van Edmond, H. P., and Paulsch, W. E. 1978. Patulin production by some fungal species in relation to water activity and temperature. *J. Food Prot.* 41: 885.

Northolt, M. D., van Edmond, H. P., and Paulsch, W. E. 1979a. Ochratoxin A production by some fungal species in relation to water activity and temperature. *J. Food Prot.* 42: 485.

Northolt, M. D., van Edmond, H. P., and Paulsch, W. E. 1979b. Penicillic acid production by some fungal species in relation to water activity and temperature. *J. Food Prot.* 42: 476.

Northolt, M. D., Verhulsdonk, C. A. H., Soentoro, P. S. S., and Paulsch, W. E. 1976. Effect of water activity and temperature on aflatoxin production by *Aspergillus parasiticus*. *J. Milk Food Technol.* 39: 170.

Notermans, S. and Heuvelman, C. J. 1983. Combined effect of water activity, pH and sub-optimal temperature on growth and enterotoxin production of *Staphylococcus aureus*. *J. Food Sci.* 48: 1832.

Ohye, D. F. and Christian, J. H. B. 1967. Combined effects of temperature, pH and water activity on growth and toxin production by *Clostridium botulinum* types A, B, and E. In *Proc. 5th Intl. Symp. Food Microbiology*, July, 1966, p. 217. Barnes and Noble, New York.

Ohye, D. F., Christian, J. H. B., and Scott, W. J. 1967. Influence of temperature on the water relations of growth of *Clostridium botulinum* type E. In *Botulism 1966, Proc. 5th Intl. Symp. Food Microbiology*, July, 1966, p. 136. Barnes and Noble, New York.

Orth, R. 1976. The influence of water activity on the spore germination of aflatoxin-, sterigmatocystin- and patulin-producing molds. *Lebensm. Wiss. Technol.* 9: 156.

Pelhate, J. 1968. A study of water requirements in some storage fungi. *Mycopathol. Mycol. Appl.* 36: 117.

Pitt, J. I. 1981. Food spoilage and biodeterioration. In *Biology of Conidial Fungi*, Vol. 2. Cole, G. T. and Kendrick, B. (Ed.). Academic Press, New York.

Pitt, J. I. and Christian, J. H. B. 1968. Water relations of xerophilic fungi isolated from prunes. *Appl. Microbiol.* 16: 1853.

Pitt, J. I. and Hocking, A. D. 1977. Influence of solute and hydrogen ion concentration on the water relations of some xerophilic fungi. *J. Gen. Microbiol.* 101: 35.

Raevuori, M. and Genigeorgis, C. 1975. Effect of pH and sodium chloride on growth of *Bacillus cereus* in laboratory media and certain foods. *Appl. Microbiol.* 29: 68.

Reeves, R. J. 1975. Behavior of *Phytophthora cinnamoni* Rands in different soils and water regimes. *Soil Biol. Biochem.* 7: 19.

Scott, W. J. 1957. Water relations of food spoilage microorganisms. *Adv. Food Res.* 7: 83.

Silverman, G. L., Munsey, D. T., Lee, C., and Ebert, E. 1983. Interrelationship between water activity, temperature and 5.5 percent oxygen on growth and enterotoxin A secretion by *Staphylococcus aureus* in precooked bacon. *J. Food Sci.* 48: 1783.

Sneh, B. and McIntosh, D. L. 1974. Studies on the behavior and survival of *Phytophthora cactorum* in soil. *Can. J. Bot.* 52: 795.

Snow, D. 1949. The germination of mold spores at controlled humidities. *Ann. Appl. Biol.* 36: 1.

Sperber, W. H. 1982. Requirements of *Clostridium botulinum* for growth and toxin production. *Food Technol.* 36(12): 89.

Tanaka, N., Goepfert, J. M., Traisman, E., and Hoffbeck, W. M. 1979. A challenge of pasteurized process cheese spread with *Clostridium botulinum* spores. *J. Food Prot.* 42: 787.

Tomkins, R. G. 1929. Studies of the growth of molds. I. *Proc. Royal Soc.* B105: 375.

Troller, J. A. 1971. Effect of water activity on enterotoxin B production and growth of *Staphylococcus aureus. Appl. Microbiol.* 21: 435.

Troller, J. A. 1972. Effect of water activity on enterotoxin A production and growth of *Staphylococcus aureus. Appl. Microbiol.* 24: 440.

Troller, J. A. 1980. Influence of water activity on microorganisms in foods. *Food Technol.* 34(5): 76.

Watson, D. H. 1985. Toxic fungal metabolites in food. *CRC Crit. Rev. Food Sci. Nutr.* 22: 177.

Young, A. B., Davis, N. D., and Diener, U. L. 1980. The effect of temperature and moisture on tenuazonic acid production by *Alternaria tenuissima. Phytopathology* 70: 607.

8

Media and Methods for Detection and Enumeration of Microorganisms with Consideration of Water Activity Requirements

Ailsa D. Hocking and John I. Pitt

Division of Food Research
Commonwealth Scientific and
Industrial Research Organisation
North Ryde, New South Wales, Australia

INTRODUCTION

The a_w of a food will have a major influence on the type of microflora capable of growing in that food and causing spoilage. It is important, therefore, to take this fact into consideration when contemplating isolation and enumeration of the significant microorganisms in the food. If the food is of high a_w (0.980–0.999), bacteria will be the main spoilage organisms, and media of high a_w, such as plate count agar, will be adequate to give an overall picture of the microbiological status of the food.

At lower a_w values (below 0.95), common spoilage bacteria will still be present, but will not be capable of growing in the food and are of less significance. If, however, the food is to be reconstituted to a high a_w, then the presence of pathogens such as *Salmonella* and *Staphylococcus* and other food-poisoning bacteria is significant, and appropriate methods for their detection must be employed.

The use of media with reduced a_w in food bacteriology is restricted to a few specific areas. In food mycology, however, reduced a_w media are important and should be used frequently, because food-spoilage fungi grow over a wide a_w range, from 0.999 to near 0.61 a_w, a fact that is often forgotten in standards that call for a simple "yeast and mold" count. Fungal spores are ubiquitous in our environment, occurring on many foods as surface contaminants. For low a_w foods, a "total yeast and mold" count on high a_w media will enumerate only the surface flora, giving no real indication of the level of potential spoilage fungi. Under these circumstances, a conventional "total yeast and mold" count can be about as useful as an aerobic total plate count when you are looking for *Clostridium*.

USE OF SALT-BASED MEDIA IN FOOD BACTERIOLOGY

There are relatively few uses for media of reduced a_w in food bacteriology. These media are all salt-based, and are used to detect and enumerate salt-requiring or salt-tolerant bacteria. Salt-based media may also be used to differentiate between species within a particular genus, such as *Vibrio* and, to a lesser extent, *Streptococcus*.

Staphylococcus aureus

Staphylococcus aureus is the most salt tolerant of the food-poisoning bacteria, growing vigorously at 0.94 a_w, but capable of growth to near 0.86 a_w at 30°C (Scott, 1953). Solid media containing up to 7.5% NaCl, such as *Staphylococcus* medium 110 and mannitol salt agar (Gilbert et al., 1969), have made use of this salt tolerance to selectively enumerate *S. aureus*. However, salt-based agar media for enumeration of *S. aureus* have fallen into disfavor partly because they involve subjective judgments on pigmentation and colony morphology. More importantly, salt-based media, including enrichment media, have been criticized for their adverse effects on stressed cells of *S. aureus* (Baird-Parker and Eyles, 1979). Better media, based on more specific selective agents such as polymixin B and potassium tellurite, are now recommended for enumeration of *S. aureus* in foods (e.g., Baird-Parker and Eyles, 1979). The most widely used medium is Baird-Parker agar (Baird-Parker, 1962; Tatini et al., 1984), which incorporates potassium tellurite, lithium chloride, glycine, and pyruvate as selective agents.

Media containing salt are still in use for selective enrichment of *S. aureus* in certain situations. Tatini et al. (1984) recommend enrichment in trypticase soy broth containing 10% NaCl, incubated at 35–37°C for 24 hr, followed by plating onto Baird-Parker agar, for the detection of low populations of *S. aureus* in processed foods.

Vibrio Species

Many *Vibrio* species have a physiological requirement for NaCl (Fig. 8.1), so salt is an important component of selective enrichment broths and plating and identification media. *Vibrio cholerae* does not have a salt requirement, and is the least salt tolerant of the vibrios. The medium most frequently recommended for selective enrichment of *V. cholerae,* alkaline peptone water, contains only 0.5–1% NaCl (ICMSF, 1978; Furniss et al., 1978).

The most common foodborne *Vibrio* species, *V. parahaemolyticus,* requires NaCl in the growth medium, and is capable of growth in up to 7–8% (w/v) NaCl (Joseph et al., 1982; Baumann et al., 1984). A variety of enrichment broths for *V. parahaemolyticus,* all containing NaCl, appear in the literature. Methods for detection range from enrichment in glucose–salt–teepol broth (GSTB; 3% NaCl) followed by subculture onto thiosulphate citrate bile salts

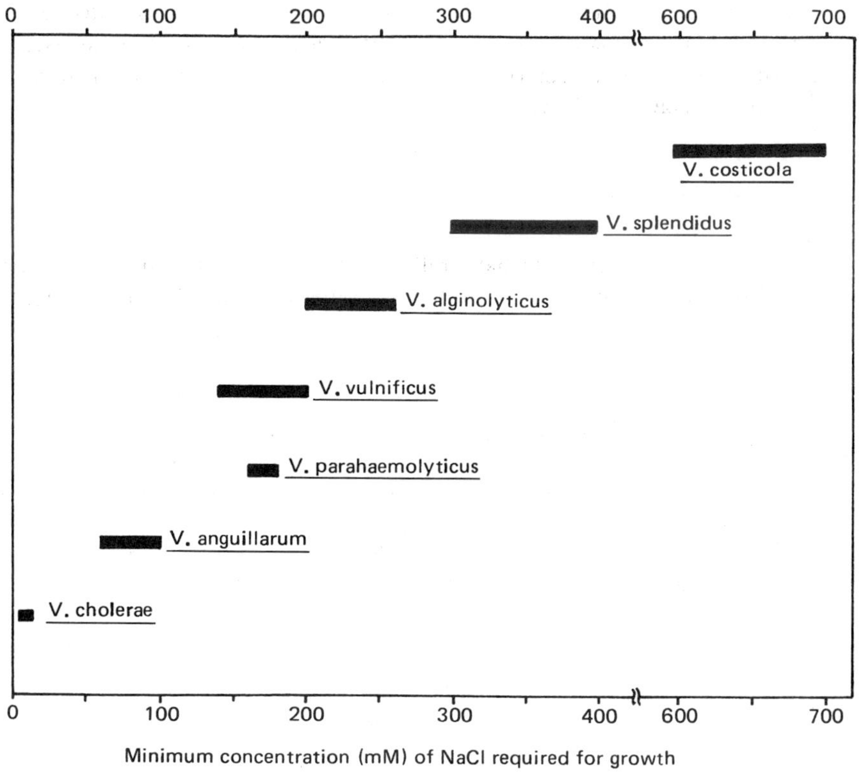

FIG. 8.1 Range of minimal NaCl concentrations required for optimal growth of selected species of *Vibrio.* (*Adapted from Baumann et al., 1984.*)

sucrose (TCBS) agar (ICMSF, 1978; Desmarchelier, 1979; Twedt et al., 1984) to trypticase soy broth (TSB) containing 7% NaCl enrichment, followed by subculture onto modified Twedt medium containing 7% NaCl (Vanderzant et al., 1974). A survey of Dutch mussels for *V. parahaemolyticus* compared five enrichment broths and concluded that enrichment in the 5% salt meat broth described by Kampelmacher et al. (1972) gave the best recovery of this organism (van der Brock et al., 1979).

Salt tolerance is an important characteristic for differentiating between *Vibrio* spp. Growth in 0%, 3%, 6%, 8%, and 10% NaCl media in conjunction with other biochemical tests can be used to distinguish between *V. cholerae, V. parahaemolyticus, V. alginolyticus, V. vulnificus, V. costicola,* and several other *Vibrio* spp. (see Table 8.1; Joseph et al., 1982; Baumann et al., 1984).

Streptococci

Media containing NaCl can be used in confirmatory tests for enterococci. The ICMSF recommended method for confirmation of a *Streptococcus* isolate as an enterococcus is to determine its ability to grow in brain heart infusion broth + 6.5% NaCl at 35–37°C after 72 hr (ICMSF, 1978).

Halophilic Bacteria

Moderately and extremely halophilic bacteria are important in foods because of their ability to grow in, and spoil, foods with high NaCl contents,

TABLE 8.1 Growth Responses of Some *Vibrio* spp. to Various Concentrations of NaCl

	Vibrio spp.				
% NaCl	cholerae	vulnificus	parahaemolyticus	alginolyticus	costicola
0	+	−	−	−	−
3	+	+	+	+	−
6	−	+	+	+	+
8	−	−	+	+	+
10	−	−	−	+	+

Adapted from Baumann et al. (1984).

such as salted and cured meats and fish, pickles, and curing brines. Halophilic bacteria are also important in some Oriental fermentation processes. For example, *Pediococcus halophilus* plays an important role in the production of soy sauce.

Extremely halophilic bacteria, the "red halophiles," belonging to the genera *Halobacterium* and *Halococcus* can grow only in media containing 2.5M (12.5% w/v) or more NaCl, and are capable of growing in saturated (5.2M) NaCl (Kushner, 1978). Although these bacteria can cause spoilage of some extremely salty commodities (e.g., salted, dried fish), the moderately halophilic bacteria, such as *Vibrio costicola, Staphylococcus xylosus,* and some *Micrococcus* spp., are more widespread, causing spoilage of curing brines and salted meats (Kushner, 1978). These bacteria grow best in the presence of 0.5–3.0M NaCl.

Salt must be incorporated into media and diluents used for isolation and enumeration of halophilic bacteria from food. For the examination of cured and pickled meats, salted, pickled and fermented vegetables, salt water fish, and shellfish, Harrigan and McCance (1976) recommend the addition of 15% (w/v) NaCl to diluents and to plate count agar. However, Gardner and Kitchell (1973) recommend 0.1% peptone water with 4% NaCl as a routine diluent for microbiological examination of cured meats. They suggest plate count agar with 2, 4, or 10% (w/v) NaCl, depending on the salt content of the sample. Similar diluents and media are recommended by Gardner (1973) for the microbiological examination of bacon-curing brines. Baross and Matches (1984) recommend phosphate buffer and Tryticase Soy Agar supplemented with NaCl equivalent to the NaCl content of the food being examined.

Isolation of extreme halophiles requires the use of media containing at least 2.5M NaCl (Gibbons, 1969). Dilution in halophilic broth and plating on halophilic agar, or enrichment in halophilic broth with incubation at 35°C for up to 12 days are recommended by Baross and Matches (1984).

ISOLATION OF BACTERIA FROM DEHYDRATED FOODS

Rehydration rate is an important a_w consideration in the recovery of bacteria, particularly pathogens such as *Salmonella,* from dried foods. Van Schothorst et al. (1979) and Andrews et al. (1983) have shown that recovery of *Salmonella* from dried milk powder is substantially increased if the sample is initially reconstituted in a small volume of diluent (1:2 or 1:2.5 sample/diluent ratio) for 1 hr instead of rehydrating to a 1:10 sample/diluent ratio. Normal enrichment or plating procedures can then be carried out. Most bacteria are sensitive to osmotic shock, and gradual rehydration minimizes this effect, particularly on sublethally damaged cells.

ISOLATING AND ENUMERATING FUNGI FROM FOODS

Choosing a Suitable Medium

Because of the wide range of a_w values over which various food spoilage fungi will grow, the choice of medium is very important, as it will determine the types of fungi isolated or enumerated. The characteristics of the foodstuff being examined should be critically assessed, and a medium should be chosen that reflects those characteristics. For high-a_w foods, such as meat, seafood, and fresh fruits, vegetables, and salads, a medium of high a_w would be suitable, but for dried and intermediate moisture foods, such as nuts, grains, spices, and confectionery, a high a_w medium will not enumerate the significant microflora.

If a medium of reduced a_w is needed, the type of solute is the second consideration. For foods high in sugar, a glucose- or glycerol-based medium may be most suitable, while for salty foods, the use of a reduced a_w medium containing some NaCl (not necessarily based entirely on salt) may be more appropriate.

Isolation or enumeration? Media for isolating fungi may not be suitable as enumeration media. This is because some fungi grow much more rapidly than others, and their spreading colonies may obscure smaller, slower growing species which may ultimately be more important in actually spoiling the food. Inhibitors such as dichloran (2,6-dichloro-4-nitroaniline) have been added to fungal enumeration media to overcome this problem (e.g., King et al., 1979; Hocking and Pitt, 1980), but these media are not suitable for isolating some of the more fastidious fungi, such as plant pathogens, or the extreme xerophiles.

Specific examples of choice of media and techniques for particular types of foods are dealt with later in this chapter.

Plating Techniques

Dilution plating. Enumeration of fungi in foods is normally achieved by dilution plating methods similar to those used in bacteriology. The most commonly recommended diluent for fungi is 0.1% peptone water, but other diluents are also suitable. Most fungal spores are not particularly susceptible to osmotic effects. Yeast cells are more sensitive than fungal spores, but are probably not as sensitive as bacteria.

For high-a_w foods, enumeration of the fungal flora by dilution plating is probably the most suitable and convenient method of establishing the mycological status of the food. The sample should be homogenized by stomaching or blending in a suitable diluent, then spread-plated onto a high-a_w fungal enumeration medium, such as dichloran rose bengal chloramphenicol (DRBC) agar (King et al., 1979), rose bengal chlortetracycline (RBC) agar (Jarvis, 1973), or oxytetracycline glucose yeast extract (OGY) agar (Mossel et al., 1970). The first two media are particularly recommended because they contain inhibitors to prevent spread of fungal colonies. These inhibitors, dichloran and rose bengal, induce compact colony development, so that a reasonable number of colonies can be distinguished on a plate. Spread-plating is recommended because with the pour-plating technique, development of colonies within the agar is delayed relative to those growing on the surface.

Many fungi from dried or intermediate moisture foods can also be enumerated by dilution plating. The sample should be rehydrated in the diluent for about 1 hr before stomaching or blending, then spread-plated on a suitable medium. Observations in our laboratory have shown that gradual rehydration can aid the resuscitation of yeast cells and fungal spores, shortening germination times and increasing subsequent growth rates.

For a more detailed description of dilution-plating techniques for fungi in foods see Pitt and Hocking (1985a).

Direct plating. In some instances, dilution plating is not the most effective way of gauging fungal infection or isolating fungi from food. For particulate foods, such as nuts or grains, direct plating can often provide more useful results (Mislivec and Bruce, 1977). Direct plating provides an estimate of the extent of infection in a commodity, which may be expressed as a percentage. Results are often not directly comparable with those obtained by dilution plating.

Direct plating is often the best way to study the degree of contamination of a commodity with a specific fungus, such as *Aspergillus flavus,* particularly if a selective medium, e.g., Aspergillus flavus and parasiticus agar (AFPA) (Pitt et al., 1983), is used. Direct plating is the only satisfactory way of isolating fastidious xerophiles, such as *Xeromyces bisporus, Eremascus* spp., and xerophilic *Chrysosporium* spp.

Most samples should be surface disinfected before plating. Soaking in a solution of sodium hypochlorite (approximately 0.4%) for 2 min, followed by a rinse of sterile water, effectively removes most contaminant spores adhering to the surface of the sample. This enables detection of hyphae that have penetrated and grown in nuts, grains, or similar commodities.

For isolation of extreme xerophiles from low-a_w foods, surface disinfection is not necessary. Samples, such as dried fruits, confectionery, fruit cake, or salted fish, can be cut into small pieces and plated directly onto a suitable medium. Alternatively, the surface of the commodity can be sampled by pressing a piece onto the agar, then removing it, leaving an impression. Any spores or mycelium transferred will form colonies in a few days (Pitt and Hocking, 1985a).

Moderately Xerophilic Fungi

Moderately xerophilic fungi are defined as fungi that are capable of growth below 0.85 a_w, but are not fastidious in their nutrient or a_w requirements. Included in the group are many members of the common food spoilage genera *Penicillium* and *Aspergillus* (especially the *Aspergillus restrictus* series), *Eurotium* spp. (the *Aspergillus glaucus* series), *Wallemia sebi,* and perhaps a few others (Fig. 8.2). These fungi are important in spoilage of a wide range of low moisture foods, especially stored grains, nuts, and spices.

Media of reduced a_w must be used to enumerate these fungi, since many of them grow slowly and compete poorly at high a_w. However, if the a_w of the medium is reduced sufficiently to allow good growth of *Wallemia, Aspergillus*

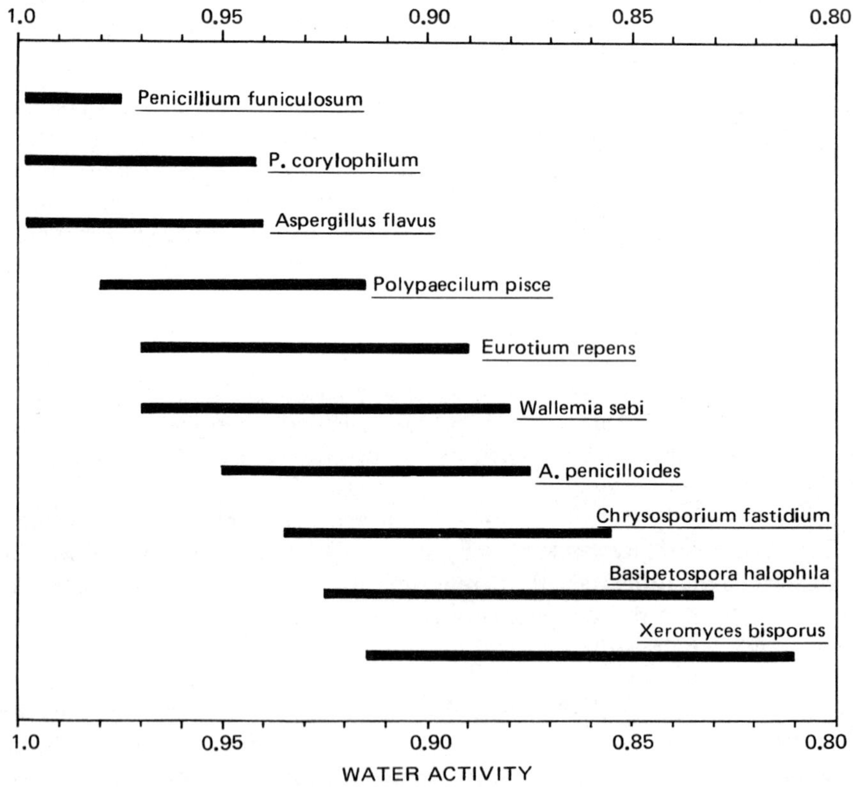

FIG. 8.2 Water activity range permitting 80% or more of the maximum growth rate of some food spoilage fungi.

penicilloides, and *A. restrictus*, *Eurotium* spp. grow luxuriantly, rapidly overgrowing the more slowly developing species. This problem has been largely overcome with the development of dichloran 18% glycerol agar (DG18) (Hocking and Pitt, 1980). DG18 agar contains 2 μg/mL dichloran to inhibit the spreading growth of *Eurotium* colonies, and 18% glycerol to reduce the a_w to 0.955. Media based on sugars are sufficiently rich to allow the *Eurotium* spp. to overcome the inhibitory effects of the dichloran, but glycerol lowers the a_w while maintaining the anti-spreading effect of dichloran. DG18 agar allows slow-growing species, such as *A. penicilloides* and *W. sebi*, to be enumerated in the presence of significant numbers of *Eurotium* colonies. Common species of *Penicillium* and *Aspergillus* also grow well on DG18 agar. The medium has an added advantage in that it encourages the development of the characteristic colored hyphae and yellow cleistothecia of *Eurotium* colonies, allowing some of the species to be differentiated on the plates.

If a high-a_w medium is used to enumerate fungi from a low-a_w commodity, the results can be quite misleading. Table 8.2 shows the differences in counts obtained from a range of low a_w samples which were plated onto DRBC (a_w 0.997) and DG18 (a_w 0.955) media. The differences in some instances were a magnitude of several $\log_{10}$ values, and the fungi that grew on the high a_w medium were often of little significance as spoilage fungi (Hocking, 1981).

Malt salt agar (MSA) (Christensen, 1946) has been used for many years to enumerate fungi from flour and related commodities. However, MSA has a number of weaknesses as an enumeration medium. Because of the reduced a_w,

TABLE 8.2 Comparison of Mold Counts Obtained on DRBC and DG18 Agars for Low-a_w Commodities

	Log total mold count per gram	
Commodity	DRBC	DG18
Semolina	5.4	6.6
Dried chillies 1	4.2	7.0
Dried chillies 2	4.9	8.2
Dried fish 1	0	4.9
Dried fish 2	5.9	6.4
Dried fish 3	4.1	7.7
Flour	5.3	6.4

Source: Hocking (1981).

Eurotium spp. spread rapidly, overgrowing slowly developing species, and sporulating heavily, causing problems with secondary colonies during the necessarily long incubation periods. Because of the high concentration of malt extract (2%), the pH of MSA is quite low (pH 4.1). If the medium is slightly overheated, the low pH and the high salt concentration cause partial breakdown of the agar gel, producing a soft, granular medium that is difficult to use for spread plating. In addition, the high salt content causes precipitation of some malt extracts, which may interfere with counting.

Fastidious Extreme Xerophiles

Fastidious extreme xerophiles are defined as those fungi which *require* reduced a_w for growth, and in addition grow poorly on media based on solutes other than sugars. There are relatively few fastidious extreme xerophiles, but they are important causes of food spoilage in high sugar commodities that, by virtue of their extremely low a_w, would normally be safe from microbial attack. Dried fruits, confectionery, fruit cakes and puddings, chocolate, and spices are susceptible to this type of spoilage if their a_w is between about 0.75 and 0.65. The species responsible are *Xeromyces bisporus,* xerophilic *Chrysosporium* spp. (*C. fastidium, C. inops, C. farinicola* and *C. xerophilum*), and the two *Eremascus* spp., *E. albus* and *E. fertilis* (Pitt, 1975). In our experience, the species most commonly causing spoilage of these products are *X. bisporus* and *C. inops. Eremascus* species are extremely uncommon. Foods with a_w values below 0.65 are generally stable, because, although *X. bisporus* is capable of growth at a_w approaching 0.61 (Pitt and Christian, 1968), growth is so slow that it would take many months for visible colonies to appear.

Fastidious extreme xerophiles either grow extremely poorly, or will not grow at all, on conventional high a_w media (see Fig. 8.2). Some of the xerophilic *Chrysosporium* spp. are able to grow weakly on high-a_w media, but *X. bisporus* will not grow on media above 0.97 a_w (Pitt and Hocking, 1977).

If a low a_w commodity, such as confectionery or dried fruit, shows signs of white mold growth, it is very likely that the fungus responsible is an extreme xerophile. Fungi of this type are usually sensitive to dilution plating, so direct plating is the method of choice. A convenient technique is to place small pieces of sample, without surface disinfection, onto a rich, low-a_w medium such as malt extract yeast extract 50% glucose (MY50G) agar (a_w 0.89; Pitt and Hocking, 1985a). An alternative method is to examine the food under a stereomicroscope and pick off pieces of mycelium or spores from the surface of the spoiled food with an inoculating needle. These pieces are placed directly onto MY50G agar at the rate of three to six inocula per plate. Colonies should develop after 1–3 wk incubation at 25°C. Quite frequently, the species mentioned above will be present in pure culture and are readily isolated by these techniques. Preliminary examination with the stereomicroscope will usually give an indication of whether this is the case; selection of growth that appears to cover the range of types seen will assist isolation of the principal fungi present by either of the above techniques.

Sometimes fastidious extreme xerophiles will be accompanied by *Eurotium* spp., which makes their isolation difficult because they grow much more slowly than *Eurotium* spp. on MY50G agar. Under these circumstances, the most suitable medium is malt extract yeast extract 70% glucose fructose (MY70GF) agar (Pitt and Hocking, 1985a), which is about 0.76 a_w. MY70GF agar is of similar composition to MY50G agar, except that equal parts of glucose and fructose are used to prevent crystallization of the medium at the concentration used (70% w/w). Growth of even the extreme xerophiles on MY70GF agar is slow, so plates should be incubated for at least 4 wk at 25°C in closed containers to prevent drying out. Once growth is apparent, small portions of the colonies should be picked off and transferred to MY50G agar to allow more rapid growth and sporulation.

Halophilic Fungi

A specific requirement for salt has not been demonstrated for any fungus. However, two species—*Basipetospora halophilia* and *Polypaecilum pisce*—have been shown to grow better in the presence of salt (Pitt and Hocking, 1985b). Both these species are common on salted, dried fish and squid from Southeast Asia, *P. pisce* being dominant on many samples of Indonesian dried fish examined in our laboratory. *Wallemia sebi*, a common spoilage fungus on salted fish from temperate regions (Frank and Hess, 1941), was rarely isolated from these tropical fish samples. This species does not have a salt requirement (Pitt and Hocking, 1977) but competes well against other xerophiles on salty foods. It is also very common on cured meats and meat products.

Two media have been developed in our laboratory for isolating halophilic fungi from salted fish. They are based on mixtures of salt and glucose. For isolation of *P. pisce*, malt extract yeast extract 5% salt 12% glucose (MY5-12) agar is recommended. *B. halophila* grows more rapidly on MY10-12 agar, which contains 10% rather than 5% salt (Pitt and Hocking, 1985a). These media have been designed as isolation media rather than enumeration media. Their high salt content makes them heat sensitive, and they may be too soft to be used for spread plating.

Yeasts in Low-a_w Commodities

Enumerating yeasts in highly viscous syrups and concentrates can be a problem, particularly if they are present in very low numbers. If fermentation is visible, yeasts will probably be present in high enough numbers to detect them by dilution plating. To minimize osmotic shock, the diluent should contain 20% sucrose or glucose.

Various media have been developed for enumeration of yeasts in syrups and fruit concentrates. Tilbury (1976) recommends Scarr's osmophilic agar (Scarr, 1959) as being simple to prepare and easy to use, but points out that the relatively high a_w (0.95) allows growth of some nonxerophilic yeasts, and occasionally bacteria as well. In our laboratory, we routinely use MY50G agar for isolation and enumeration of yeasts in low-a_w commodities. Incubation at 25°C for up to 2 wk is recommended.

Restaino et al. (1985) reported that the addition of 60% sucrose to potato dextrose agar (PDA) dramatically improved the recovery of the osmophilic yeast, *Saccharomyces rouxii,* from chocolate syrup, particularly during the log phase of growth (Fig. 8.3). Only 1–10% of the *S. rouxii* cells recovered from PDA/60% sucrose plating medium (a_w 0.92) were enumerated on standard PDA between 1 and 3 days. Restaino et al. (1985) recommended that a 60%

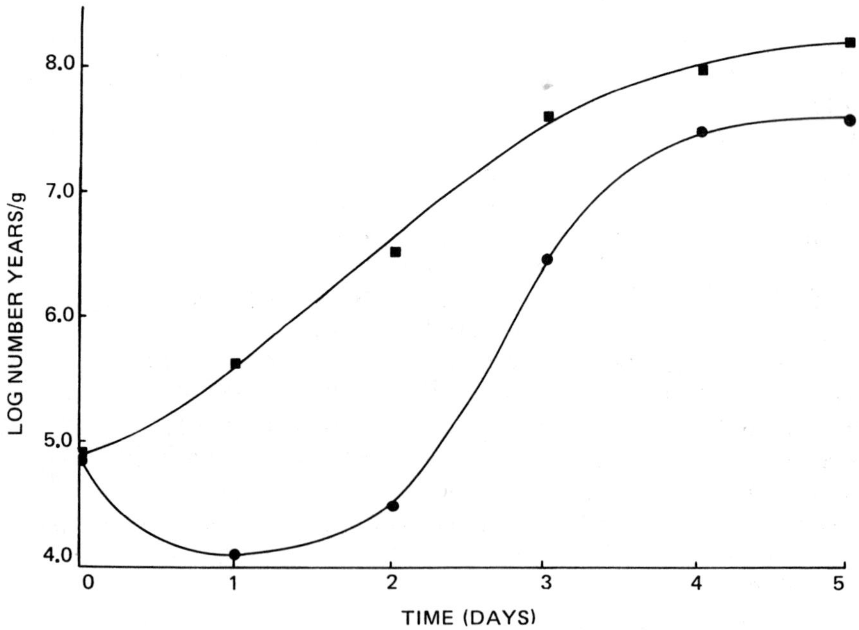

FIG. 8.3 Enumeration of *Saccharomyces rouxii* from chocolate syrup using standard potato dextrose agar (PDA; ●) and PDA with 60% sucrose (■). (*From Restaino et al., 1985.*)

sucrose–phosphate buffer diluent should be used in conjunction with the PDA/60% sucrose plating medium.

If yeasts are suspected of being present in very low numbers and the product is too viscous to be passed through a membrane filter, then the most practical method of detection is enrichment in the product itself. Dilute the product 50:50 with sterile distilled water, and then incubate for up to 4 wk. The dilution will not be sufficient to cause osmotic shock, and the increase in a_w will allow any yeasts present to grow more rapidly and thus be detected earlier.

CONCLUSION

In the isolation and enumeration of significant bacteria or fungi in foods, it is important to carefully consider the types of microorganisms capable of either spoiling that foodstuff or creating a public health risk. Appropriate methods for their recovery and enumeration can then be chosen. Gradual rehydration of dried foods is important for the recovery of pathogens such as *Salmonella* and for sublethally damaged cells.

The choice of an enumeration medium is as important in food mycology as it is in food bacteriology, because the medium used will strongly affect the types of microorganisms recovered from the food. An enumeration medium that reflects the a_w of the food will give the best indication of the load of potential spoilage microorganisms.

Formulas of selected media suitable for isolating and enumerating microorganisms from foods of reduced a_w are listed in the appendix.

ACKNOWLEDGMENT

We thank Dr. M. J. Eyles for his constructive suggestions concerning the bacteriological section of this manuscript.

APPENDIX—SELECTED MEDIA

Halophilic Agar

Casamino acids	10.0 g
Yeast extract	10.0 g

Proteose peptone	5.0 g
Trisodium citrate	3.0 g
KCl	2.0 g
$MgSO_4 \cdot 7H_2O$	25.0 g
Sodium chloride	250.0 g
Agar	20.0 g
Distilled water	1.0 liter

Dissolve ingredients in distilled water and sterilize at 121°C for 15 min. Final pH should be 7.2.

Halophilic Broth

Prepare as for halophilic agar except omit agar. This medium may be used as a diluent or as an enrichment medium for the isolation of extremely halophilic bacteria.

Dichloran Rose Bengal Chloramphenicol (DRBC) Agar

Glucose	10.0 g
Peptone	5.0 g
KH_2PO_4	1.0 g
$MgSO_4 \cdot 7H_2O$	0.5 g
Agar	15.0 g
Distilled water	1.0 liter
Rose Bengal (5% [w/v] in water, 0.5 mL)	0.025 g
Dichloran (0.2% [w/v] in ethanol, 1.0 mL)	0.002 g
Chloramphenicol	0.100 g

After addition of all ingredients, sterilize by autoclaving at 121°C for 15 min. Store prepared media away from light, which causes slow decomposition of rose bengal.

Dichloran 18% Glycerol (DG18) Agar

Glucose	10.0 g
Peptone	5.0 g

KH_2PO_4	1.0 g
$MgSO_4 \cdot 7H_2O$	0.5 g
Glycerol, A. R. grade	220 g
Agar	15.0 g
Distilled water	1.0 liter
Dichloran (0.2% [w/v] in ethanol, 1.0 mL)	0.002 g
Chloramphenicol	0.100 g

Add all ingredients except glycerol to approximately 800 mL distilled water. Steam to dissolve agar, then make up to 1 liter with distilled water. Add glycerol, giving a final concentration of 18% (w/w). Sterilize by autoclaving at 121°C for 15 min. The final a_w of this medium is 0.955.

Malt Extract Yeast Extract 50% Glucose (MY50G) Agar

Malt extract	10.0 g
Yeast extract	2.5 g
Agar	10.0 g
Distilled water	500.0 g
Glucose, A.R. grade	500.0 g

Add the minor constituents and agar to approx. 450 mL distilled water, and steam to dissolve agar. Immediately make up to 500 g with distilled water. While the solution is still hot, add the glucose all at once, and stir rapidly to prevent the formation of hard lumps of monohydrate. If lumps do form, dissolve them by steaming for a few minutes. Sterilize by steaming for 30 min. Note that this medium does not require autoclaving. Glucose monohydrate (dextrose) may be used in this medium instead of A.R. glucose, but allowance must be made for the additional water present. Use 550g dextrose and 450g basal medium. The final a_w of this medium is 0.89.

Malt Extract Yeast Extract 70% Glucose Fructose (MY70GF) Agar

Malt extract	6.0 g
Yeast extract	1.5 g
Agar	6.0 g
Distilled water	300.0 g
Glucose, A.R. grade	350.0 g
Fructose, A.R. grade	350.0 g

The procedure for making MY70GF agar is the same as that for MY50G agar. After steaming to dissolve the agar, make the solution up to 300g with distilled water, and, while still hot, add both sugars. Steam gently to dissolve if necessary. Sterilize by steaming for 30 min. Do not steam for a longer period, or autoclave, because browning reactions will render the medium toxic to fungi. MY70GF agar will take some hours to gel, because of the low proportion of water and agar. The final a_w of MY70GF agar is approximately 0.76.

Malt Extract Yeast Extract 5% Salt 12% Glucose (MY5-12) Agar

Malt extract	20.0 g
Yeast extract	5.0 g
NaCl	50.0 g
Glucose, A.R. grade	120.0 g
Agar	20.0 g
Distilled water	to 1.0 liter

Dissolve ingredients by steaming, and then sterilize by autoclaving at 121°C for 10 min. Overheating will cause softening. The final a_w of MY5-12 agar is 0.96.

Malt Extract Yeast Extract 10% Salt 12% Glucose (MY10-12) Agar

Prepare MY10-12 agar as for MY5-12 agar, but with the addition of 100g NaCl rather than 50g. Sterilize by steaming for 30 min. This medium should not be autoclaved. The final a_w of MY10-12 agar is 0.93.

Malt Salt (MSA) Agar

Malt extract	20.0 g
NaCl	75.0 g
Agar	20.0 g
Distilled water	1.0 liter

Dissolve ingredients by steaming, and then sterilize by autoclaving at 121°C for 15 min.

Scarr's Osmophilic Agar

Dissolve Difco wort agar in a 45° Brix syrup consisting of 35 parts of sucrose and 10 parts of glucose. Sterilize by autoclaving at 105°C for 20 min.

Potato Dextrose Agar (PDA)

Potatoes	250 g
Glucose	20 g
Agar	15 g
Distilled water	1 liter

Wash the potatoes, dice or slice (unpeeled), into 500 mL water. Steam or boil for 30–45 min. At the same time, melt agar in 500 mL water. Strain the potatoes through several layers of cheesecloth into the flask containing the melted agar. Squeeze some of the potato pulp through also. Add glucose, mix thoroughly, and make up to 1 liter with water if necessary. Sterilize by autoclaving at 121°C for 15 min. (*Note:* Red-skinned potatoes are not suitable for making PDA.)

REFERENCES

Andrews, W. H., Wilson, C. R., and Poelma, P. L. 1983. Improved *Salmonella* species recovery from nonfat dry milk pre-enriched under reduced rehydration. *J. Food Sci.* 48: 1162.

Baird-Parker, A. C. 1962. An improved diagnostic medium for isolating coagulase positive staphylococci. *J. Appl. Bacteriol.* 25: 12.

Baird-Parker, A. C. and Eyles, M. J. 1979. *Staphylococcus aureus.* In *Foodborne Microorganisms of Public Health Significance.* Vol. 1. Buckle, K. A., Davey, G. R., Eyles, M. J., Fleet, G. H., and Murrell, W. G. (Ed.), p. 10.1. Australian Inst. of Food Science & Technology-CSIRO Div. of Food Research-Univ. of NSW, Sydney.

Baross, J. A. and Matches, J. R. 1984. Halophilic microorganisms. In *Compendium of Methods for the Microbiological Examination of Foods,* 2nd ed. Speck, M. L. (Ed.), p. 160. American Public Health Assoc., Washington, DC.

Baumann, P., Furniss, A. L., and Lee, J. L. 1984. Genus *Vibrio.* In *Bergey's Manual of Systematic Bacteriology.* Vol. 1. Krieg, N. R. (Ed.), p. 518. Williams & Wilkins, Baltimore.

Christensen, C. M. 1946. The quantitative determination of molds in flour. *Cereal Chem.* 23: 322.

Desmarchelier, P. 1979. *Vibrio parahaemolyticus.* In *Foodborne Microorganisms of Public Health Significance.* Vol. 1. Buckle, K. A., Davey, G. R., Eyles, M. J., Fleet, G. H., and Murrell, W. G. (Ed.), p. 15.1. Australian Inst. of Food Science & Technology-CSIRO Div. of Food Research-Univ. of NSW, Sydney.

Frank, M. and Hess, E. 1941. Studies on salt fish. V. Studies on *Sporendonema epizoum* from "dun" salt fish. *J. Fish Res. Bd. Can.* 5: 276.

Furniss, A. L., Lee, J. V., and Donovan, T. J. 1978. *The Vibrios.* Public Health Laboratory Service Monograph Series No. 11. Her Majesty's Stationery Office, London.

Gardner, G. A. 1973. Routine microbiological examination of Wiltshire bacon curing brines. In *Sampling—Microbiological Monitoring of Environments.* Board, R. G. and Lovelock, D. W. (Ed.), p. 21. Academic Press, London.

Gardner, G. A. and Kitchell, A. G. 1973. The microbiological enumeration of cured meats. In *Sampling—Microbiological Monitoring of Environments.* Board, R. G. and Lovelock, D. W. (Ed.), p. 14. Academic Press, London.

Gibbons, N. E. 1969. Isolation, growth and requirements of halophilic bacteria. In *Methods in Microbiology.* Vol. 3B. Norris, F. N. and Ribbons, D. W. (Ed.), p. 169. Academic Press, New York.

Gilbert, R. J., Kendall, M., and Hobbs, B. C. 1969. Media for the isolation and enumeration of coagulase positive staphylococci from foods. In *Isolation Methods for Microbiologists.* Shapton, D. A. and Gould, G. W. (Ed.), p. 9. Academic Press, London.

Harrigan, W. F. and McCance, M. E. 1976. *Laboratory Methods in Food and Dairy Microbiology.* Academic Press, London.

Hocking, A. D. 1981. Improved media for enumeration of fungi from foods. *CSIRO Food Res. Q.* 41: 7.

Hocking, A. D. and Pitt, J. I. 1980. Dichloran-glycerol medium for enumeration of xerophilic fungi from low moisture foods. *Appl. Environ. Microbiol.* 39: 488.

ICMSF. 1978. *Microorganisms in Foods. 1. Their Significance and Methods of Enumeration,* 2nd. ed. International Commission on Microbiological Specifications for Foods, Univ. of Toronto Press, Toronto.

Jarvis, B. 1973. Comparison of an improved rose bengal-chlortetracycline agar with other media for the selective isolation and enumeration of moulds and yeasts in foods. *J. Appl. Bacteriol.* 36: 723.

Joseph, S. W., Colwell, R. R., and Kaper, J. B. 1982. *Vibrio parahaemolyticus* and related halophilic vibrios. *Crit. Rev. Microbiol.* 10: 77.

Kampelmacher, E. H., van Noorle Janson, L. M., Mossel, D. A. A., and Groen, F. J. 1972. A survey of the occurrence of *Vibrio parahaemolyticus* and *V. alginolyticus* on mussels and oysters and in estuarine waters in the Netherlands. *J. Appl. Bacteriol.* 35: 431.

King, A. D., Hocking, A. D., and Pitt, J. I. 1979. Dichloran-rose bengal medium for enumeration and isolation of molds from foods. *Appl. Environ. Microbiol.* 37: 959.

Kushner, D. J. 1978. Life in high salt and solute concentrations: halophilic bacteria. In *Microbial Life in Extreme Environments.* Kushner, D. J. (Ed.), p. 317. Academic Press, London.

Mislivec, P. B. and Bruce, V. R. 1977. Direct plating versus dilution plating in qualitatively determining the mold flora of dried beans and soybeans. *J. Assoc. Off. Anal. Chem.* 60: 741.

Mossel, D. A. A., Kleyen-Semmeling, A. M. C., Vincentie, H. M., Beerens, H., and Catsaras, M. 1970. Oxytetracycline-glucose-yeast extract agar for selective enumeration of moulds and yeasts in foods and clinical material. *J. Appl. Bacteriol.* 33: 454.

Pitt, J. I. 1975. Xerophilic fungi and the spoilage of foods of plant origin. In *Water Relations of Foods.* Duckworth, R. B. (Ed.), p. 273. Academic Press, London.

Pitt, J. I. and Christian, J. H. B. 1968. Water relations of xerophilic fungi isolated from prunes. *Appl. Microbiol.* 16: 1853.

Pitt, J. I. and Hocking, A. D. 1977. Influence of solute and hydrogen ion concentration on the water relations of some xerophilic fungi. *J. Gen. Microbiol.* 101: 35.

Pitt, J. I. and Hocking, A. D. 1985a. *Fungi and Food Spoilage.* Academic Press, Sydney.

Pitt, J. I. and Hocking, A. D. 1985b. New species of fungi from Indonesian dried fish. *Mycotaxon* 22: 197.

Pitt, J. I., Hocking, A. D., and Glenn, D. R. 1983. An improved medium for the detection of *Aspergillus flavus* and *A. parasiticus. J. Appl. Bacteriol.* 54: 109.

Restaino, L., Bills, S., and Lenovich, L. M. 1985. Growth response of an osmotolerant, sorbate-resistant *Saccharomyces rouxii* strain: Evaluation of plating media. *J. Food Protect.* 48: 207.

Scarr, M. P. 1959. Selective media used in the microbiological examination of sugar products. *J. Sci. Food Agric.* 10: 678.

Scott, W. J. 1953. Water relations of *Staphylococcus aureus* at 30°C. *Aust. J. Biol. Sci.* 6: 549.

Tatini, S. R., Hoover, D. G., and Lachica, R. V. F. 1984. Methods for the isolation and enumeration of *Staphylococcus aureus.* In *Compendium of Meth-*

ods for the Microbiological Examination of Foods, 2nd ed. Speck, M. L. (Ed.), p. 411. American Public Health Assoc., Washington, DC.

Tilbury, R. H. 1976. The microbial stability of intermediate moisture foods with respect to yeasts. In *Intermediate Moisture Foods.* Davis, R., Birch, G. G., and Parker, K. J. (Ed.), p. 138. Applied Science Publishers, Essex.

Twedt, R. M., Maddon, J. M., and Colwell, R. R. 1984. *Vibrio.* In *Compendium of Methods for the Microbiological Examination of Foods,* 2nd ed. Speck, M. L. (Ed.), p. 368. American Public Health Assoc., Washington, DC.

Van der Brock, M. J. M., Mossel, D. A. A., and Eggenkamp, A. E. 1979. Occurrence of *Vibrio parahaemolyticus* in Dutch mussels. *Appl. Environ. Microbiol.* 37: 438.

Van Schothorst, M., van Leusden, F. M., de Gier, E., Rijnierse, V. F. M., and Veen, A. J. D. 1979. Influence of reconstitution on isolation of *Salmonella* from dried milk. *J. Food Protect.* 42: 936.

Vanderzant, C., Nickelson, R., and Hazelwood, R. W. 1974. Effect of isolation-enumeration procedures on the recovery of normal and stressed cells of *Vibrio parahaemolyticus.* In *International Symposium on Vibrio parahaemolyticus.* Fujino, T., Sakaguchi, S., Sakazaki, R., and Takeda, Y. (Ed.), p. 111. Saikon, Tokyo.

9

Influences of Hysteresis and Temperature on Moisture Sorption Isotherms

John G. Kapsalis

Science and Advanced Technology Directorate
U.S. Army Natick Research, Development
& Engineering Center
Natick, Massachusetts

INTRODUCTION

Since the introduction of the term "water activity" by the microbiologist Scott (1957) on the role of a parameter other than "percent moisture" in controlling water requirements of microorganisms, and the pioneering work of Rockland (1957), Salwin (1962), and others on the protective role of a certain amount of water against oxidative changes in low-moisture foods, the moisture sorption isotherm has been used to provide answers and insights to a diversity of questions. In addition to applications by microbiologists and chemists, the sorption isotherm has been used by engineers to predict drying times and estimate the energy of dehydration, and by packaging technologists to predict the moisture transfer in multicomponent food systems enclosed in a vapor-proof container.

Successful application of the sorption isotherm, especially to new problems, depends on a basic understanding of how two central experimental conditions affect its properties and the physicochemical quantities derived from it. These are: (1) the method which has been followed in determining the moisture sorption curve, i.e., whether by adsorption or desorption, and (2) the temperature. The first of these is related to the existence of hysteresis, and the second to shifts of a_w with changing temperature.

The purpose of this chapter is to update our previous review of moisture sorption hysteresis in foods (Kapsalis, 1981) and to examine the role of temperature on the characteristics of the isotherm (equilibrium sorption) and on the thermodynamic quantities which are computed using the isotherm.

HYSTERESIS

Characteristics of Sorption Hysteresis in Foods

General considerations. Water vapor sorption hysteresis is the phenomenon according to which two different paths exist between the adsorption and desorption isotherms.

If one plots the amount of water per unit mass of solid in the ordinate and the corresponding relative vapor pressure in the abscissa, the desorption isotherm usually lies above the adsorption isotherm and a closed *hysteresis loop* is formed (Fig. 9.1).

Moisture sorption hysteresis has important theoretical and practical implications in foods. The theoretical implications range from general aspects of the irreversibility of the sorption process to the question of validity of thermodynamic functions derived from such a system. The practical implications deal with the effects of hysteresis on chemical and microbiological deterioration, and with its importance in low and intermediate moisture foods.

Due to hysteresis, a much lower vapor pressure is required to reach a certain amount of water by desorption than by adsorption. In nature, hysteresis may be considered as a built-in protective mechanism against extremities, such as loss of water due to a dry atmosphere, frost damage, and freezer burn.

The objective of this chapter is to selectively examine the effect of hysteresis and temperature on the sorption behavior of foods.

For a more extensive treatment of the subject, especially of the effect of hysteresis on chemical and microbiological deteriorations, the reader is referred to the work of Accot and Labuza (1975), Labuza et al. (1972a, b), and Labuza and Chou (1974), as well as to our earlier review (Kapsalis, 1981). Recent reports on hysteresis have been published by Bizot et al. (1985) and Johnston and Duckworth (1985). Experimental work and discussion of hysteresis in polymers has been published by Watt (1980a). The same author has published a comprehensive in-depth review of sorption phenomena in keratin

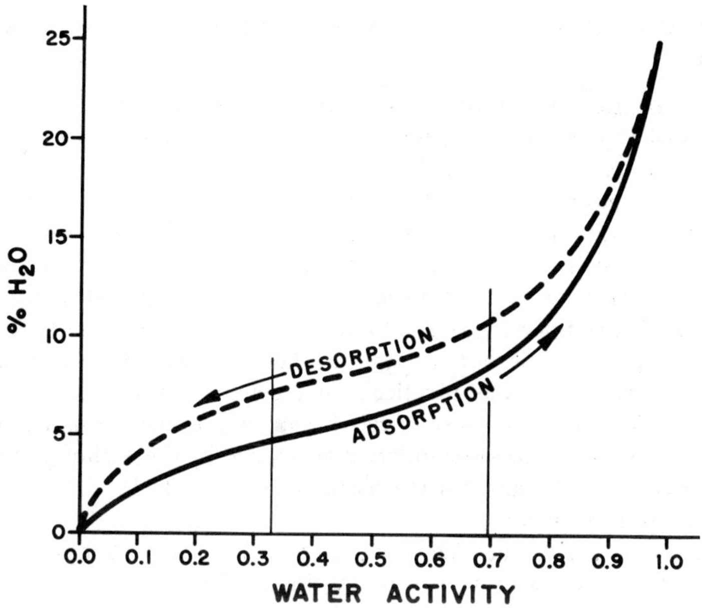

FIG. 9.1 Schematic representation of a moisture sorption hysteresis.

(Watt, 1980b). Some of the theoretical foundations of the subject have been contributed by the earlier work of, among others, Arnell and McDermot (1957), Everett (1967), McLaren and Rowen (1951), and Rao (1939a, b; 1941; 1942).

Types of hysteresis. A variety of hysteresis loop shapes can be observed in foods. Wolf et al. (1972) reported wide differences in the magnitude, shape and extent of hysteresis of dehydrated foods, depending on the type of food and the temperature. Variations could be grouped into three general types:

In high-sugar, high-pectin foods, exemplified by air-dried apple, hysteresis was pronounced mainly in the lower moisture content region, below the first inflection point of the sorption isotherm. Although the total hysteresis was large, there was no hysteresis above a_w 0.065.

In high-protein foods, exemplified by freeze-dried pork (middle), a moderate hysteresis began at high a_w, in the capillary condensation region, and extended over the rest of the isotherm to zero a_w. In both adsorption and desorption, the isotherms retained the characteristic sigmoid shape for proteins.

In starchy foods, as in freeze-dried rice, a large hysteresis loop occurred with a maximum at about a_w 0.70, which is within the capillary condensation region.

Effect of temperature on hysteresis. In the same work by Wolf et al. (1972) mentioned above, increasing temperature decreased the total hysteresis and limited the span of the loop along the isotherm. The latter change was dramatic for the apple, where the beginning of hysteresis was shifted from a_w 0.65 to a_w 0.20, and for the pork, where the beginning of hysteresis shifted from a_w 0.95 to a_w 0.60. The temperature dependence of hysteresis was in agreement with the work reported by Benson and Richardson (1955) for ethyl alcohol sorption on egg albumin, and at variance with the results on bovine serum albumin reported by Seehof et al. (1953). The latter workers observed that the amount of hysteresis was practically independent of temperature and was constant over the entire range of relative vapor pressures. This was probably due to the small temperature differential covered by the Seehof group, and to the presence in the pork (in the work of Wolf et al., 1972) of other, non-protein-absorbing species.

Iglesias and Chirife (1976b) reported that the effect of temperature on the magnitude of hysteresis varied among foods (Table 9.1). For some foods, increasing temperature decreased or eliminated hysteresis (thyme, winter savory, sweet marjoram, trout cooked, chicken cooked, chicken raw, and tapioca), while for others the total hysteresis remained constant (ginger and nutmeg) or even increased (anise, cinnamon, chamomile, and coriander).

Persistence of hysteresis upon adsorption–desorption cycles. Desorption isotherms usually give a higher water content than adsorption isotherms at the same a_w, with exceptions only in specific cases (Berlin and Anderson, 1975; Rutman, 1967). In general, the type of changes encountered upon adsorption and desorption will depend on (a) the initial state of the sorbent (amorphous versus crystalline), (b) the transitions taking place during adsorption, (c) the final a_w adsorption point, and (d) the rate of desorption. With regard to (c) and (d), if the saturation point has been reached and the material has gone into solution, rapid desorption may preserve the amorphous state due to supersaturation. (For phase transition and hysteresis, see Mellon and Hoover, 1951.)

Figure 9.2 shows a hysteresis effect, for the a_w range covered, in freeze-dried beet powder at 35°C reported by Cohen and Saguy (1983). The 3% water remaining after desorption at "dry" conditions even after prolonged storage time was attributed to water trapped (H-bonded) in amorphous sugar micro-regions, as well as water of crystallinity. This residual moisture was similar to our findings on air-dried apple (Wolf et al., 1972).

Hysteresis seems to be reproducible and quite persistent over many adsorption–desorption scans, especially at low temperatures and over relatively short

TABLE 9.1 Hysteresis Units at Different Temperatures for Various Foods

Product	Temperature (°C)	Hysteresis units (arbitrary)
Anise	25	10.1
	45	13.1
Chamomile	25	17.3
	60	36.1
Chicken, cooked	45	12.2
	60	7.3
Chicken, raw	45	7.9
	60	0
Cinnamon	25	10.7
	45	15.1
Coriander	25	15.7
	45	16.3
Ginger	25	8.0
	45	8.0
Nutmeg	25	5.9
	45	5.9
Sweet Marjoram	25	13.5
	45	0
Tapioca	25	24.1
	45	16.6
Trout, cooked	45	11.4
	60	7.0
Thyme	25	6.8
	45	0
Winter Savory	25	14.1
	45	0

Source: Iglesias and Chirife (1976b).

periods of time (Benson and Richardson, 1955; Strasser, 1969). However, exceptions to this are many. Figure 9.3 shows an example of both the effect of temperature and of recycling on the sorption isotherms of rice (Benado and Rizvi, 1985). The magnitude of the hysteresis loop decreased appreciably at the higher temperature. The effect of temperature was more pronounced on the desorption isotherms than on the adsorption isotherms. Cycling the adsorption–desorption for a second time resulted in the elimination of hysteresis and in a reversible isotherm, as shown for comparison in the 20°C isotherms of rice (Fig. 9.4).

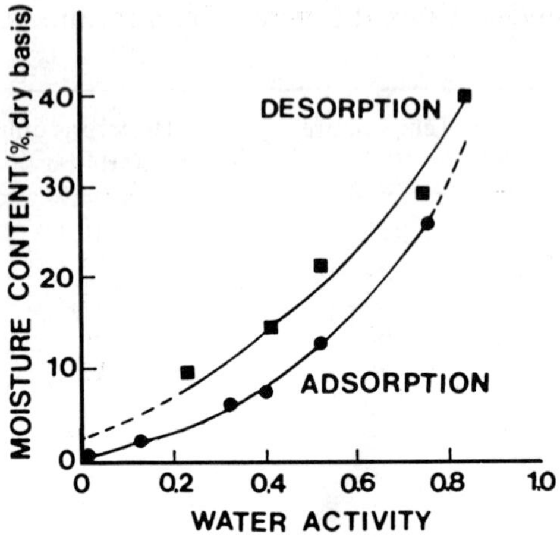

FIG. 9.2 Sorption isotherms of freeze-dried beet powder at 35°C. (*From Cohen and Saguy, 1983.*)

Rao (1942) attributed such changes of sorption phenomena resulting in the elimination of hysteresis to the elastic properties of organogels. In sorption, the capillary pores of the adsorbent become elastic and swell. Upon desorption, the removal of water causes shrinkage and a general collapse of the capillary porous structure. Alteration of structure causes elimination of hysteresis due to the abscence of capillary condensation. The key word here is "alteration." From both the structural and physicochemical points of view, one is dealing here with a modified material.

In reporting magnitudes of hysteresis loops, one should always state the procedure followed and the time allowed for equilibration. Bizot et al. (1985) reported slow drifts of desorption pseudo equilibria (−1% water, dry basis) for bulk starch samples stored over saturated salt solutions under controlled temperature (±1°C) continuing for two years, while adsorbing samples remained stable.

Elimination of hysteresis upon the second or subsequent cycles may take place for a variety of reasons. These include "mixed" hysteresis, "time-dependent hysteresis" when the experiment is carried out very slowly, change in the crystalline structure when a new crystalline form persists upon subsequent cycles (Berlin et al., 1968), swelling and increased elasticity of capillary walls resulting in a loss of power of trapping water (Rao, 1939a, b), denaturation (Chung, 1966), surface-active agents (Rutman, 1967), gelatinization (van den

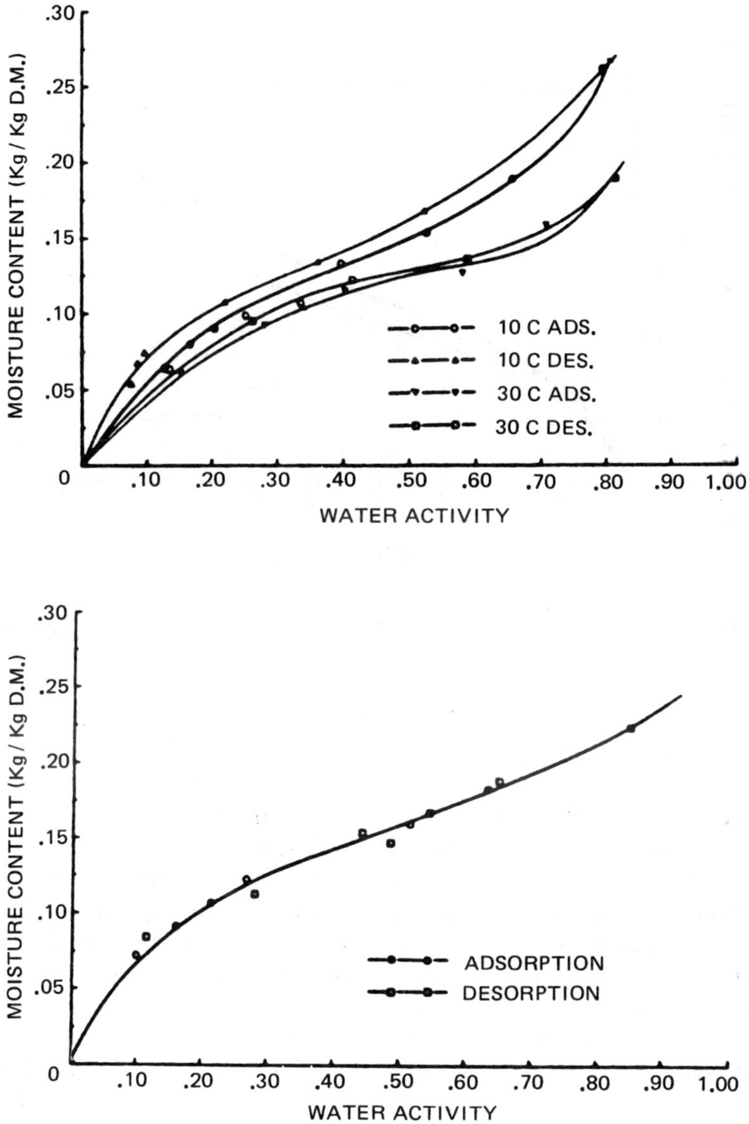

FIG. 9.3 Sorption isotherms of dehydrated rice. Upper, first adsorption–desorption cycle; lower, adsorption and desorption isotherm points at 10°C taken during the second adsorption–desorption cycle. (*From Benado and Rizvi, 1985.*)

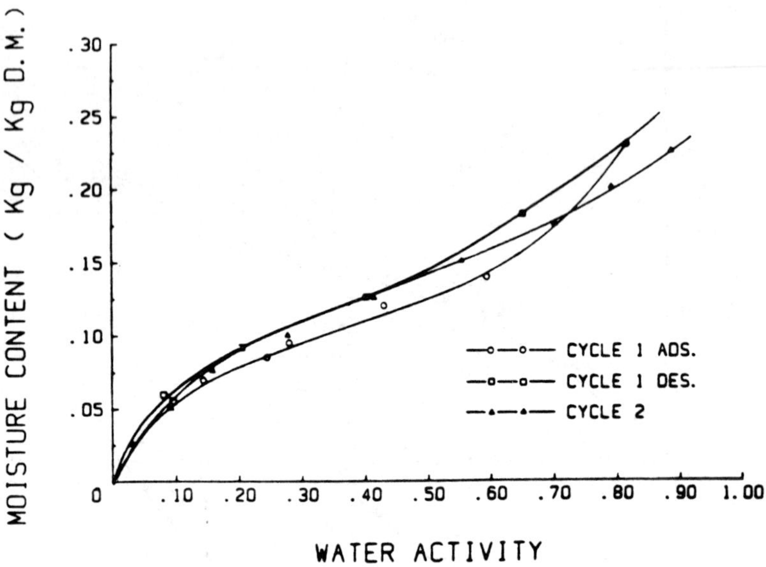

FIG. 9.4 Comparison of cycle 1 and cycle 2 isotherms of dehydrated rice at 20°C. (*From Benado and Rizvi, 1985.*)

Berg et al., 1975), and even mechanical treatment which may affect the capillary structure.

The sorption path between the boundaries and within the hysteresis loop depends on the point of the adsorption or desorption branch where the direction of sorption was reversed. In general the desorption process has a greater effect on hysteresis than the adsorption process. For example, in wool keratin where the adsorption branch seems to be more reproducible than the desorption branch, hysteresis can be eliminated by desorbing from saturation directly to low humidities.

Figure 9.5 shows schematically adsorbing–resorbing scanning curves of potato starch from the work of Bizot et al. (1985). The crossing of the loop by scanning curves has been attributed to the entrapment of water in small capillaries until the main desorption boundary is reached. At this point water is trapped mainly in the larger diameter pores (Rao, 1939b). The characteristics of the scanning paths are important in testing the predictive power of any theory of hysteresis.

In general, for most foods, hysteresis persists through a considerable part of the a_w range. The existence of high hysteresis at extremely low a_w shown in Fig. 9.6 tends to disprove the "capillary condensation" theory as the sole

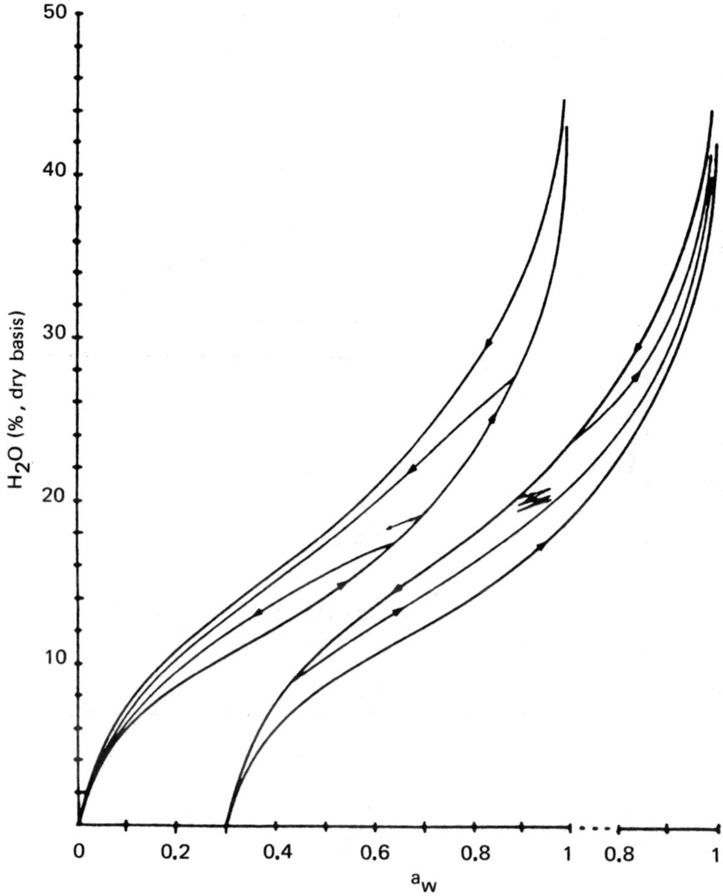

FIG. 9.5 Schematic evolutions of intermediate states for desorbing and resorbing scanning curves on native potato starch at 25°C. (*From Bizot et al., 1985.*)

causative factor (Iglesias and Chirife, 1976b). This is also applicable to materials such as starch where the network could be considered glassy or rubbery depending on the degree of hydration and where capillary condensation plays a minor role in water sorption (van den Berg, 1985).

Thermodynamic considerations. In the work of Benado and Rizvi (1985), the reversible isotherms and thermodynamic functions calculated corresponded to neither the adsorption nor the desorption branch of the isotherms

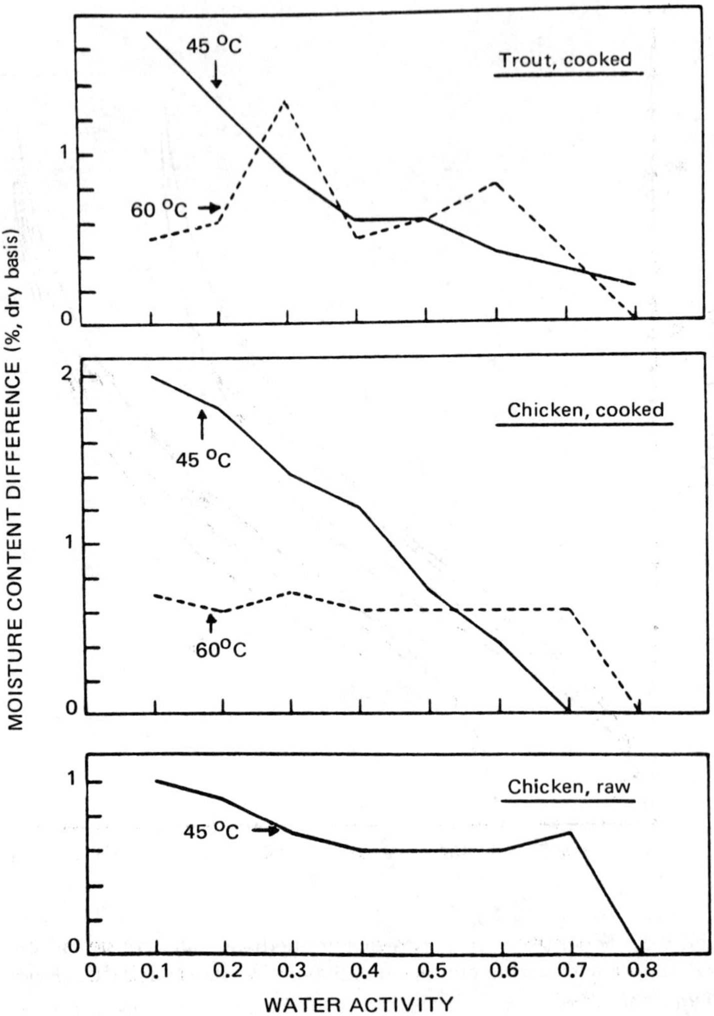

FIG. 9.6 Distribution of hysteresis relative to water activity for trout cooked, at 45 and 60°C; chicken cooked, at 45 and 60°C; and chicken raw, at 45°C. (*From Iglesias and Chirife, 1976b.*)

which exhibited hysteresis but lay in between. Thermodynamic functions calculated from the adsorption branch of the hysteresis data corresponded more closely to those calculated from the reversible isotherms. Figure 9.7 shows the entropy of sorption versus moisture content curves at 20°C for the adsorption, desorption, and reversible rice isotherms. The entropy begins to decline relatively steeply at about 8% moisture, passes through a minimum at 15% moisture, and then increases with increasing moisture, approaching the entropy of free liquid water. The decrease of entropy in the low a_w range is due to (1) lateral interactions in the adsorbed film caused by the restrictive effect of adsorbed water molecules as the available sites become saturated, and (2) structural alterations of the adsorbing food toward increased crystallinity. The rise in the entropy curve beyond the 15% moisture points indicates that the newly bound water molecules are held less strongly; they possess more degrees of freedom as a result of a gradual opening and swelling of the polymer (Bettleheim et al., 1970). Independent X-ray diffraction studies on starch by Guilbot et al. (1961) tend to support the above interpretation. These workers found discontinuity in the X-ray patterns at 9 and 15% moisture.

Changes in entropy along the isotherm were used by Kapsalis et al. (1970) and Kapsalis (1975) to illuminate changes of the structure of the adsorbent. In

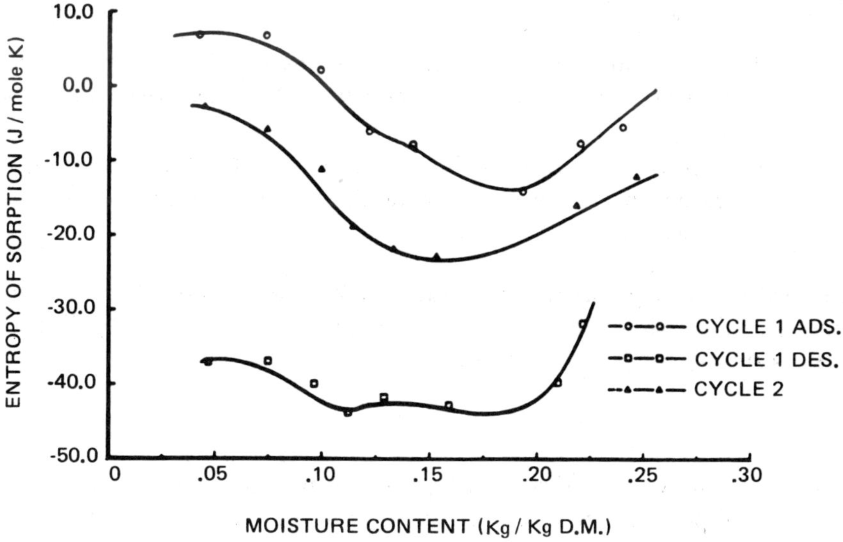

FIG. 9.7 Entropy of sorbed water at 20°C (with reference to pure liquid water) calculated from cycle 1 adsorption and desorption isotherms and cycle 2 isotherms. (*From Benado and Rizvi, 1985.*)

this work, an increase in the value of mechanical properties (compression and penetration force, moduli of elasticity, cohesiveness, etc.) within the low a_w range was attributed to increased crystallinity.

Iglesias and Chirife (1976b) determined and compared the isosteric heats of water adsorption and desorption for a large number of foods. In general, the desorption heats were significantly higher than the adsorption heats at low moisture contents. Upon further sorption, the difference rapidly decreased and practically disappeared at high moisture contents. There was no direct relationship between the observed differences in the ad–desorption heats and the distribution of hysteresis along the isotherm. The higher discrepancies between ad–desorption heat curves occurred in the range of moisture content for which the amount of hysteresis was higher, i.e., mainly in the low moisture content range.

Theories of Sorption Hysteresis

The interpretations proposed for sorption hysteresis can be classified under one or more of the following categories, based on the structure of the sorbent (Arnell and McDermot, 1957):

(a) Hysteresis on porous solids. In this category belong the theories based on capillary condensation.

(b) Hysteresis on nonporous solids. This category includes interpretations based on partial chemisorption, surface impurities, or phase changes.

(c) Hysteresis on nonrigid solids. The category deals with interpretations based on changes in structure, as these changes hinder penetration and egress of the adsorbate.

Incomplete wetting theory. Suggested by Zsigmondy (1911), the theory represents the earliest attempt at explaining hysteresis (Fig. 9.8). As all theories of capillary condensation, it is based on the Kelvin equation

$$RT \ell n\, P/P_o = -2\sigma V \cos\theta / r_m$$

where P is the vapor pressure of liquid over the curved meniscus, P_o the saturation vapor pressure at temperature T, σ the surface tension, θ the angle of contact (in complete wetting $\theta = 0$ and $\cos\theta = 1$), V the molar volume of the liquid, r_m the mean radius curvature of the meniscus defined as $2/r_m = 1/r_1 + 1/r_2$, where r_1 and r_2 are the principal radii of curvature of the liquid–vapor interface, and R the gas constant. Due to the presence of impurities (dissolved

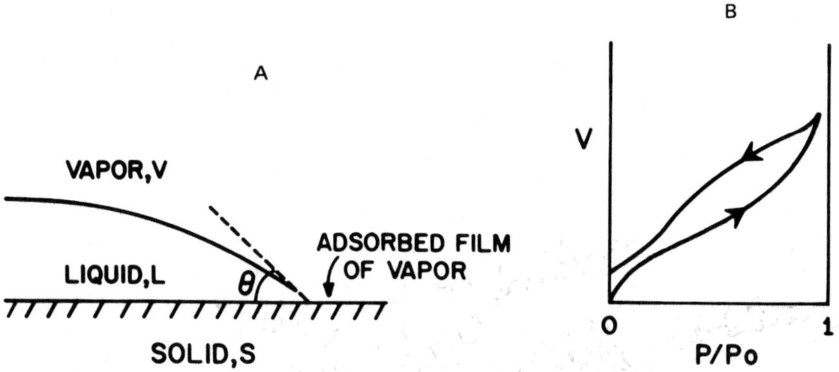

FIG. 9.8 Incomplete wetting theory of hysteresis: (A) contact angle; (B) open hysteresis.

gases, etc.) the contact angle θ of the receding film upon desorption is smaller than that of the advancing film upon adsorption. Therefore, capillary condensation along the adsorption branch of the moisture sorption isotherm will be at a higher relative vapor pressure.

The Zsigmondy theory fails to explain adsorption results at low relative vapor pressure, where in some instances it requires the value of r to be smaller than the diameter of the adsorbed molecule. It may have some application in the limited case of a hysteresis open at the lower end, in contrast to most cases in foods where the most common type of hysteresis is the closed-end, retraceable loop.

Ink bottle neck theory (Kraemer, 1931; McBain, 1935; Rao, 1941). This theory explains hysteresis on the basis of the difference in radii of the porous structure of the sorbent. The latter consists of large-diameter pores simulated by the main body of an ink bottle, equipped with narrow passages simulated by the neck of the ink bottle (Fig. 9.9).

In adsorption, condensation first takes place in the large diameter cavity at

$$P/P_o = \exp(-2\sigma V/r_2RT)$$

which results from substituting r_2 for r_m in the Kelvin equation (complete wetting, $\cos\theta = 1$). In desorption, the neck of the pore is blocked by a meniscus, which can evaporate only when the pressure has fallen to

$$P_d = P_o \exp(-2\sigma V/r_1RT)$$

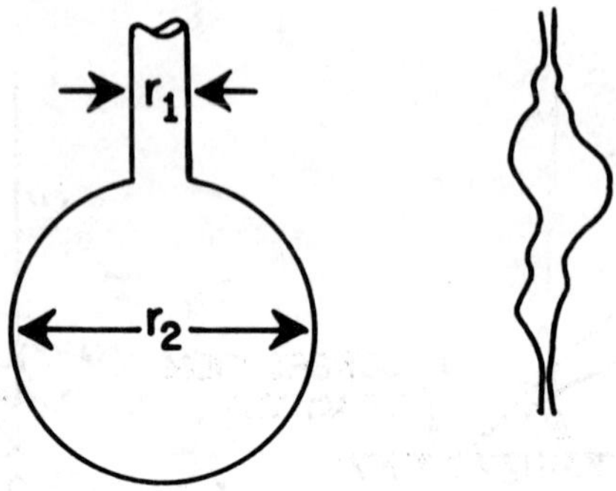

FIG. 9.9 Ink bottle neck theory of hysteresis. Left, schematic representation; right, actual pore.

at which point the whole pore empties at once. Therefore, for a given amount of water adsorbed, the pressure is greater during adsorption than during desorption.

Open-pore theory. This theory, elaborated by Cohan (1938, 1944) extends the ink bottle theory by including considerations of multilayer adsorption (Fig. 9.10). It is based on the difference in vapor pressure between adsorption P_a and desorption P_d, as affected by the shape of the meniscus. The latter is cylindrical on adsorption, where the Cohan equation applies, and hemispherical on desorption, where the Kelvin equation applies. Thus

$$P_a = P_o \exp(-2\sigma V/rRT) = P_o \exp[-\sigma V/(r_c - D)RT]$$

where r_c is the radius of the pore and D the thickness of the adsorbed film. Once condensation has taken place, a meniscus is formed and the desorption pressure P_d is given by the Kelvin equation. Assuming that wetting is complete, $\cos \theta = 1$ and

$$P_a = P_o \exp(-2\sigma V/r'RT)$$

When $r_c > 2D$, $P_a > P_d$, and hysteresis will occur. When $r_c = 2D$,

$$P_a = P_d = P_o \exp[-\sigma V/(r_c - D)RT]$$

and no hysteresis will take place (Gregg and Sing, 1967).

The pressure at which hysteresis begins should correspond to equilibrium pressure for $r_c = 2D$, i.e., for those pores that are two molecules in diameter.

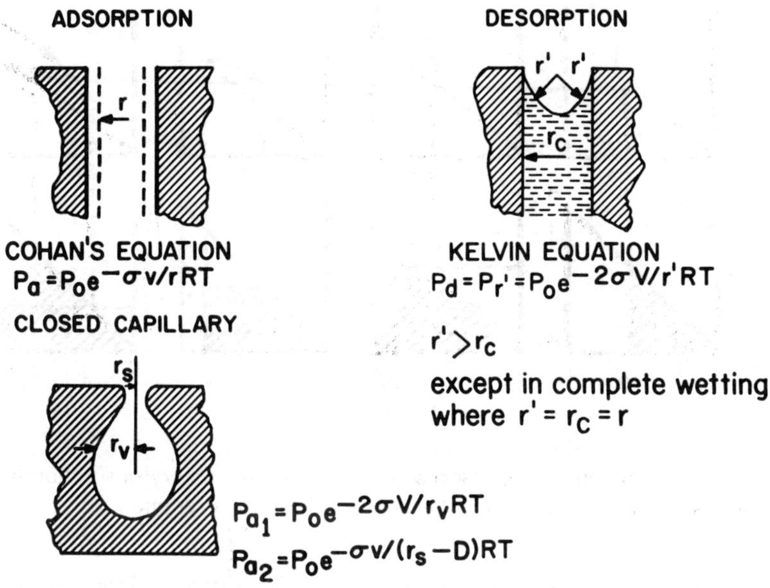

FIG. 9.10 Open-pore theory of hysteresis. (*From Cohan, 1938, 1944.*)

The domain theory. This theory has been discussed in its latest form by Everett (1967) as a reformulation of his earlier concept, which was more generalized to include the phenomena of magnetic hysteresis [see e.g., Everett and Whitton (1952) and Everett (1955)]. It is an attempt not only to explain and predict the irreversibility of the sorption process, but also to interpret sorption curves resulting by crossing the hysteresis loop through repeated scanning cycles (Fig. 9.11).

The term "domain" as first introduced with regard to magnetic hysteresis applies to a group of atoms that can exist in one of two thermodynamically metastable states. These states are separated from each other by a small but finite gap. The group of atoms should be both sufficiently small to show discrete steps in the thermodynamic curve on a small scale and sufficiently large to discount the possibility that thermal fluctuation will overcome the potential barrier separating the two states (Enderby, 1955). In Everett's theory, a pore domain is a region of pore space accessible from neighboring regions through pore restrictions. An isolated pore domain has well-defined condensation–evaporation properties involving one or more spontaneous irreversible steps. The pore domain is divided into elements of volume dV that contain either liquid or vapor. In adsorption, a liquid–vapor interface will sweep through this element of volume at a certain relative pressure x_{12}, and in desorption the same interface

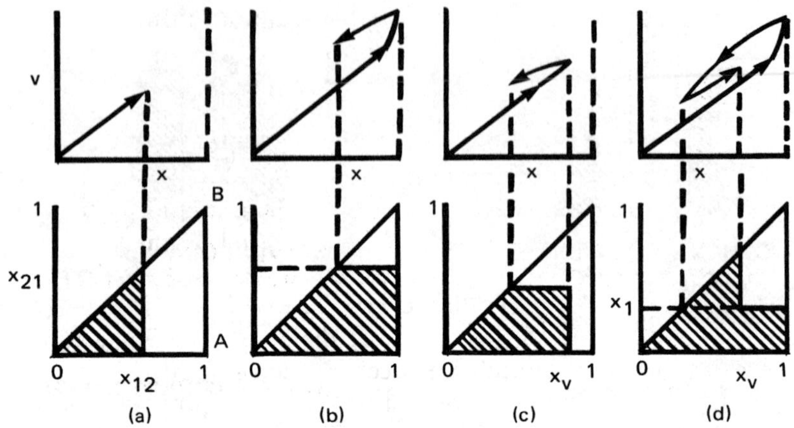

FIG. 9.11 Schematic representation of the domain theory of sorption hysteresis. (*From Everett, 1967.*) (See text for detailed description.)

will pass back through it leaving it empty at x_{21}. In an irreversible pore domain situation $x_{12} > x_{21}$. All elements of volume are classified in terms of x_{12} and x_{21}, which form the coordinates of the lower part of Fig. 9.11. Due to the inequality defined above, the points representing all elements of volume will lie in the triangle OAB. Each area $dx_{12}\,dx_{21}$ of this diagram is associated with a quantity $v(x_{12}, x_{21})$, such that $v(x_{12}, x_{21})\,dx_{12}\,dx_{21}$ is the volume of the pore domain associated with values of x_{12} and x_{21} in the ranges x_{12} to $x_{12} + dx_{12}$ and x_{21} to $x_{21} + dx_{21}$. The distribution function $v(x_{12}, x_{21})$ characterizes the properties of the pore domain. Thus, an adsorption process may be represented by the movement of a vertical line across the triangle from left to right, and at any point the state of the system may be represented by a diagram similar to the lower row of Fig. 9.11 ("domain complexion"). The diagram functions as "memory" in systems exhibiting hysteresis.

The desorption process can be described in terms of the movement of a horizontal line from top to bottom of the triangle. The properties of the system can always be expressed as the sum of integrals over triangles of the same hypotenuse OB, in a diagram of x_{21} vs x_{12} between the limits $x_{21} = x_\ell$ and $x_{12} = x_u$, as follows:

$$V(x_\ell, x_u) = \int_{x_\ell}^{x_u}\int_{x_\ell}^{x_u} v(x_{12}, x_{21})\,dx_{12}\,dx_{21}$$

In the *independent-domain* theory, each pore domain interacts with the vapor as an isolated pore. The theory fails to account quantitatively for the behavior of many systems, especially those related to capillary condensation processes. It was subsequently re-examined and broadened by Everett to consider *pore-blocking effects*. Whereas an isolated pore will fill and empty at given values of relative vapor pressures (according to the independent-domain theory), when the pore is interconnected with others, this same element may or may not undergo these changes at the same relative vapor pressures. This will depend on the way the pore is interconnected and on whether the adjacent pores are full or empty. Stated briefly, it will depend on the history of the system.

The domain theory ignores time effects. The existence of drifting "equilibria," as in potato starch, indicates that the theory is unsuitable for the analysis of scanning curves. Although a combination of the theory with relaxation processes was originally envisaged by the author, this was not accomplished (Bizot et al., 1985). For more details, the reader is referred to the original discussion by Everett (1967).

Polar and Other Interpretations of Sorption Hysteresis in Biological Materials

These interpretations do not constitute a "theory" in the literal sense of the word. Instead they are explanations that could be categorized under the main groups listed above, especially under the group relating to deformability and elastic stresses of the sorbent. Hysteresis is attributed to a deformation of the polypeptide chains within the protein molecule as the polar adsorbates (in our case water) occupy suitable positions for hydrogen bonding or ion dipole interaction. In view of our interest in biological materials, a brief discussion is appropriate.

Adsorption from the dry state by biopolymers is due to (1) side-chain amino groups, (2) end carboxylic and other groups, (3) peptide bonds, and (4) secondary structure. In general, research involving deamination or blocking of side-chain groups, (benzoylation, methylation, acetylation) has demonstrated that below 50% relative humidity the main sites of sorption are the polar side-chain groups. The contribution of the polypeptide chain becomes progressively more important at the higher humidities. For example, in wool keratin, at 80% relative humidity the peptide bonds account for almost half the adsorbed water (Watt, 1980b).

As stated earlier, hysteresis is affected more by desorption than adsorption. Experiments before and after deamination or methylation of side-chain groups in wool and benzoylation in casein (Mellon et al., 1948) did not show any appreciable changes of hysteresis. This suggests that it is the main chains of the biopolymers that are primarily responsible.

Seehof et al. (1953), on the basis of literature and their own data, supported a polar group interpretation of hysteresis where binding involves mainly the free basic groups of the protein. Table 9.2 shows a correlation of the maximum amount of hysteresis (last column) with the sum of arginine, histidine, lysine, and cystine groupings (next to last column). Besides the free basic groups of the protein molecule, sulfur linkages are also of prime importance in hysteresis (Speakman and Stott, 1936). In contrast to this work, hysteresis in casein was observed to be independent of the content of free amino groups by Mellon et al. (1948). A twofold nature of hysteresis was proposed: constant hysteresis, independent of the relative humidity desorption point, and hysteresis proportional to the amount adsorbed above the upper absorption break of the isotherm.

Van Olphen (1965) attributes sorption hysteresis in vermiculite clay to a retardation of adsorption due to the development of elastic stresses in crystallites during the initial peripheral penetration of water between the unit layers. The shift toward higher relative vapor pressure during adsorption is caused by the activation energy required to open the unit layer stacks.

Bettelheim and Ehrlich (1963) suggested that in a swelling polymer, hysteresis cannot be interpreted by capillary condensation. Rather hysteresis seems to depend on the mechanical constraints contributed by the elastic properties of the material and on the case with which the polymer swells. In one case, typified by calcium chondroitin sulfate C, the large sorptive and swelling capacity associated with polymer chains weakly bound between each other and weakly bound water led to a small hysteresis loop. In the opposite case, typified by the calcium chondroitin sulfate A, a small sorptive and swelling capacity associated with a tightly bound matrix and strongly bound water led to a large hysteresis.

TABLE 9.2 Correlation between Maximum Hysteresis and Polar Groupings in Proteins, Compiled from Literature Data

	mmol/g protein								
	Arginine 1	Histidine 2	Lysine 3	Cysteine 4		Cystine 5	RNA 6	Σ 1–6	Maximum hysteresis
Casein	0.2	0.2	0.6					1.0	1.1
TMV	0.5		0.1	0.1			2.7	3.4	3.5
Insulin	0.2	0.3	0.2	0.1		0.5		1.3	1.2
Collagen	0.5	0.1	0.3					0.9	0.8
BovSAlb	0.3	0.2	0.8		0.3			1.6	1.5
BovPAlb									1.4

Source: Seehof et al. (1953).

In the latter case, the sharp maxima in the entropy and enthalpy curves indicate strong interchain attractions.

As indicated earlier in this review, hysteresis seems to be the net result of reinforcing or competing variables between adsorbent and adsorbate. From the standpoint of the adsorbate, important factors are the hydrogen-bonding ability, the amount adsorbed, and the molar volume (Benson and Richardson, 1955). Adsorbates of greater hydrogen-bonding ability (H_2O, EtOH) develop large hysteresis loops, but those of little or no hydrogen-bonding ability (Et_2O and EtCl) give very small loops. For illustration, EtOH causes a large hysteresis loop and EtCl causes a very small one, in spite of the about equal size of the molecules and the very close dipole moments of these adsorbates. Adsorbates that are adsorbed in greater amounts cause greater hysteresis loops (greater total deformation of the protein molecule) than adsorbates adsorbed in small amounts. A typical example is water, which is adsorbed in relatively large amounts and which is strongly hydrogen bonded. Important in hysteresis is the molar volume of the adsorbate. Thus EtOH causes large hysteresis as does water, in spite of the fact that compared to water, fewer EtOH molecules are sorbed per gram of protein. In this case, the network deformation per molecule is larger for EtOH than water. On the other hand, the increasing bulk of the adsorbate molecule may result in decreased sorption due to too large local deformation, where sorption becomes energetically unfavorable, due to the increased work required by the larger deformation. For example, the amount of sorption decreases with increasing size in the R group of MeOH, EtOH, and i-C_4H_9OH.

TEMPERATURE

Effect of Temperature on the Sorption Isotherm

General considerations. Temperature affects the mobility of water molecules and the dynamic equilibrium between the vapor and adsorbed phases.

Figure 9.12 shows schematically the effects of temperature shifts on both the moisture content and a_w. If a_w is kept constant, an increase in temperature causes a decrease in the amount of sorbed water. This indicates that the food becomes less hygroscopic. As pointed out by Iglesias and Chirife (1982) this is necessitated by the thermodynamic relationship, $\Delta F = \Delta H - T\Delta S$. Since $\Delta F < 0$ (sorption is a spontaneous process) and $\Delta S < 0$ (the sorbed water molecule has less freedom), $\Delta H < 0$. Therefore, an increase of temperature represents a condition unfavorable to water sorption. An exception to this rule, as mentioned elsewhere in this review, is shown by certain sugars and other low molecular weight food constituents, which become more hygroscopic at higher temperatures because they dissolve in water.

NB

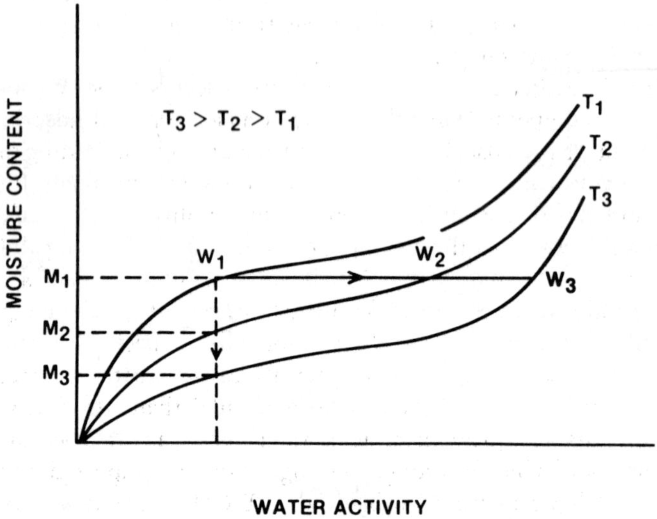

FIG. 9.12 Schematic representation of the effect of temperature on a_w and moisture content.

Temperature shifts can have an important practical effect on chemical and microbiological reactivity related to quality deterioration of a food in a closed container. As shown in Fig. 9.12, an increase of temperature causes an increase of a_w for the same moisture content. This increases the rates of reactions leading to deterioration (Labuza, 1970, 1975; Labuza and Kamman, 1983).

Figure 9.13 shows a typical example of temperature effect on the isotherms of roasted coffee (Weisser, 1985). Not all foods exhibit this type of consistent separation of isotherms at different temperatures. Figure 9.14 shows a crossing over at high a_w (at about a_w 0.78) of the 20 and 30°C adsorption isotherms of Sultana raisins, reported recently by Saravacos et al. (1986). Such a crossing-over has been reported in earlier work by Saravacos and Stinchfield (1965) on model systems of starch–glucose. It is in agreement with observations by Audu et al. (1978) on sugars, Weisser et al. (1982) on sugar alcohols, and Silverman and his associates (Silverman et al., 1983; Lee et al., 1981) for 20 and 37°C isotherms of precooked bacon. In this case, the crossing over and the progressive divergence of the isotherms with increasing a_w in the intermediate moisture range was accompanied by an increasing moisture-to-salt ratio (greater dissolution effect).

All these substances contain large amounts of low molecular weight constituents in a mixture of high molecular weight biopolymers. At lower a_w values, the sorption of water is due mainly to the biopolymers, and an increase of

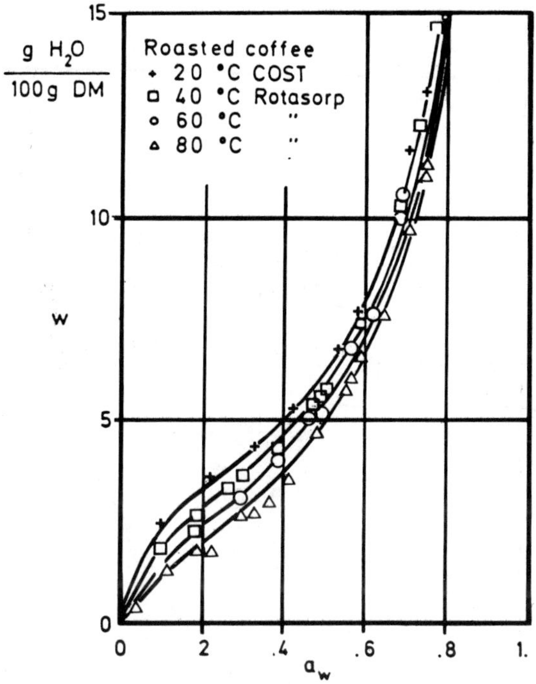

FIG. 9.13 Influence of temperature on the sorption isotherms of roasted coffee. COST and Rotasorp refer to the equipment used for the determination of isotherms. (*From Weisser, 1985.*)

temperature has the normal effect of lowering the isotherms. As the a_w is raised beyond the intermediate region, water begins to be sorbed primarily by the sugars and other low molecular weight constituents; this is indicated by the isotherm swinging upward. Dissolution (use of large amounts of water) is favored by the higher temperature (endothermic process), which offsets the opposite effect of temperature on the high molecular weight constituents. The net result is an increase of the moisture content (crossing over) of the isotherms. These relationships can be examined in terms of the sign and magnitude of the binding energy ΔH_{st}. Figure 9.15 from the work of Saravacos et al. (1986) on raisins shows the decrease of the binding energy as the temperature increases from 22 to 32°C in the low moisture region. The effect of temperature shows a crossing over of the lines at higher moisture contents due to the endothermic dissolution of fruit sugars.

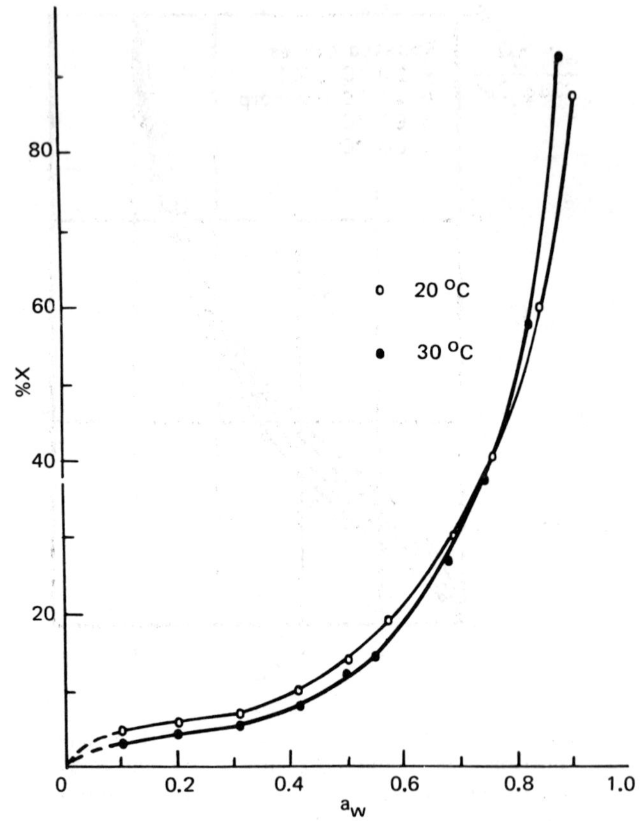

FIG. 9.14 Adsorption isotherms of Sultana raisins. % x = % moisture. (*From Saravacos et al., 1986.*)

Temperature changes affect the a_w of saturated salt solutions which are used in the determination of sorption isotherms. Labuza et al. (1984), using experimental data and thermodynamic analysis, has demonstrated that the a_w of saturated salt solutions should *decrease* with increasing temperature. If no correction is made to use the "true a_w" of the salt solution at the temperature of the study, the shape of the food isotherm will be distorted, making the shift with temperature appear larger than it actually is.

Another way by which high temperatures may affect the sorption isotherm is through irreversible changes of the food. In this case, predictive isotherm equations that are valid at lower temperatures may not be applicable at the higher temperatures (Bandyopadhyay et al., 1980).

The quantity of sugars present plays a role in whether or not crossing of

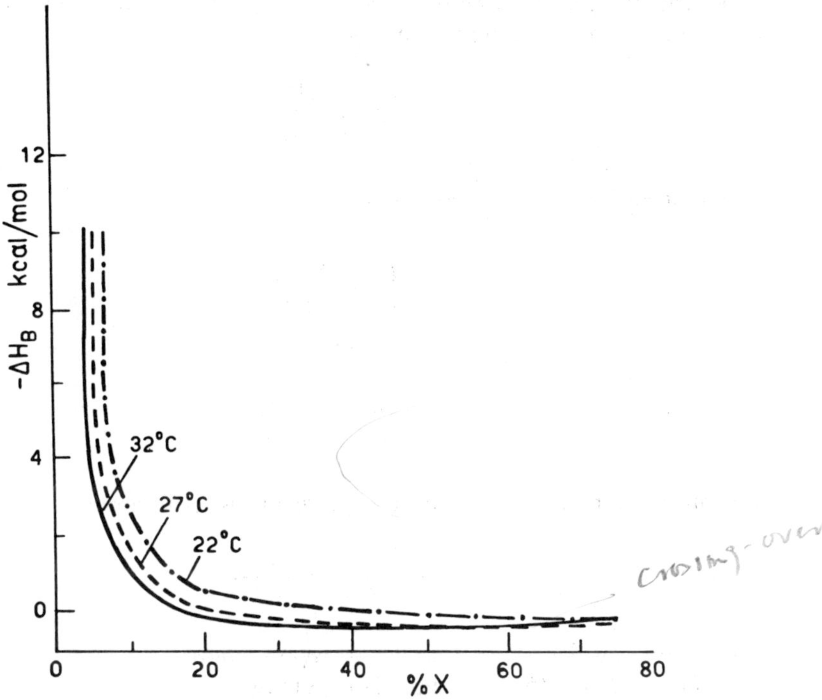

FIG. 9.15 Binding energy of sorption (ΔH_B) versus moisture content dry basis of Sultana raisins at mean temperatures 22, 27, and 32°C. % x = % moisture. (*From Saravacos et al., 1986.*)

isotherms with temperature at high water activities will take place. For example, in the work of Roman et al. (1982) on apples, no crossing was observed because of lower sugar content of apples compared with raisins studied by Saravacos et al. (1986).

Clausius–Clapeyron equation. The thermodynamic function used to express the temperature dependence of vapor pressure is the Clausius–Clapeyron equation for phase transition from an adsorbed liquid to the vapor phase.

$$\frac{dP}{dT} = \frac{\Delta H_{tot}}{T(V_{vap} - V_{liq})} \tag{1}$$

where the differential change of pressure P with temperature T is related to a total molar enthalpy change ΔH_{tot} for the phase transition and to a volume difference between the vapor and liquid.

Disregarding the volume of the liquid as negligible compared to the volume of the vapor, the equation is simplified to

$$\frac{dP}{dT} = \frac{\Delta H_{tot}}{V_{vap}T} \tag{2}$$

Substituting for $V_{vap} = \frac{RT}{P}$ results in

$$\frac{dP}{dT} = \frac{\Delta H_{tot}P}{RT^2} \tag{3}$$

or

$$\frac{d(\ell n\, P)}{dT} = \frac{\Delta H_{tot}}{RT^2} \tag{4}$$

For pure water (vapor pressure P_o) the relationship becomes

$$\frac{d(\ell n\, P_o)}{dT} = \frac{\Delta H_{vap}}{RT^2} \tag{5}$$

where ΔH_{vap} is the molar enthalpy of vaporization.

Subtracting Eq. (5) from Eq. (4) results in

$$\frac{d\left(\ell n\, \frac{P}{P_o}\right)}{d\left(\frac{1}{T}\right)} \quad \frac{\Delta H_{tot} - \Delta H_{vap}}{R} \tag{6}$$

or

$$\frac{d\left(\ell n\, \frac{P}{P_o}\right)}{d\left(\frac{1}{T}\right)} \quad \frac{\Delta H_{st}}{R} \tag{7}$$

where ΔH_{st} is the *net isosteric heat of sorption* Q_{st}.

Integration of Eq. (7), with the assumption of a constant value of Q_{st} over the a_w and temperature range considered, results in

$$\ell n\, \frac{P}{P_o} = \ell n\, W = \frac{\Delta H_{st}}{R}\frac{1}{T} + \text{constant} \tag{8}$$

where W is the water activity.

A plot of $\ell n W$ vs. $\frac{1}{T}$ will give the value for the net isosteric heat of sorption from the slope. This can also be calculated by using two temperatures (T_1, T_2) and the respective a_w (W_1, W_2) as follows:

$$\ell n \frac{W_2}{W_1} = \frac{\Delta H_{st}}{R}\left[\frac{1}{T_1} - \frac{1}{T_2}\right]. \tag{9}$$

Figure 9.16 shows a number of such plots for a meat emulsion at different moisture contents (Mittal and Usborne, 1985). The slope of the line Q'_{st} decreases as moisture content increases, indicating a decrease in the binding energy for water molecules.

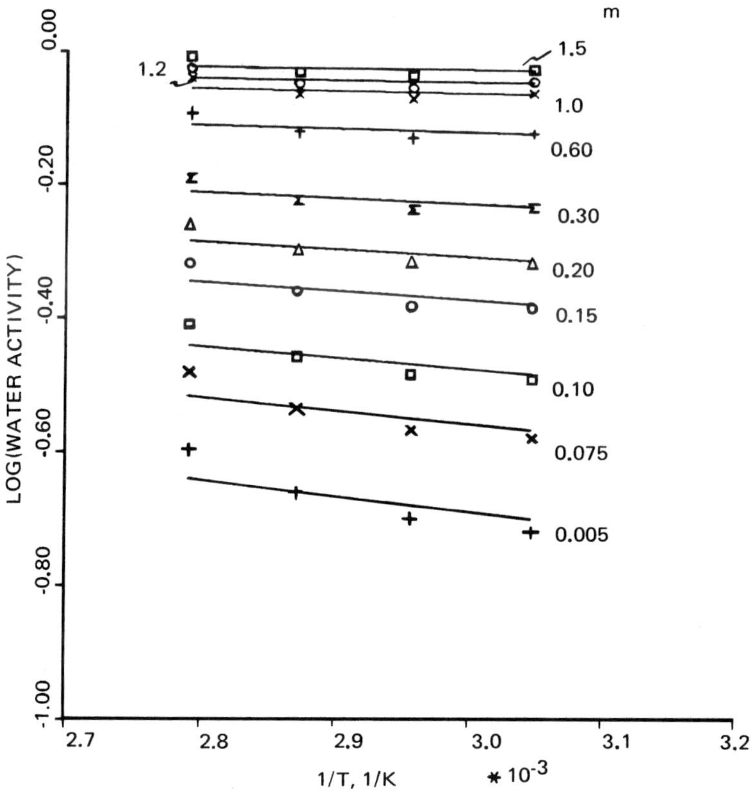

FIG. 9.16 Plots of Log_e of a_w vs. the reciprocal of the absolute temperature of a meat emulsion at various moisture contents. (*From Mittal and Usborne, 1985.*)

If Q_{st} for a particular food were known, it would be possible to predict the a_w of that food at any given temperature. Since however, standard tables for Q_{st} for different foods do not exist, prediction of a_w requires the determination of moisture isotherms for at least two temperatures.

Examples of isosteric heat curves vs. moisture content are given for high sugar foods in Fig. 9.17 and for high-protein foods in Fig. 9.18 (Iglesias and Chirife, 1976a). In Fig. 9.17, banana and pineapple show relatively similar heat of sorption curves, close to the heat of vaporization of water (horizontal line). In pineapple, the heat curve remains relatively constant as the amount of the adsorbed water increases. The behavior of grapefruit differs from the other two high-sugar items. At low moisture contents, the heat of sorption is relatively high, dropping off rapidly toward the heat of vaporization.

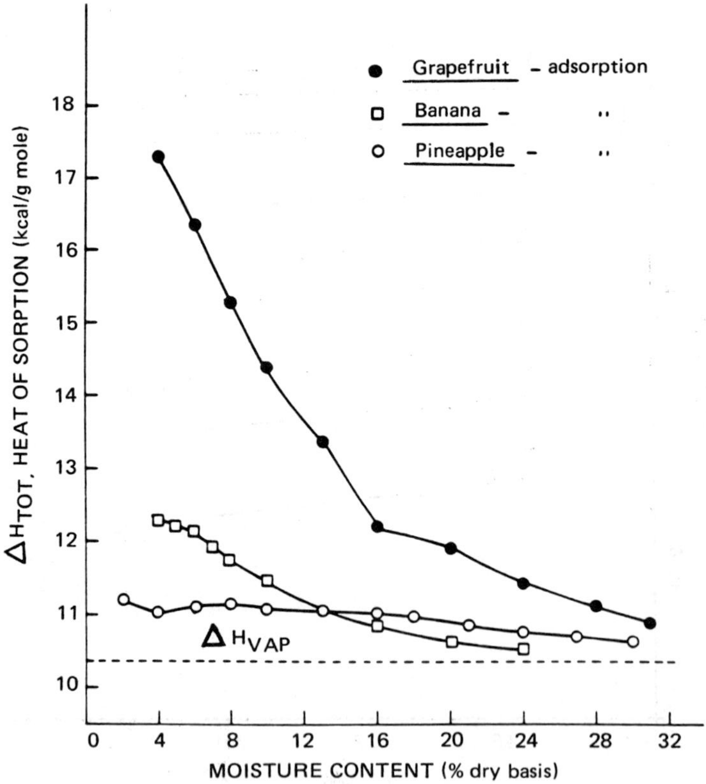

FIG. 9.17 Isosteric heat curves for water adsorption on banana, grapefruit, and pineapple. (*From Iglesias and Chirife, 1976a.*)

In high-protein foods (Fig. 9.18), there is a maximum of the adsorption heat curves at about 4.5% moisture, indicating a swelling of the polymer which exposes new sorption sites not previously available. (See also discussion of entropy changes under Thermodynamic Considerations in the first section of this review on hysteresis.) The highest adsorption heat curve of the cooked compared to the raw chicken may be due to denaturation and unfolding of the polypeptide chains, which exposes hitherto unavailable sorption sites. At high-

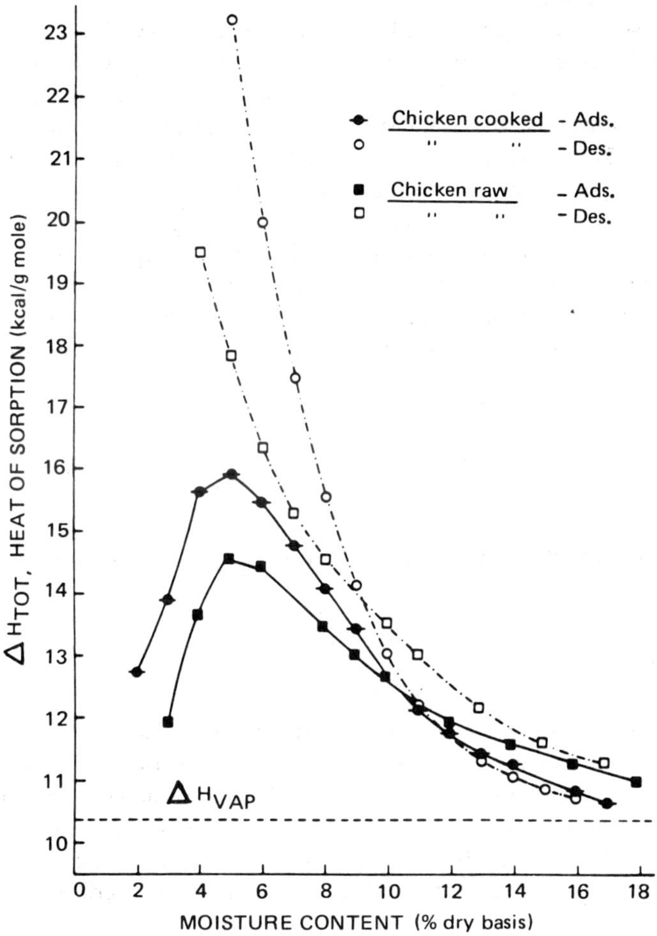

FIG. 9.18 Isosteric heat curves for water adsorption and desorption in raw and cooked chicken muscle. (*From Iglesias and Chirife, 1976a.*)

er moisture contents, swelling becomes large enough to bring the heat curves of both cooked and uncooked product close together. In desorption, an altered structure desorbs in a smooth way, resulting in the top two curves in Fig. 9.18. (See Persistence of Hysteresis upon Adsorption–Desorption Cycles.)

Clausius–Clapeyron equation applied to foods. Central to these considerations is the irreversibility of the sorption process, which has been discussed by several workers (Marshall and Moore, 1952; Everett and Whitton, 1952; Seehof et al., 1953; Gregg and Wheatly, 1957; Kingston and Smith, 1964; Franks, 1975; Watt, 1980a, b).

The presence of a persistent hysteresis indicates that the system, though reproducible, is not in true equilibrium. Although the adsorption process seems to be more reproducible, it is doubtful if either the adsorption or the desorption branch of the isotherm has greater thermodynamic significance (Amberg, 1957; Amberg et al., 1957). However, opinions among research workers vary. Recently, Bizot et al. (1985) reported that in starch "only adsorption states (carefully obtained by interval conditioning) are well defined equilibria from a thermodynamic point of view." The question of validity of thermodynamic quantities calculated from such a system was brought into sharp focus by the classical debate of La Mer with Bettelheim (La Mer, 1967).

The isosteric heats of sorption calculated from the Clausius–Clapeyron equation,

$$\frac{d\,\ell n\,P}{d(1/P)} = \frac{\overline{\Delta H}^{0}{}_{isosteric}}{R}$$

and the subsequently calculated differential entropy of sorption,

$$\overline{\Delta S}^{0} = \frac{\overline{\Delta H}^{0} - \overline{\Delta F}^{0}}{T}$$

refer to a reversible process between two defined initial and final states, which the water vapor sorption process is not. Hysteresis shows that the process is irreversible. Hence, there is an *entropy production* in the system, which indicates the degree of irreversibility, including structural changes occurring in the sorbing solid matrix. Therefore, the isosteric heat and hence the calculated "entropy of sorption" has a term $T\,\Delta S$ irreversible included and it does not strictly refer to heat transfer of the sorption process.

To evaluate the true heat effects, calorimetric experiments must be performed to measure the integral heats of sorption at different water uptakes. To compare with the isosteric heats of sorption, the differential heats of sorptions are calculated from the integral heats:

$$\left(\frac{\partial\, \Delta H}{\partial n}\right) = Q_{diff}$$

The isosteric heats of sorption now will be equal:

$$\overline{\Delta H}^{0}_{iso} = Q_{diff} + RT + T\,\Delta S_{irrev}$$

where $T\,\Delta S = Q_{irrev}$. The RT term under room temperature conditions amounts to 0.6 kcal/mole, a small correction. Thus, from a comparison of calorimetric and isosteric heats, the entropy production of the system during the sorption process can be calculated.

According to La Mer (1967) the Clausius–Clapeyron equation could be used only to compare the heat of sorption Q_{st}, calculated from the equation, with the corresponding H values obtained by direct calorimetry. This makes it possible to calculate the magnitude of the irreversibly created entropy and evolution of heat. The symbols H and S should be reserved for thermodynamic quantities calculated for a reversible process between defined states. The application of the Clausius–Clapeyron equation to sorption data should not be called "a thermodynamic analysis," unless corrections are made for loss of work and the resultant creation of entropy or heat due to hysteresis.

By combining isotherm measurements with direct calorimetric measurements, it would be possible to decide which branch of the hysteresis loop is associated with the irreversible process, if indeed irreversibility is confined to only the adsorption or desorption process. The measurements require high precision through strict and often difficult control of experimental variables, due to the small value of Q_{irrev} (Everett and Whitton, 1952).

From a practical standpoint, the interest in foods usually lies in *relative, comparative* changes of the heat of sorption associated with variables of dehydration and other processes. Within these confines, the Clausius–Clapeyron equation is useful in engineering practice. After examining the literature on irreversibility, Iglesias and Chirife (1976b) concluded that the heat changes involved in irreversible processes were probably small compared to the overall energy changes, and therefore could be neglected in a general qualitative description.

Another consideration affecting the heat of sorption values calculated from the Clausius–Clapeyron equation is the linearity of ΔH_{vap} and ΔH_{st} with temperature. Saravacos and Stinchfield (1965) found that the linearity of plots of $1/T$ vs. equilibrium vapor pressure was limited to certain temperature ranges. For example, in freeze-dried potato there was a change of slope below 20°C. The equilibrium vapor pressure in beef and peach followed the Clausius–Clapeyron equation from −20 to 20°C, and from 20 to 50°C. The slopes of these plots (and therefore the heats of adsorption) were higher at temperatures above 20°C than at lower temperatures.

With regard to the role of the temperature range (zone of isotherm), Iglesias and Chirife (1976a), using literature data, examined the aggregate error in the isosteric heat of sorption of Eq. (9), which is inherent in plotting, curve drawing, and interpolation, i.e., the error of graphical manipulation (not the error involved in the measurement of isotherm). They reported an error of 2.5% in the determination of W_1 and W_2, depending on the zone of the isotherm considered. This amounts to a relative mean square error of their ratio of 3.5%. The precision of Q_{st} depends on the factor $T_1T_2/(T_2 - T_1)$; for the calculations based on the 25–45°C isotherms the minimum error in the isosteric heat was

$$Q_{\eta}^{st} \pm 329 \text{ cal/g mole}$$

For the 45–60°C isotherms the error was

$$Q_{\eta}^{st} \pm 492 \text{ cal/g mole}$$

With regard to the accuracy and precision involved in the measurement of the sorption isotherm, the reader is referred to the significant contributions of the recent collaborative 32-laboratory European project "COST 90 and 90 bis on Physical Properties of Foodstuffs" (Spiess and Wolf, 1983; Wolf et al., 1985). Table 9.3 shows the results of standard deviation $\bar{s}$, mean repeatability $\bar{r}$ (for tests performed at short intervals at one laboratory by one operator with

TABLE 9.3 Mean Precision Data of Sorption Measurements of Real Food Materials and of Microcrystalline Cellulose

Product	Std. dev. $\bar{s}$	Mean repeatability g H_2O/100g DM $\bar{r}$	Reproducibility $\bar{R}$
Cod	0.67	0.18	2.08
Casein	0.63	0.12	1.94
Whey protein	1.08	0.23	3.33
Maltodextrin	1.24	0.21	3.87
Pectin	1.18	0.16	5.64
Beef	0.52	0.28	1.66
Soy protein	1.26	0.21	4.06
MCC	0.25	0.17	0.72

Source: Wolf et al. (1985).

the same equipment), and reproducibility $\overline{R}$ (for tests performed at different laboratories, which implies different operators and different equipment).

It was concluded that the COST-Projects 90 and 90 bis have "demonstrated that applying sufficient care good accuracy and good precision can be obtained" (Wolf et al., 1985).

With regard to alterations of food material, when considering application of the Clausius–Clapeyron equation, it is obvious that anything that affects the sorption isotherm affects the thermodynamic quantities derived from it. This includes the prehistory of the sample, such as various preheat treatments, blanching, freezing, and drying (Heldman et al., 1965; Saravacos, 1967; Iglesias and Chirife, 1976c, d; Hayakawa et al., 1978; Bolin, 1980). These treatments may change the polar and other groups that bind water, along with changes in the capillary and other configurations of the food structure. The result may be changes over a certain range or the total span of the sorption isotherm. For example, in the work of Greig (1979) denaturation of native cottage cheese whey had no effect on the sorption isotherm at low relative humidities but significantly increased sorption at high relative humidities. San José et al. (1977) found that the drying method (freeze and spray drying) of lactose-hydrolyzed milk did not affect the adsorption isotherms, but had a profound effect on the desorption isotherms. The impact on the adsorption vs. desorption branches within different a_w ranges will depend on changes of capillarity, swelling, crystallinity, and other characteristics of structure.

Moisture sorption isotherm equations. Temperature is part of any model that attempts to describe and predict a_w data. A large number of equations have been proposed for fitting water sorption isotherms of foods (see critical reviews by Chirife and Iglesias, 1978; van den Berg and Bruin, 1981; Lomauro et al., 1985). Success has been variable. In general, none of these models or equations can reproduce the sorption isotherm over the whole range of a_w for all foods.

Recently, interest in this area has been stimulated by the use of the three-parameter equation of Guggenheim–Anderson–de Boer (G.A.B.) shown in Figs. 9.19 and 9.20. This equation is an extension and a general representation of the B.E.T. equation, taking under consideration the modified properties of the sorbed water in the multilayer region. It has been found useful, especially by European workers, for fitting isotherm data up to very high a_w, e.g., in the work of van den Berg (1981, 1984) up to about a_w 0.9. Detailed reports of the theoretical context, meaning of the different terms, and procedures for solving the equation have been published (van den Berg, 1984; Wolf et al., 1985; Lomauro et al., 1985; Weisser, 1985; Schar and Ruegg, 1985).

For our purpose, it will suffice to say that the G.A.B. equation is probably best for fitting sorption isotherm data of a majority of food products, describ-

$$W = \frac{W_m C K a_w}{(1 - K a_w)(1 - K a_w + C K a_w)} \tag{1}$$

or written explicitly in water activity:

$$a_w = \frac{2 + (W_m/W - 1)C - \{[2 + (W_m/W - 1)C]^2 - 4(1 - C)\}^{0.5}}{2K(1 - C)} \tag{2}$$

W = water content
W_m = water content at "monolayer"
C = G.A.B. sorption constant related to monolayer properties
K = G.A.B. sorption constant related to multilayer properties

FIG. 9.19 The Guggenheim–Anderson–de Boer (G.A.B.) sorption equation.

ing the temperature effect over a range of at least 40 degrees (van den Berg, 1985). For example, Weisser (1985) observed that the G.A.B. equation with its temperature-dependent coefficients was suitable for describing the effect of temperature on the sorption behavior of several food components, e.g., apple pectin, soy and whey proteins, and sodium caseinate, in the temperature range 25–80°C. On the other hand, Saravacos et al. (1986) found that, although the experimental data of Sultana raisins fitted well the 3-parameter G.A.B. (and also Halsey) equations, the best fit was provided by the 5-parameter D'Arcy–Watt equation, which provides for both physical adsorption and dissolution of sugars. The final choice of an equation will depend on a compromise between the desired closeness of fit and convenience with regard to the number of parameters involved.

Influence of subfreezing temperatures on water activity. Relative to the large volume of literature reports on isotherm data above freezing temperatures, information on a_w below freezing is limited. Until recently this field has been in some uncertainty due to early reports that vapor pressures of animal tissues over the temperature range −26 to −1°C ranged from 13 to 20% lower than those of pure ice at the same temperatures (Dyer et al., 1966; Hill and Sunderland, 1967). The lower vapor pressures were attributed to the formation of solid solutions. However, later work by Storey and Stainsby (1970), Fennema (1978, 1981), Mackenzie (1975), and other researchers demonstrated that the vapor pressures of frozen biological materials were equal to the vapor pressure of ice at the same temperatures. Water activity values at sub-

$$C = c \exp\left(\frac{H_m - H_n}{RT}\right)$$

$$K = k \exp\left(\frac{H_p - H_n}{RT}\right)$$

c, k = entropic accommodation factors
H_m = molar sorption enthalpy of the monolayer
H_n = molar sorption enthalpy of the multilayer
H_p = molar enthalpy of evaporation of liquid water
R = ideal gas constant
T = absolute temperature

FIG. 9.20 Constants of the Guggenheim–Anderson–de Boer (G.A.B.) equation.

freezing temperatures can be calculated rather than measured, according to the equation

$$a_w = p_w/p_{sw}$$

where a_w is the water activity, p_w the vapor pressure of water generated by the sample, and p_{sw} the vapor pressure of pure supercooled water (not of ice!) at the same temperature. Figure 9.21 shows a Clausius–Clapeyron type of plot for temperatures below and above freezing, from the work of Fennema (1978). The plots are linear with the slopes being different for temperatures below and above freezing. The a_w at any given temperature from −1 to −15°C can be read from the solid line. The solid line at subfreezing temperatures is valid for muscle tissues and probably for all biological materials including foods (Mackenzie, 1975). For example, the a_w of a relatively high moisture sample (such as muscle) at an initial freezing point of −2°C would decrease from about 0.99 at 20°C to 0.98 at −2°C, and then to 0.86 at −15°C. Other products at different initial freezing points would have their own line plots intersecting the −1–−15°C line, each at a different point.

The central meaning of the above plot is this: In the presence of an ice phase, the a_w values depend on the temperature, irrespective of the solute(s) present and irrespective of the kind of food. The a_w values above freezing depend on both temperature and sample composition.

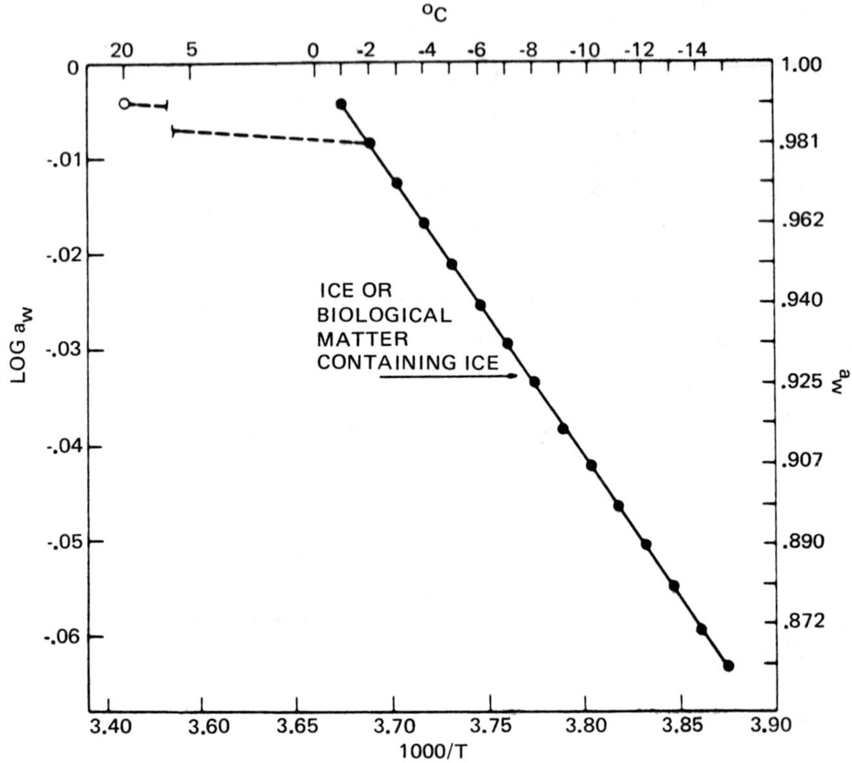

FIG. 9.21 Relationship between a_w and temperature for samples above and below freezing. (*From Fennema, 1978, 1981.*)

FUTURE WORK

Listed in Table 9.4 is an outline for future work on hysteresis in foods. This work relates experiments on rheological measurements (stress relaxation), water sorption measurements (isotherms), and thermodynamic measurements (calorimetry). The entropy production, discussed earlier in this chapter, is an expression of the irreversibility of the sorption process. It will be of both theoretical and practical interest to interpret hysteresis in foods employing the above three-pronged approach. Stress relaxation parameters (such as the elastic component, the viscous component, and the relaxation time, derived from the Maxwell model) can be used to quantify the mechanical rearrangement of structure. Application of direct calorimetry (possibly microcalorimetry) will

TABLE 9.4 Outline for Future Work on Hysteresis in Foods

Stress relaxation curves at different water activity increments on adsorption and desorption
Hysteresis measurements
Thermodynamics
  Calorimetric measurements of integral heat of sorption to determine Q_{diff}

$$\left(\frac{\partial\, \Delta H}{\partial n}\right) = Q_{diff}$$

  Isosteric heat of sorption from isotherms to determine entropy production

$$\overline{\Delta H^{0}_{iso}} = Q_{diff} + RT + T\, \Delta S_{irrev}$$

  Relate difference in stress relaxation between ad-desorption to hysteresis
  Interpret results thermodynamically

necessitate strict control of experimental variables in view of the small quantities of heat involved. The choice of foods will be important, model systems of proteins, carbohydrates, etc., being appropriate for providing (among other advantages) geometrical control for the rheological work.

ACKNOWLEDGMENT

I am indebted to Professor George Saravacos at Rutgers University for sharing with me his most recent work on the sorption equilibria of Sultana raisins, and for his helpful suggestions concerning this manuscript.

REFERENCES

Accot, K. M. and Labuza, T. P. 1975. Microbial growth response to water sorption preparation. *J. Food Technol.* 10: 603.

Amberg, C. H. 1957. Heats of adsorption of water vapor on bovine serum albumin. *J. Am. Chem. Soc.* 79: 3980.

Amberg, C. H., Everett, D. H., Ruiter, L., and Smither, F. W. 1957. The thermodynamics of adsorption and adsorption hysteresis. In *Surface Activity,* Vol. 2. Schulman, J. H. (Ed.), p. 3. Butterworth, London.

Arnell, J. C. and McDermot, H. L. 1957. Sorption hysteresis. In *Surface Activity,* Vol. 2. Schulman, J. H. (Ed.), p. 113. Butterworth, London.

Audu, T. O. K., Loncin, M., and Weisser, H. 1978. Sorption isotherms of sugars. *Lebensm. Wiss. U.-Technol.* 11: 31.

Bandyopadhyay, S., Weisser, H., and Loncin, M. 1980. Water adsorption isotherms of foods at high temperature. *Lebensm. Wiss. U.-Technol.* 13: 182.

Benado, A. L. and Rizvi, S. S. H. 1985. Thermodynamic properties of water on rice as calculated from reversible and irreversible isotherms. *J. Food Sci.* 50: 101.

Benson, S. W. and Richardson, R. L. 1955. A study of hysteresis in the sorption of polar gasses by native and denatured proteins. *J. Am. Chem. Soc.* 77: 2585.

Berlin, E. and Anderson, B. A. 1975. Reversibility of water vapor sorption by cottage cheese whey solids. *J. Dairy Sci.* 58: 25.

Berlin, E., Anderson, B. A., and Pallansch, M. J. 1968. Water vapor sorption properties of various dried milks and wheys. *J. Dairy Sci.* 51: 1339.

Bettelheim, F. A. and Ehrlich, S. H. 1963. Water vapor sorption of mucopolysaccharides. *J. Phys. Chem.* 67: 1948.

Bettelheim, F. A., Block, A., and Kaufman, L. J. 1970. Heats of water vapor sorption in swelling biopolymers. *Biopolymers* 9: 1531.

Bizot, H., Buleon, A., Mouhous-Riou, N., and Multon, J. L. 1985. Some facts concerning water vapor sorption hysteresis on potato starch. In *Properties of Water in Foods.* Simatos, D. and Multon, J. L. (Ed.). Martinus Nijhoff Publishers, Dordrecht.

Bolin, H. R. 1980. Relation of moisture to water activity in prunes and raisins. *J. Food. Sci.* 45: 1190.

Chirife, J. and Iglesias, H. A. 1978. Equations for fitting water sorption isotherms of foods, Part 1—A Review. *J. Food Technol.* 13: 159.

Chung, D. S. 1966. Thermodynamic factors influencing moisture equilibrium of cereal grains and their products. Ph.D. thesis, Kansas State Univ., Manhattan.

Cohan, L. H. 1938. Sorption hysteresis and the vapor pressure of concave surfaces. *J. Am. Chem. Soc.* 60: 433.

Cohan, L. H. 1944. Hysteresis and the capillary theory of adsorption of vapors. *J. Am. Chem. Soc.* 66: 98.

Cohen, E. and Saguy, I. 1983. Effect of water activity and moisture content on the stability of beet powder pigments. *J. Food Sci.* 48: 703.

Dyer, D. F., Carpenter, D. K., and Sunderland, J. E. 1966. Equilibrium vapor pressure of frozen bovine muscle. *J. Food Sci.* 34: 196.

Enderby, J. A. 1955. The domain model of hysteresis, Part 1, Independent domains. *Trans. Faraday Soc.* 51: 835.

Everett, D. H. 1955. A general approach to hysteresis, Part 4, An alternative formulation of the domain model. *Trans. Faraday Soc.* 51: 1551.

Everett, D. H. 1967. Adsorption hysteresis. In *The Solid-Gas Interface.* Flood, E. A. (Ed.), p. 1055. Marcel Dekker, Inc., New York.

Everett, D. H. and Whitton, W. I. 1952. A general approach to hysteresis. *Trans. Faraday Soc.* 48; 749.

Fennema, O. 1978. Enzyme kinetics at low temperature and reduced water activity. In *Dry Biological Systems.* Crow, J. H. and Clegg, J. S. (Ed.), p. 297. Academic Press, New York.

Fennema, O. 1981. Water activity at subfreezing temperatures. In *Water Activity: Influences on Food Quality.* Rockland, L. B. and Stewart, G. F. (Ed.). Academic Press, New York.

Franks, F. 1975. Water, ice and solutions of simple molecules. In *Water Relations of Foods.* Duckworth, R. B. (Ed.). Academic Press, New York.

Gregg, S. J. and Sing, K. S. W. 1967. In *Adsorption, Surface Area and Porosity.* Academic Press, New York.

Gregg, S. J. and Wheatley, K. H. 1957. The heat and the entropy of adsorption of benzene vapor on alumina. In *Proc. 2nd Int. Congr. Surface Activity.* Butterworth, London.

Greig, R. I. W. 1979. Sorption properties of heat denatured cheese whey protein. Part II. Unfreezable water content. *Dairy Ind. Int.* June: 15.

Guilbot, A., Charbonniere, R. Abadie, P., and Drapon, R. 1961. The contribution of water to the fine structure of macromolecular starch chains. *Staerke* 13: 204.

Hayakawa, K. I., Matas, J., and Hwang, P. 1978. Moisture sorption isotherms of coffee products. *J. Food Sci.* 43: 1026.

Heldman, D. R., Hall, C. W., and Hedrick, T. I. 1965. Vapor equilibrium relationships of dry milk. *J. Dairy Sci.* 48: 845.

Hill, J. E. and Sunderland J. E. 1967. Equilibrium vapor pressure and latent heat of sublimation for frozen meats. *Food Technol.* 21: 1276.

Iglesias, H. A. and Chirife, J. 1976a. Isosteric heats of water sorption on dehydrated foods. Part I. Analysis of the differential heat curves. *Lebensm.-Wiss. U.-Technol.* 9(2): 116.

Iglesias, H. A. and Chirife, J. 1976b. Isosteric heats of water vapor sorption on dehydrated foods. Part II. Hysteresis and heat of sorption comparison with BET theory. *Lebensm.-Wiss. U.-Technol.* 9: 123.

Iglesias, H. A. and Chirife, J. 1976c. Prediction of the effect of temperature on water sorption isotherms of food material. *J. Food Technol.* 11: 109.

Iglesias, H. A. and Chirife, J. 1976d. Equilibrium moisture content of air-dried beef. Dependence on drying temperature. *J. Food Technol.* 11: 565.

Iglesias, H. A. and Chirife, J. 1982. *Handbook of Food Isotherms*. Academic Press, New York.

Johnston, K. A. and Duckworth, R. B. 1985. The influence of soluble components on water sorption hysteresis in foods. In *Properties of Water in Foods*. Simatos, D. and Multon, J. L. (Ed.). Martinus Nijhoff Publishers, Dordrecht.

Kapsalis, J. G. 1975. The influence of water on textural parameters in foods at intermediate moisture levels. In *Water Relations of Foods*. Duckworth, R. B. (Ed.). Academic Press, New York.

Kapsalis, J. G. 1981. Moisture sorption hysteresis. In *Water Activity: Influences on Food Quality*. Rockland, L. B. and Stewart, G. F. (Ed.). Academic Press, New York.

Kapsalis, J. G., Walker, J. E., Jr., and Wolf, M. 1970. A physico-chemical study of the mechanical properties of low and intermediate moisture foods. *J. Texture Studies* 1: 464.

Kingston, G. L. and Smith, P. S. 1964. Thermodynamics of adsorption in capillary systems. Part I. Origin of irreversibility. *Trans. of the Faraday Soc.* 60: 705.

Kraemer, E. O. 1931. In *A Treatise on Physical Chemistry*. Taylor, H. S. (Ed.), Ch. XX, p. 1661. D. Van Nostrand Co., New York.

Labuza, T. P. 1970. Properties of water as related to the keeping quality of foods. *Proc. 3rd Intl. Congr. Food Sci. and Technol.*, p. 618. Institute of Food Technologists, Chicago, IL.

Labuza, T. P. 1975. Interpretation of sorption data in relation to the state of constituent water. In *Water Relations in Food,* Duckworth, R. B. (Ed.). Academic Press, New York.

Labuza, T. P. and Chou, H. E. 1974. Decrease of linoleate oxidation rate due to water at intermediate water activity. *J. Food Sci.* 39: 112.

Labuza, T. P. and Kamman, J. F. 1983. Reaction kinetics and accelerated tests simulation as a function of temperature. In *Computer-Aided Techniques in Food Technology*. Saguy, I. (Ed.), p. 71. Marcel Dekker, New York.

Labuza, T. P., Kaane, A., and Chen, J. Y. 1984. Effect of temperature on the moisture sorption isotherms and water activity shift of two dehydrated foods. *J. Food Sci.* 50: 385.

Labuza, T. P., Cassil, S., and Sinskey, A. J. 1972a. Stability of intermediate moisture foods. 2. Microbiology. *J. Food Sci.* 37: 160.

Labuza, T. P., McNally, L., Gallapher, D., Hawkes, J., and Hurdato, F. 1972b. Stability of intermediate moisture foods. 1. Lipid oxidation. *J. Food Sci.* 37: 154.

La Mer, V. K. 1967. The calculation of thermodynamic quantities from hysteresis data. *J. Colloid Interface Sci.* 23: 297.

Lee, R. Y., Silverman, G. J., and Munsey, D. T. 1981. Growth and enterotoxin A production by *Staphylococcus aureus* in precooked bacon in the intermediate moisture range. *J. Food Sci.* 46: 1687.

Lomauro, C. J., Bakshi, A. S., and Labuza, T. P. 1985. Evaluation of food moisture sorption isotherm equations. Parts I and II. *Lebensm.-Wiss. U.-Technol.* 18: 110.

Marshall, P. A. and Moore, W. J. 1952. Sorption of ammonia by silk fibroin. *J. Am. Chem. Soc.* 74: 4779.

Mackenzie, A. P. 1975. The physico-chemical environment during freezing and thawing of biological materials. In *Water Relations of Foods*. Duckworth, R. B. (Ed.). Academic Press, New York.

McBain, J. W. 1935. An explanation of hysteresis in the hydration and dehydration of gels. *J. Am. Chem. Soc.* 57: 699.

McLaren, A. D. and Rowen, J. W. 1951. Sorption of water vapor by proteins and polymers: a review. *J. Polymer Sci.* 7: 289.

Mellon, E. F. and Hoover, S. R. 1951. Hygroscopicity of amino acids and its relationship to the vapor phase adsorption of proteins. *J. Am. Chem. Soc.* 73: 3879.

Mellon, E. F., Korn, A. H., and Hoover, S. R. 1948. Water adsorption of proteins. II. Lack of dependence of hysteresis in casein on free amino groups. *J. Am. Chem. Soc.* 70: 1144.

Mittal, G. S. and Usborne, W. R. 1985. Moisture isotherms for uncooked meat emulsions of different composition. *J. Food Sci.* 50: 1576.

Rao, K. S. 1939a. Hysteresis in the sorption of water on rice. *Current Sci.* 8: 256.

Rao, K. S. 1939b. Hysteresis loop in sorption. *Current Sci.* 8: 468.

Rao, K. S. 1941. Hysteresis in sorption. I–VI. *J. Phys. Chem.* 45: 500.

Rao, K. S. 1942. Disappearance of the hysteresis loop. The role of elasticity of organogels in hysteresis in sorption. Sorption of water on some cereals. *J. Phys. Chem.* 45: 517.

Rockland, L. B. 1957. A new treatment of hygroscopic equilibria: Application to walnuts (Juglans regia) and other foods. *Food Res.* 22: 604.

Roman, G. N., Urbicain, M. J., and Rotstein, E. 1982. Moisture equilibrium in apples at several temperatures. Experimental data and theoretical considerations. *J. Food Sci.* 47: 1484.

Rutman, M. 1967. The effect of surface active agents on sorption isotherms of model systems. M.S. thesis, Massachusetts Inst. of Technology, Cambridge.

Salwin, H. 1962. The role of moisture in deteriorative reactions of dehydrated foods. In *Freeze Drying of Foods*. Fisher, F. R. (Ed.). National Academy of Sciences-National Research Council, Washington, DC.

San José, C., Asp, N. G., Burvall, A., and Dahlgvist, A. 1977. Water sorption in lactose hydrolyzed milk. *J. Dairy Sci.* 60: 1539.

Saravacos, G. D. 1965. Freeze-drying rates and water sorption of model food gels. *Food Technol.* 19: 193.

Saravacos, G. D. 1967. Effect of the drying method on the water sorption of dehydrated apple and potato. *J. Food Sci.* 32: 81.

Saravacos, G. D. and Stinchfield, R. M. 1965. Effect of temperature and pressure on the sorption of water vapor by freeze-dried food materials. *J. Food Sci.* 30: 779.

Saravacos, G. D., Tsiourvas, D. A., and Tsami, E. 1986. Effect of temperature on the water adsorption isotherms of sultana raisins. *J. Food Sci.* 51: 381.

Schär, W. and Rüegg, M. 1985. The evaluation of G.A.B. constants from water vapor sorption data. *Lebensm.-Wiss. U. Technol.* 18: 225.

Scott, W. J. 1957. Water relations of food spoilage microorganisms. In *Adv. Food Res.* 7: 83. Academic Press, New York.

Scott, V. N. and Bernard, D. T. 1983. Influence of temperature on the measurement of water activity of food and salt systems. *J. Food Sci.* 48: 552.

Seehof, J. M., Keilin, B., and Benson, S. W. 1953. The surface areas of proteins, V. The mechanism of water sorption. *J. Am. Chem. Soc.* 75: 2427.

Silverman, G. J., Munsey, D. T., Lee, C., and Ebert, E. 1983. Interrelationship between water activity, temperature and 5.5 percent oxygen on growth and enterotoxin A secretion by *Staphylococcus aureus* in precooked bacon. *J. Food Sci.* 48: 1783.

Speakman, J. B. and Stott, C. J. 1936. The influence of drying conditions on the affinity of wool for water. *J. Text. Inst.* 27: T186.

Spiess, W. E. L. and Wolf, N. R. 1983. The results of the COST 90 project on water activity. In *Physical Properties of Foods*. Jowitt, D., et al. (Ed.), p. 65. Applied Science Publ., London.

Storey, R. M. and Stainsby, G. 1970. The equilibrium water vapor pressure of frozen cod. *J. Food Technol.* 5: 157.

Strasser, J. 1969. Detection of quality changes in freeze-dried beef by measurement of the sorption isobar hysteresis. *J. Food Sci.* 34: 18.

van den Berg, C. 1981. Vapor sorption equilibria and other water starch interactions: a physicochemical approach. Doctoral thesis, Agricultural Univ., Wageningen, Netherlands.

van den Berg, C. 1984. Description of water activity of foods for engineering purposes by means of the G.A.B. model of sorption. In *Engineering and Food,* Vol. 1, p. 311. McKenna, B. M. (Ed). Elsevier Applied Science Publ., London.

van den Berg, C. 1985. Development of B.E.T.-like models for sorption of water on foods, theory and relevance. In *Properties of Water in Foods.* Simatos, D. and Multon, J. L. (Ed.). Martinus Nijhoff Publishers, Dordrecht.

van den Berg, C. and Bruin, S. 1981. Water activity and its estimation in food systems: theoretical aspects. In *Water Activity: Influence on Food Quality.* Rockland, L. B. and Stewart, G. F. (Ed.). Academic Press, New York.

van den Berg, C., Kaper, F. S., Weldring, A. G., and Wolters, I. 1975. Water binding by potato starch. *J. Food Technol.* 10: 589.

Van Olphen, H. 1965. Thermodynamics of interlayer adsorption of water in clay. I. Sodium vermiculite. *J. Colloid Sci.* 20: 822.

Watt, I. C. 1980a. Adsorption-desorption hysteresis in polymers. *J. Macromol. Sci.-Chem.* A14(2): 245.

Watt, I. C. 1980b. Sorption of water vapor by keratin. *J. Macromol. Sci.-Rev. Macromol. Chem.* C18(2): 169.

Weisser, H. 1985. Influence of temperature on sorption isotherms. ICEF 4, Paper No. 130, Edmonton, Alberta, Canada.

Weisser, H., Weber, J., and Loncin, M. 1982. Water vapor sorption isotherms of sugar substitutes in the temperature range 25 to 80°C. Inter. *Zeits. Lebens. Technol. Verfahrenstechnik* 33: 89.

Wolf, W., Spiess, W. E. L., and Jung, G. 1985. Standardization of isotherm measurements (COST-Project 90 and 90 bis). In *Properties of Water in Foods.* Simatos, D. and Multon, J. L. (Ed.). Martinus Nijhoff Publishers, Dordrecht.

Wolf, M., Walker, J. E., and Kapsalis, J. G. 1972. Water vapor sorption hysteresis in dehydrated foods. *J. Agr. Food Chem.* 20: 1073.

Zsigmondy, R. 1911. Structure of gelatious silicic acid. Theory of dehydration. *Z. Anorg. Chem.* 71: 356.

10

Critical Evaluation of Methods to Determine Moisture Sorption Isotherms

Walter E. L. Spiess and Walter Wolf

Federal Research Centre for Nutrition
Karlsruhe, Federal Republic of Germany

INTRODUCTION

Water vapor sorption by solids depends upon many factors, among which chemical composition, physical–chemical state of ingredients (such as the degree of denaturation or cross-linking), and physical structure are the most important. These parameters determine substantially the quantity of sorbate absorbed and the kinetics of the sorption process. The various forms of water bindings (Table 10.1) make it, at the present time, nearly impossible to predict the course of sorption isotherms or to precalculate the equilibrium water content for a given water vapor pressure in the product environment. It is possible only for some well-specified product groups to indicate, by way of estimation, those ranges in which equilibrium water content for a certain water vapor pressure can be expected. So whenever a correlation between water content and vapor pressure for a complex food system is required, it has to be determined by experiment. But even if equilibrium data are determined in this way, attention has to be paid to the fact that the values obtained are to

TABLE 10.1 Sorption by Solid Sorbents

No interaction sorbent-/sorbate molecules		
monolayer settlement	adsorption	residual binding forces of surface molecules
multilayer settlement	capillary condensation	capillary forces
Interaction sorbent-/sorbate molecules		
chemical reaction	chemisorption	primary valencies (ionic bond)
formation of cryst. hydrate	absorption	lattice forces
formation of solutions	absorption	molecular power field
struct. change swelling	absorption	secondary valencies (hydrogen bond)

some extent influenced also by the kind of measurement applied. In general, comparable data are obtained only by measuring methods that are nearly identical.

PROBLEMS OF WATER VAPOR SORPTION ISOTHERM MEASUREMENT

In view of the relationship

$$X = f(a_w)_{T=const} = f(P_D/P_{Do})_{T=const}$$

where

X = equilibrium water content of sorbent (dry base)
a_w = water activity of sorbent
P_D/P_{Do} = relative water vapor pressure in the sorbent environment
T = temperature of sorbent and sorbent environment,

it becomes obvious that the measurements of water vapor sorption isotherms include determination of the partial water vapor pressure and temperature in the sample environment, and measurement of the equilibrium water content within the sample (Table 10.2).

TABLE 10.2 Measuring Methods for Water Vapor Sorption Isotherms

Gravimetric methods	
continuously	discontinuously
evacuated systems dynamic systems static systems	
Manometric, hygrometric methods	
continuously	discontinuously
directly indirectly	
Special methods	

Source: Gal (1967).

In the case of partial water vapor pressure measurement one has to differentiate between partial pressure above the sorbate source and the vapor pressure in the sample environment. Determination of the equilibrium water content requires in most cases that the water content be determined before and after the sorption process. For this purpose different measuring methods have been elaborated, depending upon the product concerned. The equilibrium water content in the sample, however, is influenced not only by the method of water determination, but also by the experimental arrangement of the sorption process including selection of equilibration times and partial pressure differences, ad- and desorption, etc.

In the following the most important factors influencing the results of the sorption process will be discussed briefly and details of a standardized measuring method will be presented.

Sorption Process

In the majority of devices available to determine sorption isotherms, the sorption process essentially determined by mass- and heat transport processes in the sorbent (Table 10.3) consists in principle of four mass- and three heat transfer steps which have to be regarded separately: release of sorbate from the

TABLE 10.3 Mass and Heat Transport Coefficients (D = Diffusion Coefficient; λ = Thermal Conductivity)

Transport mechanism	Product		Gas pore size d/m[a]		
	High WC[c]	Low WC[c]	10^{-9}	10^{-6}	10^{-1}
Mass transport $D/m^2/s$	5.10^{-9}	5.10^{-10}	10^{-7}	10^{-5}	10^{-4}
Heat transport $\lambda/W/m^2\ K$	0.7			0.02[b]	

[a]Diameter of pore, equipment response.
[b]At atmospheric pressure.
[c]WC = water content.

sorbate source, transport of sorbate from sorbate source to product surface, sorption of sorbate by sorbent surface, and, finally, the transport process within the sorbent to those places where the water molecules are bound. Incorporation of water molecules into the product matrix may lead to formation of hydrates and/or solutions, or gels by swelling, depending upon the kind and composition of the product. Heat transport processes take the opposite direction: Heat released during the sorption process is transported to the sorbent surface, where it is discharged into the environment of the sorbent and in some cases transmitted to the sorbate source.

As far as heat discharge from the sorbent is concerned, two extreme cases exist, depending on the water activity difference between sorbate source and sorbent. If this difference is small, the released heat of sorption is low and the sorption process takes place under largely isotherm conditions. If the difference is large, the quantity of heat of sorption released may lead to a considerable rise in sample temperature and hence to nonisothermal, i.e., uncontrolled conditions.

Measurement of Partial Water Vapor Pressure

To determine the partial water vapor pressure over a sorbate source or in the sorbent environment, either manometers or hygrometers can be used. Partial pressure manometers, usually unsuitable for routine measurements, are applied mainly if very accurate data are required. Such manometers show an accuracy of ± 0.1% relative humidity. Another method for direct measurement of the water vapor pressure is the dew-point method, which, after some technical improvements, can be used as a routine laboratory method; its accuracy is ± 0.1% relative humidity.

For routine measurements, several electric hygrometers have been developed in recent years. These devices, which measure water vapor pressure indirectly through material properties that depend on water content of the environmental air, are as accurate as the dew-point method. Because of their mode of action, manometric and hygrometric methods are predominantly used to measure individual equilibrium points, i.e., to determine the vapor pressure in the sample environment.

If complete sorption isotherms are required, it is advisable to use systems that provide the required partial water vapor pressure in the sorption containers. This is accomplished by admission of conditioned air or through aqueous solutions with a given constant water vapor pressure. In the latter case, saturated and nonsaturated solutions of electrolytes and nonelectrolytes can be used. Mostly saturated salt solutions and sulfuric acid solutions are used. Problems arising during sorption measurements by use of salt solutions have shown that the use of saturated salt solutions, which are easier to handle in general than sulfuric acid solutions, require skill and experience. Accuracy of the vapor pressure adjustment, which depends very strongly on the salt used and on the sorption temperature, varies between 0.05 and 2% relative humidity. The accuracy of data obtained over sulfuric acid solutions can be ± 0.1% relative humidity over a larger temperature range (15–60°C). At temperatures above 40°C interactions between the H_2SO_4 and sorbent molecules may take place.

If the water vapor pressure is adjusted by conditioned air (two-temperature system, two-pressure system, mixture of two air streams) an accuracy of ± 0.1% relative humidity can be achieved as well. These three systems involve greater sophistication and expense. The air-mixture system and the two-pressure system are primarily used for dynamic sorption measurements, whereas the two-temperature system has been the method of choice for static measurements.

Measurement of Water Content

If complete sorption isotherms are required, it is advisable to measure either pure ad/absorption- or pure desorption isotherms. For this it is necessary to condition samples either to the lowest measuring point of the ad/absorption isotherm or to the highest point of the desorption isotherm. To achieve better reproducibility, it is recommended in the case of ad/absorption isotherms to condition samples at a relative vapor pressure of $P_D/P_{Do} = 0$. Such a pretreatment of samples, which must not affect the nature of the product, has the additional advantage that the sorbent, at the beginning of the sorption measurement, is available in water-free condition. Its weight may serve as basis for the water content calculations after completion of the sorption processes. In some cases, however, this method of evaluation of the water content is not satisfactory because predrying even in vacuum may not produce the same results as an official, more destructive method of water determination. It is therefore necessary to control the water content after completion of the sorption process by an appropriate standard method. In case the results of the two methods differ, the sorption isotherm should be calculated on the basis of the results obtained by the standard method.

In the case of measurement of desorption isotherms it is advisable, for reasons of accuracy, to determine the water content only after completion of sorption measurements.

Practical Implementation of Sorption Measurements and Interpretation of Experimental Results

The need for sorption data on food and food components was the reason for the development of a variety of equipment of different design and principle and of various types of measuring devices. Corresponding to the requirements, equipment may comprise simple desiccators or sophisticated measuring devices with automatic weight and temperature control and end-point determination. The great variety of equipment developed for industry and research may be regarded as an indication of the unsatisfactory situation prevailing and the need for improved equipment which would permit precise measurement of individual equilibrium conditions and complete isotherms. At present, reproducibility and repeatability of sorption isotherm measurements are very unsatisfactory. In recent years scarcely any novel equipment has been developed. Efforts have been concentrated on improving individual elements, such as hygrometers and balances. Some progress has been achieved by standardization of the entire measuring process.

Since it is very difficult to discuss available equipment with respect to achievable accuracies, some problems involved in the measurement of sorption data shall be examined on the basis of the principles outlined before. These discussions will concentrate primarily on static (gravimetric, discontinuous) methods used most frequently in industry and research.

Independent of the fact that the main resistance to mass transport arises from the product itself, one should make sure that all connective elements between sorbate source and sorbent be as short as possible and their diameters be as wide as possible to allow unhindered vapor transport from sorbate source to sorbent. Ten centimeters may be regarded as a guideline value for the maximum distance between sorbate source and sorbent. The ratio of area of the smallest equipment diameter to the visible sorbent surface should not exceed 5:1. The sorbent should be placed into weighing bottles with a large diameter to maximize heat discharge from the sorbent. The sorbent thickness should not exceed ¼ of the container diameter ($h_s \simeq$ ¼ d). Container height should be approximately equal to the container diameter ($h_c = d$).

During static measurements, the desirable environmental vapor pressure is adjusted in an optimal way by aqueous salt solutions. It is advantageous to use saturated solutions. The solutions should be prepared from analytical grade salts. Care must be taken to ensure that there is sufficient salt in the solid phase and sufficient excess liquid in the container. The salts must not degrade under the heat influence or become contaminated.

Great care is required for thermostatizing the measuring equipment as sample temperature must be constant during the entire measuring process. Only very slight temperature fluctuations between sorbent and environmental atmosphere are permissible. In routine measurements temperature deviations should not exceed ± 0.2°C. Measurements for scientific purposes or reference measurement must be within ± 0.02°C.

At least 10 points should be measured at different environmental vapor pressures for complete isotherms. It is advisable to measure these points on the same equipment at the same time. The salts or any other sorbate source should be selected in such a way that the measuring points are evenly distributed over the relative vapor pressure range $0 \leq P_D/P_{Do} \leq 1$.

A balance capable of determining changes in water content of ± 0.1% absolute for routine measurements, and of ± 0.01% absolute for reference methods is desirable. If the balance is not integrated in the system, it should be placed in the immediate vicinity to minimize environmental influence.

Precision recommendations for equipment used for sorption measurement are listed in Table 10.4.

Product

Samples should be representative of the product and should be comminuted as finely as possible. In cases where the product fine structure is ex-

TABLE 10.4 Precision Requirements of Equipment for Sorption Measurement

Physical property	Routine methods	Reference methods
Temperature (°C)	±0.2	±0.02
Rel. humidity (%)	±1.0	±0.1
Equil. moisture content (%)	±0.1	±0.01

pected to influence the water quantity absorbed, such structural elements should be preserved during comminution. For measurement of pure ad/absorption isotherms it is in general necessary to condition the samples before the actual sorption process. For ad/absorption measurements the product is completely predried, provided that drying does not change its properties. For foods that contain no volatile components except water, drying in high vacuum at temperatures below 50°C over a period of several days is recommended. The product should be spread in a uniform layer in weighing bottles.

The relative water vapor pressure in the sample environment can be adjusted to the desired value either directly in one or in several steps. If the environmental vapor pressure is adjusted in one step (integral sorption), the rate of water absorption is determined by transport mechanisms on the basis of concentration gradients in the sorbent. Initial water uptake is followed by a rearrangement of the water molecules in the sorbent to obtain a stress-free matrix. During the rearrangement step it is possible that water is released so that the quantity absorbed passes a maximum until it stabilizes at the final equilibrium water content.

In stepwise adjustment of the environmental vapor pressure (interval sorption), water uptake and rearrangement phases are parallel and the eventual sorption maximum is always less distinct than with integral sorption. In the case of integral sorption, the equilibrium water content, after having passed its maximum, is in general higher than the value recorded after interval sorption. Information as to whether the two values approximate after very extended equilibration times has not been available.

General Remarks

Sorption isotherms obtained according to the guidelines presented above still need some supplementary information and interpretation, particularly for practical use, as the brief discussion of integral and interval sorption processes has shown. A selection of important information is compiled in Table 10.5.

Details of the product state (particle size and configuration) are of special significance mainly when sorption data are used in drying and storage techniques. In these cases it should be noted that isotherms obtained from comminuted material have to be regarded as extreme values which are reached at the center of larger particles or bulks only after extended equilibration times.

In evaluating the correctness of sorption data, the temperatures applied and temperature constancy during measurements deserve particular attention. This includes also temperature fluctuations in the environment to which the product is exposed during manipulations such as weight control. Factors to be taken into account in evaluating temperature influences include the temperature

TABLE 10.5 Catalogue of Important Criteria for Evaluation of Sorption Data

A. **Test material**	
General product information	
Designation	Kind, variety, lot
Origin	Country, supplier, year of harvest, degree of ripening, age
Composition	Protein, fat, carbohydrates, ash, etc., in composite food percentage of ingredients. Physical-chemical state
Pretreatment	Brief description of processing methods. Drying: process, drying degree and rate, conditions of pressure, temperature and time
Specific experimental information	
Particle size and configuration	Comminution method, particle size distribution, geometrical structure
Sampling	Random sample, representative average sample
Sample size	Weight, volume, layer thickness
B. **Method**	
Control methods	
Determination of product water content	Precise description (if possible official or ISO method), accuracy
Control of relative humidity	Generation of vapor pressure (e.g., salt solution, H_2SO_4 solution, waterbath, etc.), measuring devices; tolerance
Control of temperatures	Absolute values; tolerance
Sorption apparatus	
Type of apparatus	Precise description of apparatus
Weight sensing element	Balance type, weighing accuracy
a_w sensing element	Method and precision of a_w adjustment and control
Gas flow pattern	Diffusion; forced gas movement
Exposure time of sample	Integral time of exposure to humid atmosphere
Kind of isotherm	Ad-, desorption, integral or interval sorption, initial state
C. **Report of experimental data**	
Water activity a_w; equilibrium moisture content (dry basis); physical state (e.g., free flowing, slightly caked, firmly caked, etc.)	

dependence of vapor pressure adjustment above the sorbate source and in the product.

To represent sorption data for interpolations or for inclusion into computer models of drying processes, closed functions are used; the Guggenheim–Anderson–de Boer (GAB) equation is one of the relations most frequently applied in this respect. Before such equations are used, one has to make sure that they fit the peculiarities (irregularities) of the sorption data to be described.

REFERENCE SYSTEM

In order to allow a reliable judgment of sorption measurements and sorption data, a reference system is recommended. In recent years a group of European research laboratories developed such a system and is still trying to improve it and extend the field of its application. The reference system consists essentially of a reference material, simple standard equipment with handling procedures, and recommendations for the evaluation of results.

Reference Material

As reference material, microcrystalline cellulose Avicel PH 101 (MCC) manufactured by the FMC Company was selected. The material must be:

1. Stable in its crystalline structure, which is responsible for sorption behavior, in a temperature range −18–80°C, with minimal changes in its sorption characteristics; when exposed to temperatures above 100°C the sorption properties may change. Thermal degradation is expected to begin at temperatures above 120°C.
2. Stable in its sorption properties after 2–3 repeated ad- and desorption cycles, even in cases where adsorption was extended to complete humidification of the material. However, when ad- and desorption cycles are repeated more than three times, significant changes in the sorption behavior have to be expected.

Furthermore it can be stated that:

1. All sorption data obtained, including results published in the pertinent literature, resulted in a sigmoid shape of the sorption isotherm when plotted in the usual way.
2. The water content can be determined by means of alternative methods (e.g., gravimetric and titrimetric) if the results are comparable.

Another advantage is that the substance, according to information supplied by the manufacturer, is available as a biochemical analytical agent in constant quality.

Standard Equipment and Handling Procedure

The sorption device consists of sorption containers, Petri dishes on trivets, and weighing bottles in which the sample material is exposed to the humid atmosphere in the container.

The sorption containers are simple preserving jars (1L volume) which can be tightly sealed against water vapor by means of rubber seal rings and glass covers (Fig. 10.1).

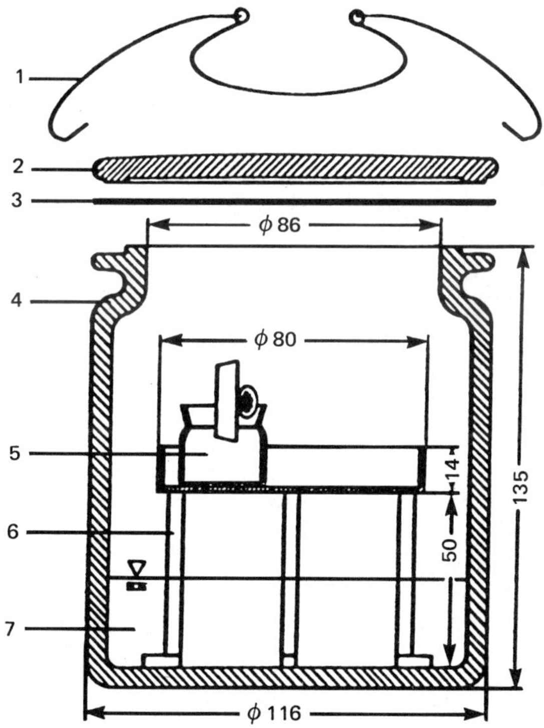

FIG. 10.1 Standardized sorption apparatus, sorption container: (1) locking clamp; (2) lid, glass; (3) rubber seal ring; (4) sorption container, glass; (5) weighing bottle with ground-in stopper; (6) Petri dish on trivet; (7) saturated salt solution.

The Petri dishes made of Duran® smeltable glass allow sufficient space to accommodate five weighing bottles within the sorption container over the surface of the salt solution.

As sample flasks low-shape weighing bottles, according to DIN 12 605, with ground-in stopper have proven satisfactory. (Dimensions of the bottle: 25 mm × 25 mm).

For temperature control sorption containers must be placed in a thermostated cabinet. Rectangular thermostats are especially suitable (Fig. 10.2). During the sorption experiments the thermostat should be covered with a plastic foam lid to prevent heat exchange by radiation. In case the thermostat

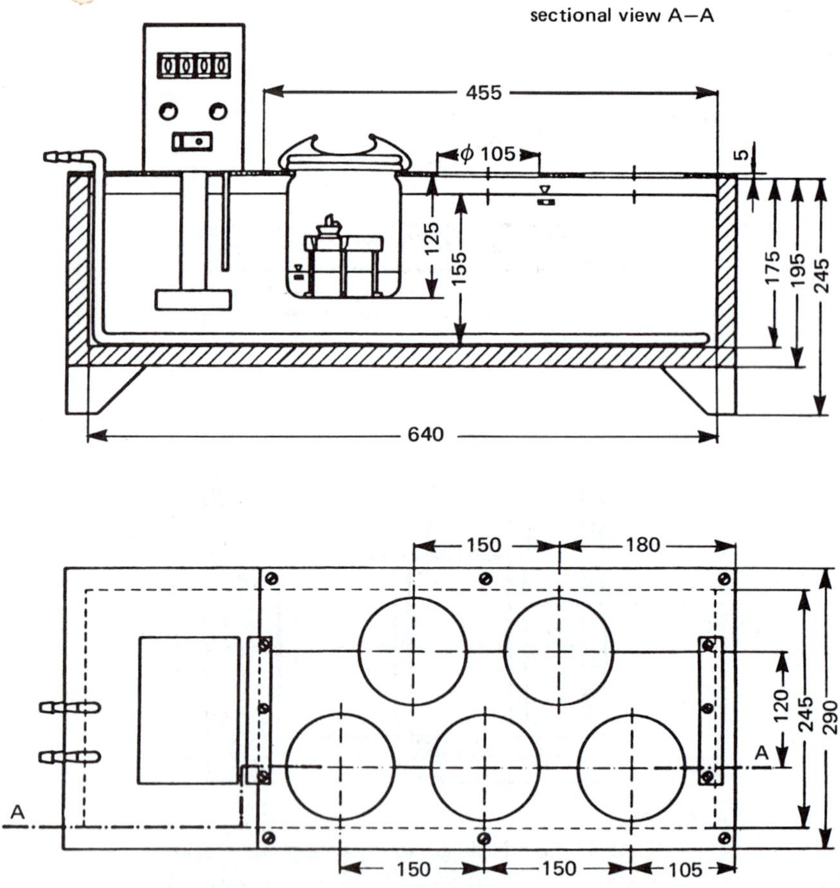

FIG. 10.2 Standardized sorption apparatus. Rectangular thermostat prepared for the accommodation of five sorption containers.

is not protected against heat losses by radiation, temperature differences in the sorption containers may occur to such an extent that the precision of the measurement is significantly reduced.

For practical sorption isotherm measurements the following equipment and materials should be available:

1. Sorption apparatus, consisting of a rectangular thermostat; 5–10 sorption containers; and 5–10 Petri dishes on trivets.
2. Twenty-five weighing bottles (25 mm × 25 mm) with ground-in stopper.
3. Desiccator, about 250 mm diameter, with perforated ceramic bottom plate (to accommodate at least 25 weighing bottles), bottom space filled up to a height of 10 mm with P_2O_5.
4. Laboratory drying cabinet.
5. Analytical balance.
6. Salts to regulate the relative humidity in the sorption containers.

The salt solutions should be prepared in the following way: fill the salt quantity as indicated in Table 10.6 into the sorption container; with stirring add the quantity of distilled cold water indicated in Table 10.6; close the sorption container and allow to stand 1 wk; stir the solutions briefly once a day. *Note:*

TABLE 10.6 Preparation of Recommended Saturated Salt Solutions at 25°C

Salt	Relative humidity (%)	Quantity: Salt (g)	Quantity: Water (mL)
LiCl	11.15	150	85
CH_3 COOK	22.60	200	65
$MgCl_2$	32.73	200	25
K_2CO_3	43.80	200	90
$Mg(NO_3)_2$	52.86	200	30
NaBr	57.70	200	80
$SrCl_2$	70.83	200	50
NaCl	75.32	200	60
KCl	84.32	200	80
$BaCl_2$	90.26	250	70

NaBr and LiCl show a strong tendency to clump. Thus it is recommended to comminute the clumps with a metal spatula.

Methodology.
Preparation of weighing bottles.

Clean with chromosulphuric acid.

Rinse thoroughly with distilled water.

Dry for 3 hr at 105°C.

Cool in the desiccator over P_2O_5 approximately 1 hr.

Apply cover only after the weighing bottle has cooled down.

Weigh the bottles on an analytical balance (W1).

Determination of dry matter.

Fill about 400 mg MCC into each bottle, placing lids on the bottles as shown in Fig. 10.1.

Predry for 3 hr at 100°C. Dry for 7 days in the desiccator over P_2O_5.

Close weighing bottles in the desiccator as quickly as possible. To avoid systematic errors close the bottles in random order (not in numerical order).

Weigh the bottles on an analytical balance (W2).

Sorption measurement (temperature 25 ± 0.1°C).

Open a series of five weighing bottles (five replicates) and place them into the corresponding sorption container. (The lids should be placed on the bottles as shown in Fig. 10.1). Seal the container (air-tight).

Weigh after a period of 4 days during which the substance is allowed to reach equilibrium (W3).

If possible, the sorption isotherm should be measured completely (i.e., at all a_w values indicated in Table 10.6). In a thermostat having space for only five sorption containers, salts No. 1, 3, 5, 7, and 9 should be used in the first test series, and salts No. 2, 4, 6, 8, and 10 in the second.

Explanations and calculations.

a_w = water activity of the saturated salt solutions used.

W1 = weight of the empty, dry weighing bottle (tara).

W2 = weight of the dry sample plus weight of the weighing bottle.

W3 = weight of the sample at equilibrium plus weight of weighing bottle.

DM = dry matter in sample = W2 − W1.

X = water content of sample = $100 \cdot (\mathrm{W3} - \mathrm{W2})/(\mathrm{W2} - \mathrm{W1})$.

$\bar{X}$ = mean value of the water content (five measurements).

s = standard deviation of the five measurements.

At high water vapor pressures in the sorbostat when microbial growth can be expected, a fungicide should be placed on a special support in the containers to avoid microbial contamination of the sorbent. Phenyl mercury acetate (C_6H_5-HgOCO-CH_3 (highly toxic!) serves the purpose best without affecting the sorption process. With some qualifications (not for fatty products) thymol can be used.

Reference Sorption Isotherm of MCC

Sorption measurements for MCC (results of an international collaborative study with 32 participating laboratories) were summarized, i.e., for the 10 relative partial water vapor pressures and the relevant statistical data. The mean values were used to construct the most probable sorption isotherm for MCC (Fig. 10.3).

The standard deviations of the equilibrium water contents vary from s = 0.19 to s = 0.31g H_2O/100g DM. The mean standard deviation and the means of the precision data were calculated. The mean standard deviation within the range covered by the experiments was $\bar{s}$ = 0.25g H_2O/100g DM; the mean repeatability was $\bar{r}$ = 0.17g H_2O/100g DM; and the mean reproducibility was $\bar{R}$ = 0.72g H_2O/100g DM.

For a more convenient and precise use of the MCC sorption isotherm (e.g., interpolation) the GAB-model was applied to the data and an explicit equation, which fits the experimental data well, was derived:

$$X = \frac{X_m C k a_w}{(1 - k a_w)(1 - k a_w + C k a_w)}$$

where

X = water content on a dry basis, and

C, k, X_m = GAB constants to be determined by regression analysis (see Fig. 10.3)

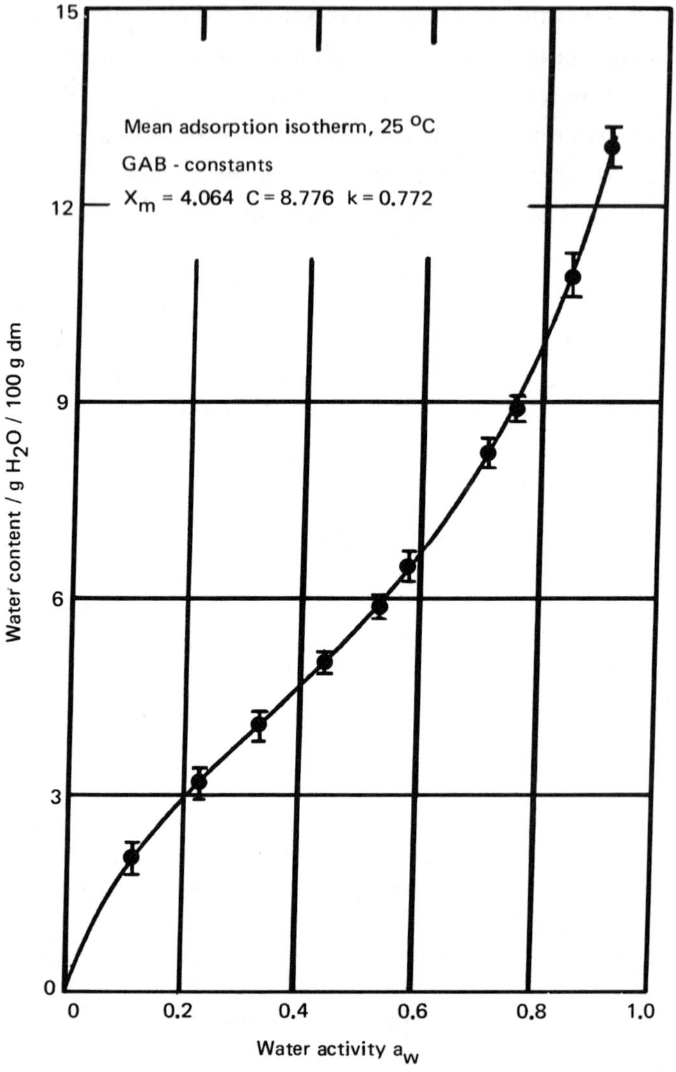

FIG. 10.3 Mean adsorption isotherm of MCC. (Results of a collaborative study with 32 participating laboratories within the framework of COST 90 project.)

Recommendations for Measurement of Water Vapor Sorption Isotherms

For the practical implementation of sorption measurements the following steps are recommended:

1. Measure the adsorption isotherm of MCC (Avicel) according to the standardized procedure (five replicates).
2. Compare the results with the mean value and precision data of the collaborative study.
3. If agreement is poor, search for errors and repeat the measurements.
4. If good agreement is obtained, initiate work with the material to be investigated.

Whether the agreement obtained is good or poor has to be decided with the aid of the precision data (reference value, repeatability, reproducibility).

If n determinations performed by one laboratory under repeatability conditions produce a mean value $\bar{X}$ which is to be compared to reference value X_r, then good agreement, on a 95% probability level, is obtained if the difference $/\bar{X} - X_r/$ is equal to or smaller than a so-called critical difference D_{cr}, the value of which is defined by the term:

$$D_{krit} = |\bar{X} - X_r| = \frac{1}{\sqrt{2}} \cdot \sqrt{R^2 - r^2\frac{n-1}{n}}$$

For the precision data obtained in the collaborative study:

$r = 0.17\text{g } H_2O/100\text{g } DM$
$R = 0.72\text{g } H_2O/100\text{g } DM$
$n = 5$ replicates in the laboratory to be judged

the critical difference has the value:

$$D_{cr} = 0.498$$

In a random case, for example, with $X_r = 5.61\text{g } H_2O/100\text{g } DM$ (equilibrium water content of MCC for $a_w = 0.5$, according to the sorption isotherm elaborated in the study) mean values of $\bar{X}$ which are within the interval of $5.11 \leq \bar{X} \leq 6.11$ are in good agreement with the recommended sorption isotherm.

Mean values falling outside of the interval are of poor agreement and should be considered questionable. In most cases in which the difference $/\bar{X} - X_r/$ is larger than the critical difference D_{cr}, systematic errors in the experimental procedure are very likely.

Where the mean value $\bar{X}$ of five measurements differs from the reference value X_r by not more than the critical difference, the standard deviation s obtained has to be compared to the repeatability standard deviation s_r as established in the collaborative study for the experimental procedure in question. The repeatability standard deviation s_r is defined as

$$s_r = \frac{r}{2.83}$$

where the factor 2.83 represents the 95% probability level. If another probability is requested, this factor changes to 2.32 for a 90% and to 3.65 for a 99% probability level.

For a judgment of the precision obtained, the following estimation is recommended: If s is equal to or smaller than s_r, the experimental procedure tested can be considered good; if s is larger than s_r, the variability of the test is too high; in this case more careful measurements are necessary.

If good agreement of the data with MCC as test material is obtained, measurements of real food material can be started. The following procedure is recommended:

1. Evaluate the necessary desorption (predrying) conditions for setting the zero water content or a water content considerably below the equilibrium water content of the lowest a_w value used.
2. Determine the time to reach equilibrium conditions.
3. Select appropriate standard procedures for the determination of the equilibrium water content.

CONCLUSION

The present brief survey of practical procedures for measuring sorption isotherms is intended to help obtain insight into problems which may arise. The recommended procedures for sorption isotherm measurements have been made in a form that allows those who have had little experience in this field to obtain reasonable results after a short training period.

REFERENCES

The survey has been based essentially on the following communications as well as some of our own unpublished data. For additional references see Wolf et al. (1985).

Gal, S. 1967. *Die Methodik der Wasserdampf-Sorptionsmessungen.* Springer Verlag, Berlin, Heidelberg, New York.

Gal, S. 1975. Recent advances in techniques for the determination of sorption isotherms. In *Water Relations of Foods.* Duckworth, R. B. (Ed.). p. 139. Academic Press, London.

Gal, S. 1983. The need for, and practical applications of, sorption data. In *Physical Properties of Foods.* Jowitt, R., Escher, F., Hallström, B., Meffert, H. F. T., Spiess, W. E. L., and Vos, G. (Ed.), p. 12. Applied Science Publishers, London, New York.

Rüegg, M., Blanc, B., and Lüscher, M. 1979. Hydration of casein micelles: kinetics and isotherms of water sorption of micellar casein isolated from fresh and heat-treated milk. *J. Dairy Res.* 46: 325.

Spiess, W. E. L. and Wolf, W. R. 1983. The results of the COST 90 project on water activity. In *Physical Properties of Foods.* Jowitt, R., Escher, F., Hallström, B., Meffert, H. F. T., Spiess, W. E. L., and Vos, G. (Ed.), p. 65. Applied Science Publ., London, New York.

Watt, I. C. 1980. Sorption of water vapor by keratin. *J. Macromolecular Sci.—Rev. Macromolecular Chem.* C 18(2): 169.

Watt, I. C. 1983. The theory of water sorption by biological materials. In *Physical Properties of Foods.* Jowitt, R., Escher, F., Hallström, B., Meffert, H. F. T., Spiess, W. E. L., and Vos, G. (Ed.), p. 27. Applied Science Publ., London, New York.

Weisser, H. 1980. Die Technik der Sorptions- und Wasseraktivitätsbestimmung. In *Hochschulkurs Ausgewählte Themen der Lebensmittelverfahrenstechnik "Wasseraktivität."* Loncin, M. (Ed.). Institut für Lebensmittelverfahrenstechnik, Universität Karlsruhe.

Wolf, W., Spiess, W. E. L., and Jung, G. 1985. *Sorption Isotherms and Water Activity of Food Material.* Science and Technology Publ., Hornchurch, England.

11

Applications of Nuclear Magnetic Resonance

Shelly J. Richardson and Marvin P. Steinberg

University of Illinois
Urbana, Illinois

INTRODUCTION

The extreme importance of water content to the overall stability and acceptability of foods has been known for many years. The actual content of water in a food has been shown to be an imprecise indicator of stability (Franks, 1982); rather it is the "nature," "state," or "availability" of that water that determines its eventual deterioration.

Investigators throughout the past 30 years (van den Berg and Bruin, 1981) have proposed that the thermodynamic "state" of the water, referred to as its chemical activity (a_w), was the critical factor in determining product stability. Even though the concept of a_w has grown to be the "most popular measure of biological viability and also of technological quality of food products" (Franks, 1982), it still leaves much to be desired. For instance, the minimum a_w for growth of microorganisms is most often reported as a range of values because the precise value depends on several additional parameters, such as thermodynamic equilibrium (e.g., sorption hysteresis concerns) (van den Berg and

Bruin, 1981; Franks, 1982), environmental conditions (e.g., temperature, pH, Eh) (Leistner and Rodel, 1976), and chemical agents utilized to adjust the a_w (Gould and Measures, 1977). Thus, a measure of the "availability" of water is still needed.

One of the most successful techniques employed to investigate this "availability" of water in biological systems is nuclear magnetic resonance (NMR) spectroscopy (Fuller and Brey, 1968; Steinberg and Leung, 1975; Nagashima and Suzuki, 1984; Richardson et al., 1986a). NMR provides a rapid, sensitive, direct, and most importantly a noninvasive, nondestructive determination of not only the quantity of water present, but also the structure and dynamic characteristics of water (Berendsen, 1975) in complex systems, such as foods. The vast majority of methods for both quantitation of total water (e.g., oven and vacuum moistures, Karl Fisher titration and distillation methods) and determination of dynamic and structural information [e.g., such as Differential Scanning Calorimetry (DSC)] are invasive and destructive to the sample. Several problems arise when employing invasive or destructive methods. For example, in oven drying the nature of the sample is changed by an increase in temperature and co-current decrease in moisture content to obtain the desired measure of moisture content. Another example is DSC, where the measurements obtained during sample heating are used to make *inferences* regarding the state of the water in the sample under standard conditions. Therefore, these data provide only an indirect measure of the state of the water in the system. Auxiliary problems with destructive methods include the inability to reuse the sample for replicate measurements with the same method or for determination of additional properties employing other methods and the imprecise nature of these methods.

The noninvasive character of NMR coupled with its ability to characterize as well as quantitate water serves to emphasize its extensive usefulness in investigating water in foods, on both a theoretical and applied basis. Therefore, the objectives of this review are to discuss several NMR techniques available for study of water and to present studies illustrating various applications of NMR to the investigation of water in biological systems, especially as related to foods and food constituents. Due to the rapid growth in this field, it will not be possible to elaborate on the topics presented. However, the reader will be directed to references that supplement this text.

NMR TECHNIQUES FOR WATER

Theory

Nuclear magnetic resonance (NMR) spectroscopy is based on the measurement of resonant radio frequency adsorption by nuclear spins in the presence of an applied magnetic field (B_0). Only certain nuclei possess a nonzero spin value. A nucleus that has this property has a magnetic moment, μ_N, and can interact with applied magnetic fields. Such nuclei show a uniform or zero energy level in the natural state (no B_0). However, when placed in a magnetic field, they interact with this field and distribute themselves among several energy levels; the number of levels is equal to $2I + 1$ where I is the spin value. For water, there exist three nuclei that possess this spin property; the proton (^1H; $I = 1/2$), deuterium (^2H or D; $I = 1$) and oxygen-17 (^{17}O; $I = -5/2$). For the ^1H nucleus in liquid water, two energy levels result from this magnetic field interaction. The energy levels correspond to the spins aligned along ($-1/2$) and against ($+1/2$) B_0. The spins which orient against B_0 have a higher energy. The population of nuclei at each energy level is dictated by the Boltzmann distribution.

The ^2H and ^{17}O nuclei with $I > 1/2$ not only interact with applied magnetic fields but possess an additional property, a quadrupole moment that allows the nuclei to interact with the electric fields produced by neighboring electrons and nuclei (Campbell and Dwek, 1984). This quadrupolar interaction is often very strong and therefore dominates the ^2H and ^{17}O NMR behavior, as will be discussed below. The detailed theory underlying NMR spectroscopy is given in several references (Abragam, 1961; Dwek, 1973; Berliner and Reuben, 1980; Laszlo, 1983; Campbell and Dwek, 1984; Jelinski, 1984).

There are two basic types of NMR spectroscopy: the wide-line or frequency-sweep method and the pulsed or transient response method. The wide-line measurement is analogous to tuning a piano note by note, striking each note and listening to the response. The pulsed measurement is comparable to all the keys being struck at once and extracting the response of each note from the total sound (Campbell and Dwek, 1984).

Both types of NMR measurements have been used to study water in biological systems (Toledo et al., 1968; Shanbhag et al., 1970; Civan and Shporer, 1975; Okamura et al., 1978; Halle et al., 1981; Lioutas et al., 1986; Richardson et al., 1986a). However, essentially only the pulsed technique is used today because it is more versatile and powerful (Campbell and Dwek, 1984). For example, in the pulsed method three water nuclei (^1H, ^2H, and ^{17}O) can be probed, whereas in the wide-line method only the ^1H nucleus has been probed. Therefore, further discussion in this section will be limited to pulsed NMR.

In pulsed NMR, a large permanent B_0 field is applied to the sample. The B_0 field should be as homogeneous as possible so as to apply a uniform magnetization to the sample. To improve the effective B_0 field homogeneity, the tube containing the sample is often spun about its vertical axis (Campbell and Dwek, 1984). A pulse is applied by a rotating magnetic field (B_1) at right angles to B_0. This causes the spins of all nuclei being probed to become aligned in a single direction. This rotating field is generated through a tuned coil and generally consists of a burst or pulse of radio frequency energy. This

B_1 radio frequency pulse rotates the sample magnetization by an angle which is dependent upon its length and strength, i.e., a 90° pulse rotates the sample magnetization 90° from B_0. The B_1 pulse is thus defined in terms of angles of magnetic rotation from B_0. After the pulse has been applied, the spin system tends to return to its equilibrium position in the B_0 field through dephasing (i.e., losing its alignment generated by the B_1 field) via spin–lattice (or longitudinal) and spin–spin (or transverse) relaxation processes. The spin–lattice relaxation, T_1, is defined as the time constant that characterizes the rate at which the z vector component of the magnetization returns to its equilibrium value. The spin–spin relaxation, T_2, is defined as the time constant that characterizes the rate of decay of the magnetization in the x-y plane.

After the equilibrium magnetization is reestablished (i.e., the T_1 relaxation process has occurred), the radio frequency pulse may be repeated. The induced voltage generated by the spin dephasing process is monitored in the x-y plane by a resonant radio frequency coil. The result is a plot of voltage intensity against time called the free induction decay (FID). This FID is a complex superposition of signals from all the relaxing nuclei. However, the FID signal can be resolved using the mathematical method known as Fourier transformation; this transforms the FID into the desired frequency-domain spectrum of intensity against frequency, which is the usual manner of reporting the data.

Recent Advances

The steady advancement of NMR spectroscopy has extended and enhanced its capabilities for investigating water dynamics and structure in biological systems. Examples are high-field (Brevard and Granger, 1981; Markley and Ulrich, 1984) solids NMR (Maciel, 1984), two-dimensional NMR (Peemoeller et al., 1981), quadrupole relaxation processes (Hubbard, 1970; Bull et al., 1979; Westlund and Wennerstrom, 1982), models for interpretation of magnetic resonance data (Zimmerman and Brittin, 1957; Woessner, 1962, 1974; Hallenga and Koenig, 1976; Hsi and Bryant, 1977; Walmsley and Shporer, 1978; Halle and Wennerstrom, 1981; Halle et al., 1981; Shirley and Bryant, 1982; Finney et al., 1982; Kumosinski and Pessen, 1982), pulse sequencing and decoupling techniques.

High-field NMR (large B_0) improved peak resolution and allowed the field dispersion dependence of the relaxation process to be explored (Hallenga and Koenig, 1976; Bryant, 1978; Halle et al., 1981; Finney et al., 1982). The dependence of the ^1H NMR differential relaxation rate on magnetic strength in wheat flour suspensions was reported by Richardson et al. (1986a). This dependence was explored by comparing ^1H NMR relaxation at 20 and 360 MHz with the Ostroff-Waugh multipulse sequence (discussed below) at several flour concentrations (Fig. 11.1). The 20- and 360-MHz ^1H NMR data at the low flour

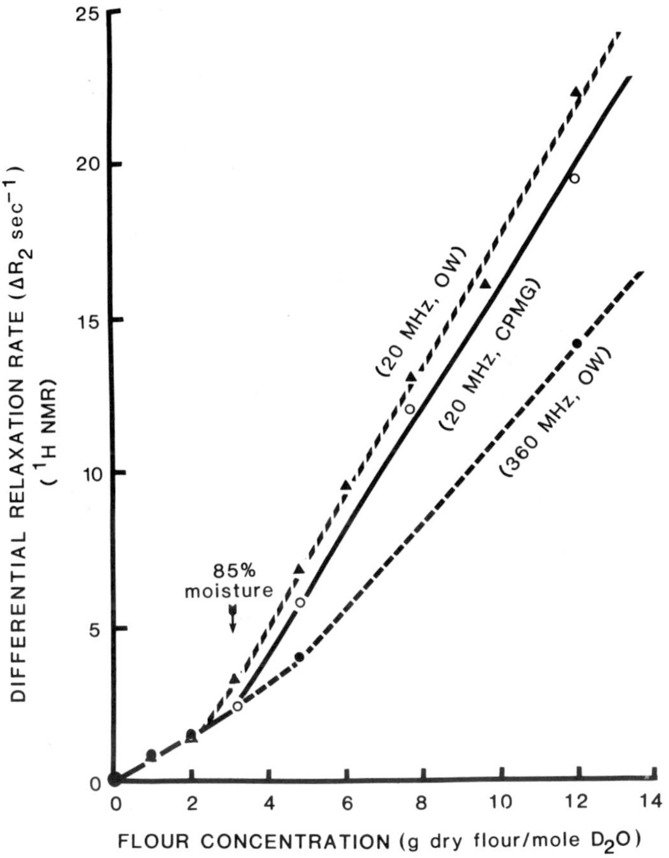

FIG. 11.1 Dependence of the differential proton NMR relaxation rates on flour concentration in D_2O at two frequencies and two multipulse sequences (OW and CPMG). (*From Richardson et al., 1986a.*)

concentrations were virtually the same. However, at higher flour concentrations, above approximately 3g dry flour/mole D_2O, the 20-MHz data showed a larger differential relaxation rate than the 360-MHz data. This difference between the two frequencies at the higher flour concentrations was attributed to cross-relaxation (defined below), but the difference was not as large as would be expected if cross-relaxation was the dominant NMR relaxation mechanism. A variety of pulse sequence techniques (Hahn, 1950; Carr and Purcell, 1954; Meiboom and Gill, 1958; Ostroff and Waugh, 1966) have been developed to reduce the effect of magnetic inhomogeneities on the measurement of T_2. An example is the 90°-τ–180°-τ pulse sequence, where τ is the time between pulses. This sequence generates an echo at time 2τ. This echo has a decreased sensitivity to field inhomogeneity effects (Campbell and Dwek, 1984).

The Carr–Purcell–Meiboom–Gill (CPMG) (Meiboom and Gill, 1958) multipulse sequence is used for liquid samples. However, in the case of highly viscous or solid samples, the CPMG sequence may be less useful because strong dipolar interactions can dominate the spin echo behavior (Baianu et al., 1978; Lioutas, 1984). The pulse technique, proposed by Ostroff and Waugh (1966), (OW) was shown to be adequate at such high solids concentration as well as at the low concentration range (Lioutas, 1984; Lioutas et al., 1986). This work was done on hydrated lysozyme ranging in concentration from 1 to 98.5% in deuterium oxide. Richardson et al. (1986a) applied both the CPMG and OW pulse sequences at 20-MHz ^1H resonance frequency to study the mobility of water in wheat flour suspensions (Fig. 11.1). Two linear regions and a transition at approximately 85% moisture, which was the pattern observed throughout the study, was obtained for both multipulse sequences. Both the CPMG and OW multipulse sequences yielded comparable differential relaxation rates (ΔR_2) in the "liquid" range. However, as previously discussed, the CPMG sequence may be less useful for high-viscosity liquids or semisolids because strong dipolar interactions can dominate the spin echo behavior. Therefore, the OW sequence has the advantage that it can be used over the entire wheat flour concentration range from dilute suspension to dough systems (Richardson et al., 1986a).

These NMR spectroscopy advancements have led to a focus on two major questions: 1) which nucleus should be probed, and 2) which model should be adopted for data interpretation. These questions will be discussed individually below.

Nucleus to be probed. As discussed by Richardson et al. (1986a), the majority of the NMR studies investigating the water component have been done using low field ^1H NMR. However, there are problems with the interpretation of ^1H NMR relaxation data from macromolecular systems both at low and high fields. The major concern is with the relative influence of the various relaxation mechanisms that contribute to the line width of the water peak in the ^1H NMR spectra of such complex systems (Kalk and Berendsen, 1976; Edzes and Samulski, 1978; Koenig et al., 1978; Peemoeller et al., 1986). They stated that the two relaxation mechanisms contributing significantly to the water proton line broadening are cross-relaxation and proton exchange.

Cross-relaxation is caused by dipole–dipole interaction between the protons of the water and the protons of the macromolecule. This occurs when the dipole moment of a water proton couples with the dipole moment of a macromolecular proton and transfers out its magnetization to the macromolecular proton (Fig. 11.2). This causes the coupled dipole moments to cross-flip or

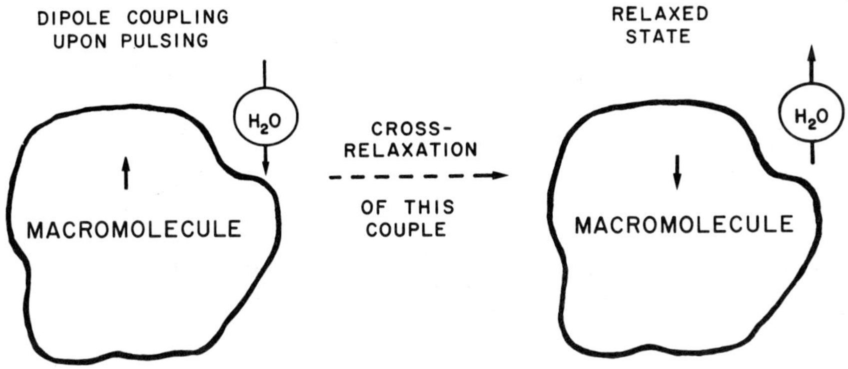

FIG. 11.2 Schematic representation of cross-relaxation. Solid arrows represent the proton magnetic dipole moments for the macromolecule and the water components. Coupled moments cross-flip as the water proton transfers its magnetization energy to the macromolecular proton.

cross-relax without the physical exchange of protons. The consequence is that relaxation rates of all the coupled protons tend to become equal (Berendsen, 1975). Also, the phenomenon of cross-relaxation decreases the apparent relaxation time and results in shorter T_1 or T_2. Most important, this causes the water to appear to be bound to a greater degree than it actually is.

Proton exchange is the physical exchange of protons between distinct states of water, i.e., "bound" and "free" water (Fig. 11.3). It is most often referred to as "chemical exchange." The magnitude of this effect depends on at least three factors: 1) the rate of exchange ($1/t$, Fig. 11.3) between sites, 2) the fraction of nuclei at the "bound" site, and 3) the intrinsic relaxation rates of the nuclei in the "bound" and "free" states (Berliner and Reuben, 1980). For a detailed discussion on how each of these factors affects the observed relaxation rate see Berliner and Reuben (1980).

For example, if we assume that 1) there are only two populations of water, "bound" and "free," and 2) their protons exchange at a rate greater than that of the NMR frequency (i.e., t is very short), then the observed relaxation rate is a weighted average over both water populations. The situation described above is known as the "isotropic two-state model with fast exchange" (Zimmerman and Brittin, 1957; Derbyshire, 1982). This model is discussed in detail below.

In a similar manner, chemical exchange can also occur between these water states and prototropic residues on the macromolecule, i.e., $-OH$, $+NH_3$, and $-NH_2$ (Halle et al., 1981). However, in most cases, the exchange between water states ("bound" and "free") dominates since the water is usually present

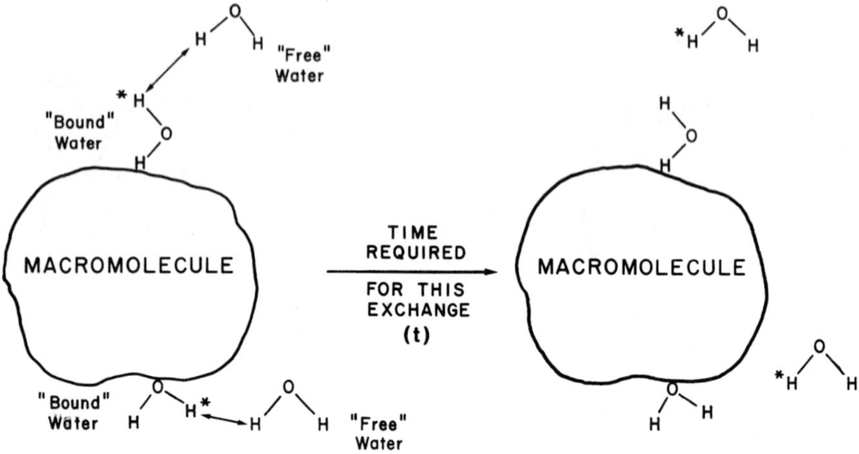

FIG. 11.3 Schematic representation of chemical exchange between water molecules. There is a physical exchange between a free water proton (H) and a proton of water bound to the macromolecule (*H). Time *t* is the time required for this physical exchange to occur.

in molar concentrations that are much larger than the macromolecular concentrations.

Because of these concerns with ^{1}H NMR, recent interest in measuring water binding and hydration of macromolecules has been directed to both ^{2}H and ^{17}O NMR (Halle and Wennerstrom, 1981; Halle et al., 1981; Laszlo, 1983; Lioutas, 1984; Richardson et al., 1986a). Probing the ^{2}H and ^{17}O nuclei eliminates the cross-relaxation experienced with the ^{1}H nucleus because the former are quadrupolar nuclei, and their relaxation process is dominated by electric field gradient interactions (Halle et al., 1981). However, ^{2}H nuclei do experience chemical exchange.

The advantages of measuring ^{17}O relaxation are discussed by Halle et al. (1981). A concern with ^{17}O NMR is that the ^{17}O relaxation is influenced by proton exchange broadening, especially for a narrow pH range around neutrality (Rabideau and Hecht, 1967; Halle et al., 1981). This proton exchange broadening is caused by a coupling between the ^{17}O and ^{1}H dipole moments. It is decreased by the use of deuterium oxide (D_2O) as a solvent (Lioutas, 1984; Richardson et al., 1986a) and can be eliminated through proton decoupling (Earl and Niederberger, 1977). Therefore, since ^{17}O NMR measurements directly monitor the molecular motions of water molecules, the oxygen nucleus is the best probe for studying water binding, molecular motions of water, and water–protein interactions in solutions (Halle et al., 1981; Lindman,

1983) and suspensions (Lioutas, 1984; Lioutas et al., 1986; Richardson et al., 1986a, b).

Model for data interpretation. A major difficulty in studying the dynamics of water associated with food components and systems is the extreme complexity encountered (van den Berg and Bruin, 1981). Food systems are heterogeneous, multicomponent assemblies which include two or more of the following components: water, protein, lipids, carbohydrates (simple and complex), vitamins, minerals, and/or salts. This complexity is exacerbated by a multitude of other factors, such as interaction among components (Bull and Breese, 1970; Gal, 1975; Arakama and Timasheff, 1982a, b; Chinachoti and Steinberg, 1984, 1985), growth and harvest conditions, postharvest treatment (Potter, 1978), processing conditions and environmental conditions (e.g., temperature, pH), to name a few. Therefore, a mathematical model is needed to analyze the relaxation of the water in these complex systems so as to obtain an understanding of the relation between the water and other components.

The two relaxation parameters usually studied are the spin–lattice relaxation time, T_1, and the spin–spin relaxation time, T_2. These relaxation times can be employed to calculate the rotational correlation time, τ_c. There is no general solution to the relationship between T_1, T_2, and τ_c. Specific mathematical relationships must be derived for each particular mechanism of spin relaxation and each nucleus probed (Kuntz and Kauzmann, 1974). For simple theoretical cases (e.g., two protons interacting with each other through a fixed distance and tumbling isotropically), the complex quantum mechanics have been elucidated (Kubo and Tomita, 1954; Solomon, 1955). However, as the number of factors under consideration increases (e.g., rotational modes, intermolecular interactions, anisotropic reorientation) and as the system under investigation becomes more intricate (e.g., heterogeneous samples, macromolecular bi-phasic systems), the complexity of the relationship between T_1, T_2, and τ_c is dramatically increased (Mantsch et al., 1977; Walmsley and Shporer, 1978; Bull et al., 1979; Halle and Wennerstrom, 1981; Kumosinski and Pessen, 1982). Therefore, to investigate the behavior of water in complex systems, simplifying assumptions along with basic interpretive modeling must be introduced (Bull et al., 1979).

Finney et al. (1982) stated that a major consequence of this problem of complexity is that the interpretation of much of the NMR data is strongly model-dependent—so much so that a change in the model may result in a very different picture of the system under examination. However, this is also true of other techniques used to investigate water–protein interactions, such as X-ray diffraction, neutron scattering, and infrared and Raman spectroscopy (Finney et al., 1982). Thus, great care must be taken when drawing conclusions from experimental data.

The majority of the NMR data, until the last five years, was interpreted in terms of a two- (or more) state fast-exchange model (Finney et al., 1982). When the exchange time between states is short in relation to the NMR relaxation times of each state, the relaxation behavior shows a single relaxation time:

$$T_{obs}^{-1} = \sum_i P_i T_i^{-1} \tag{1}$$

where T_{obs}^{-1} is the observed relaxation time, T_i^{-1} is the relaxation time of the i_{th} state, and P_i is the probability that the nucleus is found in that state (Cooke and Kuntz, 1974). Now, if we assume that there are fundamentally two possible water states, bound and free, Eq. (1) simplifies to the standard two-state model, with fast exchange (Derbyshire, 1982):

$$T_{obs}^{-1} = P_B T_B^{-1} + P_F T_F^{-1} \tag{2}$$

where T_B^{-1} is the relaxation rate of bound water, P_B is the probability of water being bound, T_F^{-1} the relaxation rate of the free water, and P_F is the probability of water being free. Since P_F equals $(1 - P_B)$ we can cast Eq. (2) into a linear form:

$$T_{obs}^{-1} = P_B(T_B^{-1} - T_F^{-1}) + T_F^{-1} \tag{3}$$

Thus, the two-state model, with fast exchange, predicts a linear relationship between the observed relaxation rate (T_{obs}^{-1}) and the probability of the water being in the bound state (P_B), where P_B can be related to the concentration of substance in solution. This relationship generally holds for dilute solutions (Derbyshire, 1982); however, departures from linearity often occur at higher concentrations (Woodhouse, 1974; Halle et al., 1981; Richardson et al., 1986a). Three mechanisms to account for this nonlinearity have been proposed. Derbyshire (1982) invoked two mechanisms: a change in hydration number and a change in the relaxation rate of the bound water (T_B^{-1}). The third mechanism, presented by Kumosinski and Pessen (1982) for dilute protein solutions, attributes the nonlinearity to charge repulsion or charge fluctuations as predicted by the Kirkwood and Shumaker (1952) theory.

One of the major applications of the third model is the use of activities in place of concentrations when dealing with systems strongly deviating from ideality. The activity of a substance, a, in solution is related to its concentration by the activity coefficient, γ:

$$a = \gamma c \tag{4}$$

where c is the concentration in grams of substance/mole water. The activity coefficients can be obtained from the virial expansion of osmotic pressure as a function of concentration:

$$d \ln \gamma / dc = 2B_0 + 3B_2 c + \ldots \tag{5}$$

where the B parameters are the virial coefficients. The application of virial coefficients to the nonideality of macromolecules in solution is discussed by Richards (1980) and Bates (1982). According to the theory underlying the third model, plotting the NMR relaxation rate against activity of the substance will yield a linear plot. The method for calculation of the virial coefficients is given by Kumosinski and Pessen (1982) and Richardson et al. (1986a, b). Application of this model to β-lactoglobulin A is illustrated in Fig. 11.4 (Kumosinski and Pessen, 1982). Plotting the relaxation rate against concentration yields the curved lines. However, plotting the relaxation rate against activity results in a linear relation, where the activity coefficient corrects for intermolecular interactions. The Kumosinski Model has been applied to a

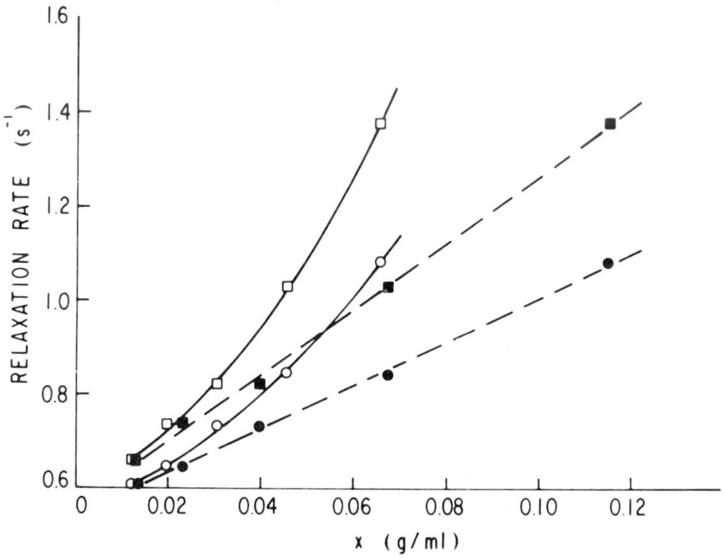

FIG. 11.4 Dependence of water proton relaxation rates, longitudinal (circles) and transverse (squares), on both concentration and activity. Dependence is curvilinear with concentration (open) but linear with activity (closed). (*From Kumosinski and Pessen, 1982.*)

variety of other systems including lysozyme with and without salt (Kumosinski and Baianu, 1986), wheat flour suspensions (Richardson et al., 1986a), and corn starch and amylopectin suspensions in both H_2O and D_2O (Richardson et al., 1986b); the latter is shown under Applications.

Despite the concern with deviation from linearity at higher concentrations, the two-state, fast-exchange model has been used by several investigators to interpret their data (Child and Pryce, 1972; Cooke and Wien, 1973; Woodhouse, 1974; Hansen, 1976; Oakes, 1976; Lioutas, 1984). It is often the model of choice due both to its simplicity and its ability to yield reasonable values for the correlation time, τ_c, under extreme narrowing conditions (Halle et al., 1981; Lioutas, 1984; Richardson et al., 1986a, b). It must be emphasized that all the bound water in this model is assumed to be equivalent and have identical molecular properties, which are different, however, from those of the free, bulk water. Derbyshire (1982) presented an extensive discussion on the concerns associated with this assumption.

An extension of the two-state model is the three-state model proposed by Fuller and Brey (1968) for water sorbed on serum albumin. This model suggests three different states of water; the first state involves direct binding of water molecules to available polar groups on the protein, the second state involves the formation of a hydration shell around the protein (the structure of which is markedly different than that of liquid water), and the third state is very similar to that of liquid water. Several other investigators have applied the three-state model to their data with varying definitions of "boundness" attributed to each of the states (Brey et al., 1968; Cooke and Kuntz, 1974; Ablett et al., 1976; Hsi et al., 1976; Grigera and Mascarenhas, 1978). Other models recently developed (Finney et al., 1982) are briefly discussed below.

A cross-relaxation model was proposed by Bryant and co-workers (Hsi and Bryant, 1977; Bryant and Shirley, 1980a, b). The results of this model, as generalized from Bryant's lysozyme data, indicate that "the perturbed solvent region is not extensive, and the rotational correlation time of the water at the protein surface is shorter than nanoseconds" (Finney et al., 1982). This indicates that the water at the interface is indeed far from "ice-like" and much closer to bulk-water than previous thought. This phenomenon of cross-relaxation has been investigated by several other researchers (Kalk and Berendsen, 1976; Koenig, 1980; Bryant and Shirley, 1980a, b; Peemoeller et al., 1984).

An alternative model set forth by Hallenga and Koenig (1976) proposed a hydrodynamic interaction mechanism between solute (e.g., protein) and solvent water molecules. This mechanism regards the water molecules (surrounding the solute molecules) as participating in an anticorrelated reverse rotation about the solute molecule in a manner that would conserve angular momentum. Employing this interpretation, Hallenga and Koenig (1976) estimated a correlation time for this water of 10^{-9} sec. In a further study by

Koenig et al. (1978), diluting the protons with deuterons indicated that there is also a contribution from cross-relaxation.

There exist several other relaxation models which have been applied to a variety of systems as follows:

1. Anisotropic water motion (Walmsley and Shporer, 1978; Halle et al., 1981; Shirley and Bryant, 1982; Peemoeller et al., 1986).
2. Two-step model: a fast anisotropic reorientation superimposed on a more extensive slow motion (Halle and Wennerstrom, 1981).
3. Combined model of anisotropic motion and distribution of correlation times (Shirley and Bryant, 1982).
4. Bimodal distribution of correlation times (Peemoeller et al., 1986).

A further discussion of some of these models, as related to the applications of NMR, follows.

APPLICATIONS OF NMR

The NMR techniques and associated models described above as well as extensions of these models (Steinberg and Leung, 1975; Lillford et al., 1980; Nagashima and Suzuki, 1981, 1984; Callaghan et al., 1983a, b; Lang and Steinberg, 1983) have been applied to the study of water dynamics and structure in a variety of systems from pure water to cellular biological systems to complex macromolecular food systems (Cooke and Wien, 1973; Hansen, 1976; Leung et al., 1976; Lillford et al., 1980; Franks, 1982; Leung et al., 1979, 1983; Richardson et al., 1986a).

To clearly yet briefly cover a large number of investigations, this review will be divided into the following sections: basic investigations, applied investigations, and quality control applications.

Basic Investigations

A large number of NMR investigations have centered on the basic research aspects of water associated with macromolecular systems (Fuller and Brey, 1968; Koenig et al., 1975, 1978; Kalk and Berendsen, 1976; Walmsly and Shporer, 1978; Halle and Wennerstrom, 1981; Halle et al., 1981; Kumosinski and Pessen, 1982; Shirley and Bryant, 1982; Lioutas et al., 1986; Peemoeller et al., 1986). The majority of these studies have focused on protein.

Bryant and co-workers have reported several investigations on crystals and powders of hydrated lysozyme (Hsi et al., 1976; Hsi and Bryant, 1977; Hilton et al., 1977; Koenig et al., 1978; Bryant, 1978; Bryant and Shirley, 1980b; Shirley and Bryant, 1982; Bryant et al., 1982; Borah and Bryant, 1982; Bryant and Jarvis, 1984). Early investigations proposed a two-phase, slow chemical exchange model to interpret the observed nonexponential T_1 relaxation behavior (Bryant, 1978). However, this model engendered several difficulties, such as extensively long proton lifetimes in the vicinity of the protein. This difficulty and others are discussed by Bryant and Shirley (1980a, b). They attempted to avoid these difficulties by employing a cross-relaxation model (Hilton et al., 1977; Bryant, 1978; Koenig et al., 1978; Bryant and Shirley, 1980a, b). Results of applying their cross-relaxation model indicated that only a small region of water is disturbed due to the protein and that the reorientation correlation time of this disturbed water is shorter than nanoseconds. This water was only two orders of magnitude slower than free water. Thus, their model of water of hydration is very different from the "ice-like" concept presented by previous workers (Frank and Evans, 1945; Klotz, 1958).

In a later study, Shirley and Bryant (1982) included anisotropic motion of the water and distribution of activation barriers for reorientation to their cross-relaxation analysis.

Peemoeller et al. (1984) studied cross-relaxation at the lysozyme–water interface using a ^1H NMR line-shape-relaxation correlation approach which employed selective inversion of the ^1H magnetization. The results suggested that at least three different correlation times were necessary to characterize the water mobility in wet lysozyme.

Fullerton et al. (1986) employed a ^1H NMR titration method to study the motional properties of water molecules in conjunction with globular proteins. The method was specifically applied to the lysozyme–water system. Four distinct water fractions were reported: (1) superbound water, correlation time 10^{-6} sec; (2) polar-bound water, 10^{-9} sec; (3) structured water, 10^{-11} sec; and (4) bulk water, 10^{-12} sec. The bound plus structured water fractions extend to 1.4g H_2O/g lysozyme and approximately two to three layers from the surface of the lysozyme macromolecule. The structural water is related to nonisotropic water rotation in conjunction with hydrophobic patches on the lysozyme. The bulk water exhibited motions typical of free water and corresponded to water amounts greater than 1.4g H_2O/g lysozyme.

Halle et al. (1981) employed ^{17}O NMR, which avoids the problem of cross-relaxation, to measure both the T_1 and T_2 relaxation rates in seven aqueous protein (including lysozyme) solutions. The data were successfully analyzed in terms of a fast-exchange, two-state model with local anisotropy. The results indicated that the proteins are hydrated by approximately two water layers which have a reorientation rate less than one order of magnitude slower than that of bulk water.

Lioutas et al. (1986), using both ^{17}O and ^{2}H high-field NMR, investigated lysozyme hydration as well as the relationship between a_w and NMR water mobility. Good agreement was found between the hydration numbers determined by ^{17}O NMR and those calculated based on water sorption isotherm analysis. Correlation times from the ^{17}O NMR data at 20°C were determined for three water types: tightly bound water, 41.4 ps, weakly bound water, 27.2 ps, and "multilayer" or trapped water, 17.0 ps. These values are in good agreement with the concept presented by other workers that the mobility of lysozyme bound water is only one or two orders of magnitude less than that of bulk water.

Peemoeller et al. (1986) measured T_1 and T_2 as a function of temperature (219–309 K) at two frequencies (5 and 30.6 MHz) in lysozyme powder hydrated with deuterium oxide. The deuteron relaxation results were compatible with a water molecule dynamics model based on either a bimodal distribution of correlation times or anisotropic motion. However, based on comparisons with previous proton data (Peemoeller et al., 1984), the anisotropic motion model was suggested as providing a more reasonable description of the molecular motion of the water.

Other basic NMR studies and materials include T_1 and T_2 ^{1}H NMR measurements on a clay–water system (Woessner, 1974; Woessner, 1977), ^{2}H NMR measurements of lecithin-water systems at different compositions and temperatures (Ulmius et al., 1977), and ^{17}O and ^{2}H NMR measurements of water orientation and motion in phospholipid bilayers (Tricot and Niederberger, 1979).

Despite the large number of investigations carried out, there still remain several unresolved questions (Bryant, 1978; Koenig, 1980) and much controversy over the nature or state of water in macromolecular systems. Two of the major issues concern the model(s) to be employed for data interpretation (Finney et al., 1982; Kumosinski and Pessen, 1982; Shirley and Bryant, 1982), and the lifetime and nature of the local water and/or water–protein structure (Bryant and Shirley, 1980a, b; Bryant and Halle, 1982). To resolve these issues, further research is required; e.g., ^{17}O NMR T_1 measurements should be made as a function of temperature and/or frequency at various concentrations (Baianu, 1986; Peemoeller et al., 1986). These T_1 experiments will determine the exchange rate of the bound and free water by locating the position of the T_1 minima.

Applied Investigations

Nuclear magnetic resonance techniques, both wide-line and pulsed, have been applied to study the water component of a variety of systems. The purpose of this section is to present an overview of these studies. The discussion will be divided into three parts: wide-line NMR, pulsed NMR, and relation of NMR to physical properties.

Wide-line NMR. The earliest applied NMR studies on water employed wide-line NMR. These studies were based on two different techniques (Brosio et al., 1983): (1) Differentiation of bound and free water by their freezing points; the free water freezes at 0°C and ice does not give a signal so the signal can be related to the amount of bound water present (Toledo et al., 1968), and (2) Differentiation of bound and free water by their different relaxation times; at a high power of radio frequency the measured signal can be attributed to bound water only (Shanbhag et al., 1970; Mousseri et al., 1974; Okamura et al., 1978).

The use of wide-line NMR has decreased due to the development of pulsed NMR techniques; therefore, only a brief review is presented here. Steinberg and Leung (1975) reviewed the subject in more detail. Toledo et al. (1968), employing a PA-7 Varian wide-line NMR, found that flour samples of different moisture contents showed the same amount of liquid water at a given subfreezing temperature. This unfrozen water was defined as bound water. It was concluded that the amount of water bound by a given weight of dry matter is independent of the total moisture content. Shanbhag et al. (1970) developed a NMR calibration curve for the moisture content of wheat flour. By plotting NMR units (at RF 0 decibels)/g dry matter against g water/g dry matter they obtained two linear regions with a break at 0.5g water/g dry matter. They defined this break as the bound water capacity (BWC) of the flour, i.e., total water that gives an NMR signal at RF 0.

Basler and Lechert (1973) reported a similar experiment on starch gels at 295 K. The results showed that the determination of bound water was a pure effect of RF-saturation, which could be calculated from T_1 and T_2.

In an effort to examine the concept and physical significance of BWC, Leung and Steinberg (1979) determined the BWC of 14 food materials. From their results, which were similar to those of Shanbhag et al. (1970), they concluded that the mobility of the water was material-dependent and correlated with the steepness of the slope of the calibration curve for each material; the steeper the slope, the more strongly the water is immobilized by the material.

Duckworth and Kelly (1973), employing wide-line NMR, investigated the ^1H magnetic resonance absorption of hydrated colloids, potato starch, and agar, in the presence and in the absence of various test solutes, such as urea, sucrose, ethyl urea, and hexamine. In general they found that the size of the resonance absorption signal was unaffected by the addition of a solute at low moisture contents. However, above a particular moisture content for each solute–absorbent combination, an increased signal was observed. They attributed this increase in signal to the mobilization of the solute (i.e., sucrose) itself in the water present. Duckworth et al. (1976), extending the Duckworth and Kelly (1973) ^{1}H NMR investigation, reported that the mobilization point and pattern of glucose in potato starch was influenced by the form of the glucose added to the system. Duckworth (1981) suggested that the ^{1}H NMR mobilization point of the solute coincides with the intersection point of water sorption isotherms for the pure colloid itself and for the colloid with the added solute. The latter is discussed on page 271.

Pulsed NMR. Several pulsed NMR techniques have been developed to study various aspects of water in complex systems; examples include water binding, water characterization, water mobility, and water diffusion. These are discussed individually below.

Water binding. Water binding of several macromolecules as determined by ^{1}H pulsed NMR was carried out at 10 MHz by Leung et al. (1976). Both T_1 and T_2 were measured. T_1 values were plotted as a function of moisture content and, for each of the macromolecules studied, a minimum was observed at 0.15–0.25g water/g dry matter. The T_2 relaxation curves exhibited nonexponential behavior and were resolved into two exponential decay curves with relaxation times of T_{21} for the bound fraction and T_{22} for the mobile fraction. The amount of water in this bound fraction showed a remarkable consistency (0.194 ± 0.011g water/g dry matter) among all six of the macromolecules studied. Callaghan et al. (1983b) investigated four different wheat starch pastes by the ^{1}H NMR freezing method of Derbyshire and Duff (1973). They reported an average bound water value of 0.35g water/g dry matter for all four pastes tested.

Water-binding values of soy protein concentrate and ovalbumin using two NMR methods was reported by Hansen (1976). In the first method, water-binding values were obtained by measuring the ^{1}H NMR T_2 as a function of water content for the two proteins and plotting the results according to the isotropic two state model with fast exchange. The plot consisted of two linear segments; the intersection was taken as the bound water value. These were 0.26 and 0.20g water/g solids for the soy protein concentrate and the ovalbumin, respectively.

In the second method, the values were obtained by measuring the water NMR ^{1}H signal amplitude as a function of temperature. The assigned water-binding values corresponded to the amount of water which does not freeze at temperatures below 0°C, where the amount of water is determined from the amplitude of the water NMR ^{1}H signal. The values obtained for the soy protein concentrate and the ovalbumin at −50°C were 0.26 and 0.33 g water/g solids, respectively. Further discussion comparing the two methods, to each other and to other methods, is given by Hansen (1976).

Brosio et al. (1983) employed low resolution (20 MHz) pulsed ^{1}H NMR to study water binding by powdered milk. The spin-echo decay curve was nonexponential. This nonexponentiality was accounted for by the presence of three different relaxation times T_{2S} (solid), T_{2B} (bound water), and T_{2F} (free water). From the three-component analysis, both the relaxation time values and the relative abundance were obtained for milk samples at concentrations from 0.10–1.80 mL water/g dry matter. Unlike the results obtained by Leung et al. (1976), the water-binding values here were not homogeneous and increased from 0.11 to 0.20g bound water/g dry matter as total water increased from 0.25 to 0.90 mL water/g dry matter.

Water characterization. Lang and Steinberg (1981), using the Smith (1947) isotherm equation, showed a marked difference in sorption behavior between macromolecules (e.g., starch and casein) and solutes (e.g., sucrose and sodium chloride). They reported that each of these components sorbed a different "state" of water over a broad a_w range. The water associated with the macromolecule (polymer) was termed "polymer" water, while that associated with the solute was termed "solute" water. Lang and Steinberg (1983) related this isotherm differentiation of water states to pulsed ^1H NMR T_1 and T_2 measurements to characterize the differences between the states, i.e., "polymer" and "solute" waters. Under the instrument parameters used, their results indicated a distinct difference in the NMR signal response for the different water states. For example, Fig. 11.5 shows the first minus second pulse signals with increasing variable time delay (VTD) for samples of starch, sucrose, and a 90:10 mixture of starch:sucrose. All samples were equilibrated to a_w 0.91. Under the range of VTD used, the starch gave a constant signal; in contrast, sucrose showed a decreasing signal with increasing VTD. A mixture of starch and sucrose gave a signal intermediate between the two individual components, close to that for the starch.

A low field (10 MHz) pulsed ^1H NMR instrument was used by Chinachoti and Steinberg (1986) to quantitatively determine polymer and solute waters in a starch–sucrose model system. An equation for each state of water was developed from T_1 relaxation data. When these equations were checked against calculations based on isotherm data for the model systems and for each component (Lang and Steinberg, 1981; Chinachoti and Steinberg, 1984), 45° lines with correlation coefficients > 0.97 were obtained.

Nagashima and Suzuki (1981, 1984) presented a pulsed ^1H NMR method to characterize bound water in a variety of systems, such as amino acids, beef, condensed milk, and starches. The method consists of measuring the free induction decay (FID) amplitude with decreasing (or increasing) temperature. The results were reported by one or both of the following procedures: (1) the variation of the FID amplitude was recorded as a function of temperature, and (2) the FID curve was resolved into two components characterized by long and short relaxation times, which correspond to the liquid and solid state

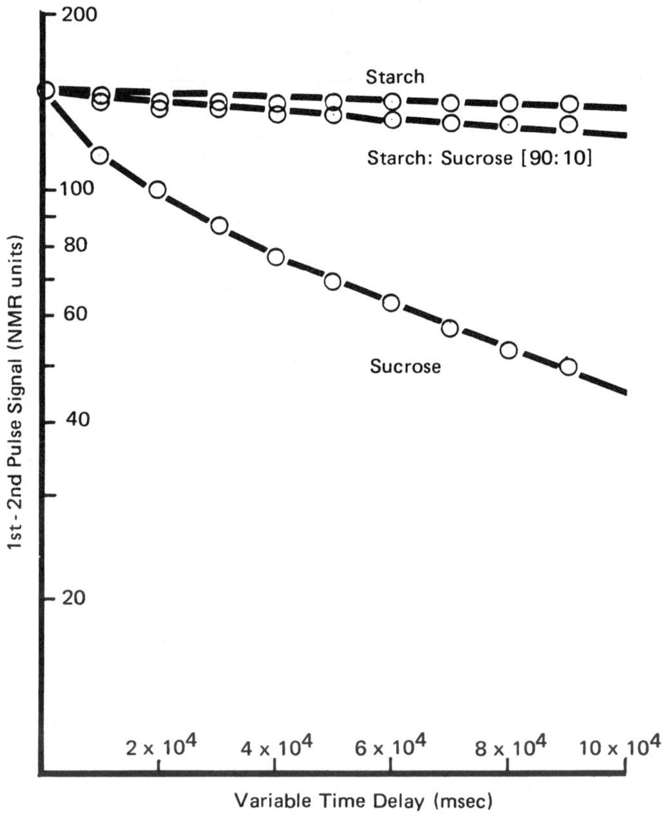

FIG. 11.5 NMR characterization of water on starch, sucrose, and a starch:sucrose mixture, each at a_w 0.91. (*From Lang and Steinberg, 1983.*)

proton populations, respectively. Next, the population of liquid state protons was converted to the unfrozen water (UFW) content of the sample (in g water/g dry matter) at each temperature. Finally, the UFW was recorded as a function of temperature. For both procedures, the resultant plot was referred to as a freezing curve when the measurements were taken as the temperature was decreasing and as a thawing curve when the temperature was increasing. These authors presented an extensive number of these freezing and thawing curves and, where appropriate, discussed their relation to each other (e.g., individual amino acids) as well as their relation to other properties of the system, such as hysteresis (e.g., egg yolk), gelatinization, and retrogradation (e.g., waxy corn starch). Such a freezing curve for native, gelatinized, and retrograded waxy maize starch is shown in Fig. 11.6.

Water mobility. The major application of pulsed NMR has been for the elucidation of the mobility of water. The mobility and binding aspects of water are often interrelated. A specific model of interpretation may be applied to the mobility data (i.e., T_1 and T_2 relaxations) to extract a value for the amount of bound water. Therefore, where applicable, the bound water values calculated from the mobility data are also reported in this section.

This review of the water mobility studies is organized according to the number of component(s) being investigated: single sorbent systems, two-sorbent systems, and multi-sorbent systems.

Single Sorbent Systems—Starches: Because starch is a major food hydrocolloid (Luallen, 1985), NMR has been extensively applied to several aspects of starch hydration (Richardson et al., 1986c). Most NMR studies of the starch–water interaction have been carried out by ^1H NMR, measuring T_1 and/or T_2 (Hennig and Lechert, 1974; Lechert, 1976; Lechert and Hennig, 1976; Leung et al., 1976; Kashkina et al., 1979; Lechert et al., 1980; Nakazawa et al., 1980, 1983; Lechert and Schwier, 1982; Lang and Steinberg, 1983; Callagan et al., 1983b) and by ^2H NMR (Tait et al., 1972a, b; Lechert and

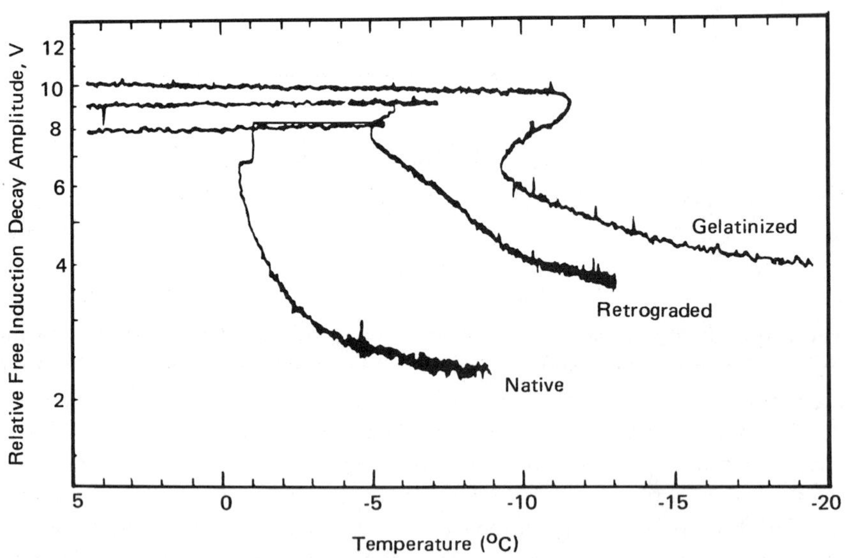

FIG. 11.6 Relation of NMR response to freezing of 50% solutions of native, gelatinized, and retrograded waxy corn starch. (*From Nagashima and Suzuki, 1981.*)

Hennig, 1976; Hennig and Lechert, 1977; Lechert et al., 1980; Lechert and Schwier, 1982).

Leung et al. (1976) found that T_1 showed single exponential, i.e., single phase, behavior against water content and/or a_w. The T_2 relaxation, on the other hand, exhibited two-phase behavior. This indicates the existence of two water fractions with different mobilities, where the exchange between the different fractions is slow compared to their relaxation times (Zimmerman and Brittin, 1957). This "two water fraction" behavior was also reported by Hennig and Lechert (1974) and Hennig (1977). In contrast, an investigation by Kashkina et al. (1979) reported results indicating a multiphase adsorption of water on the starch, where each phase differed in its degree of mobility.

Callaghan et al. (1983b), in studying wheat starch pastes, employed carbon-13 NMR to investigate the mobility of the starch and ^1H NMR for the water. The ^1H NMR relaxations (T_1 and T_2) were characterized by a single time constant indicating that a two-state model with fast exchange may be adequately accounting for such observations. This two-state model was also proposed by two other investigators; Lelievre and Mitchell (1975) for wheat starch–water suspensions up to 40% starch based on T_2 measurements, and Nakazawa et al. (1980) for gelatinized nonglutinous and glutinous rice starch based on both T_1 and T_2 measurements. Nakazawa et al. (1980) obtained correlation times for bound water and developed a model for calculating the bound water fraction of starch.

Further insight into the mechanism of mobility of the bound water can be obtained by the ^2H resonance on starches (Lechert et al., 1980). Two specific motions are possible: isotropic or anisotropic reorientation. Deviations from isotropic motion cause a splitting of the ^2H resonance line which is a measure of the anisotropy of the motion. Several studies have been designed to investigate these motions in starch (Lechert and Hennig, 1976; Hennig, 1977; Hennig and Lechert, 1977; Lechert et al., 1980; Lechert and Schwier, 1982; Schwier and Lechert, 1982). The overall conclusion from these studies was that different starches show different degrees of water reorientation rates. For example, the "bound" water in potato starch showed distinct anisotropic water motion, while that in wheat starch showed only weak anisotropic motion and that in corn starch showed isotropic motion (Lechert et al., 1980). The explanation given for the isotropic motion in corn starch was that the relative residence time in the ordered state is too short, which may be caused by the starch structure itself or by impurities in the starch, such as protein (Lechert et al., 1980).

Due to the recent problems with interpretation of ^1H NMR data, interest in measuring water mobility in starch, as will be seen for other components as well, has shifted from ^1H NMR to ^{17}O and ^2H NMR (Halle et al., 1981; Richardson et al., 1986b, c). Richardson et al. (1986b, c) studied water mobility in corn starch suspensions and powders ranging in concentration from 10 to 96% corn starch solids by high-field ^{17}O and ^{2}H NMR. They reported four different regions of water mobility: Region I, 10–40% solids; Region II, 40–60% solids; Region III, 60–92% solids; and Region IV, 92–96% solids. The dependence of the ^{17}O and ^{2}H NMR transverse relaxation rates (R_2 (sec^{-1}) = $1/T_2$) in deuterium oxide with increasing corn starch concentration is presented in Fig. 11.7 for Regions I through III and in Fig. 11.8 for Regions III and IV. Region III begins in Fig. 11.7 and continues in Fig. 11.8. The standard isotropic two-state model with fast exchange was used to interpret the data for the high moisture regions (Fig. 11.7) by means of a physical model and a chemical activity model. According to the first, plots of relaxation rate against concentration indicated three regions of decreasing mobility. According to the chemical activity model, the relaxation rate was linear with chemical activity over the entire concentration range (Fig. 11.9) as predicted by this model (Kumosinski and Pessen, 1982). The concentration plots for the low moisture regions (Fig. 11.8) showed two regions of mobility, trapped (Region III) and

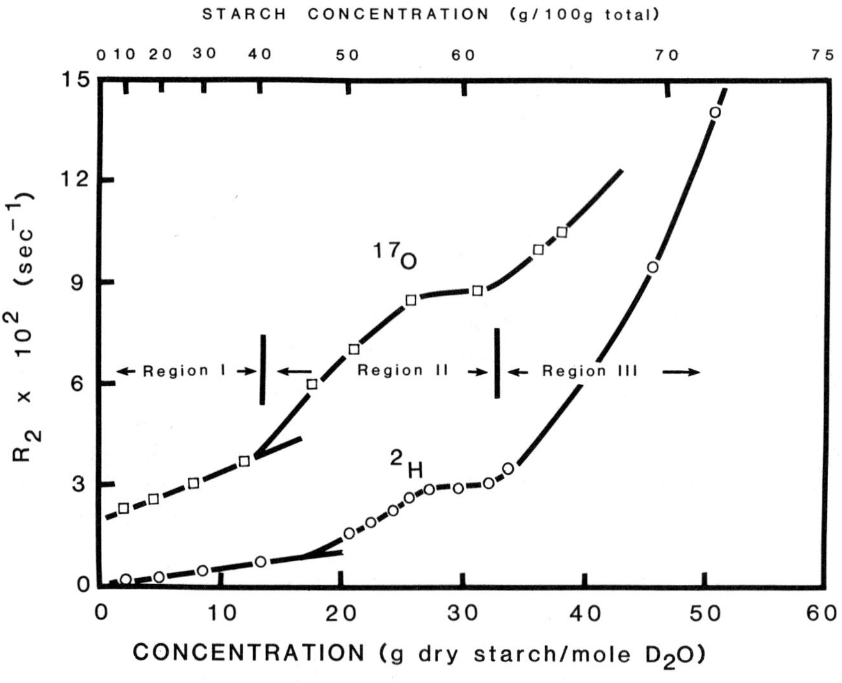

FIG. 11.7 Dependence of NMR transverse relaxation rate (R_2) on corn starch concentration in D_2O for ^{17}O and ^{2}H nuclei. (*From Richardson et al., 1986b.*)

monolayer (Region IV) water. A similar study investigating amylopectin has also been reported (Richardson et al., 1985b).

Mora-Gutierrez and Baianu (1986), in studying modified starches, probed all three water nuclei, ^{1}H, ^{2}H, and ^{17}O. The NMR transverse relaxation rates, R_2, were plotted as a function of the inverse of moisture content (g solid/g water). Maltodextrin Lo-Dex 5, for example, exhibited linear behavior for all three nuclei. The ^{1}H and ^{2}H NMR showed a marked departure from this linearity at approximately 10 and 30% solids, whereas the ^{17}O NMR remained linear throughout the concentration range studied, 1–44% solids. They discussed the possible mechanisms for the departures from linearity as well as the differences between responses of the three nuclei.

Single Sorbent Systems—Sugars: Many of the NMR studies on monosaccharides, such as glucose and sucrose, have been coupled with dielectric relaxation data (Tait et al., 1972c; Allen and Wood, 1974; Suggett, 1976). In the study by Tait et al. (1972c) employing ^{17}O NMR, it was observed that the

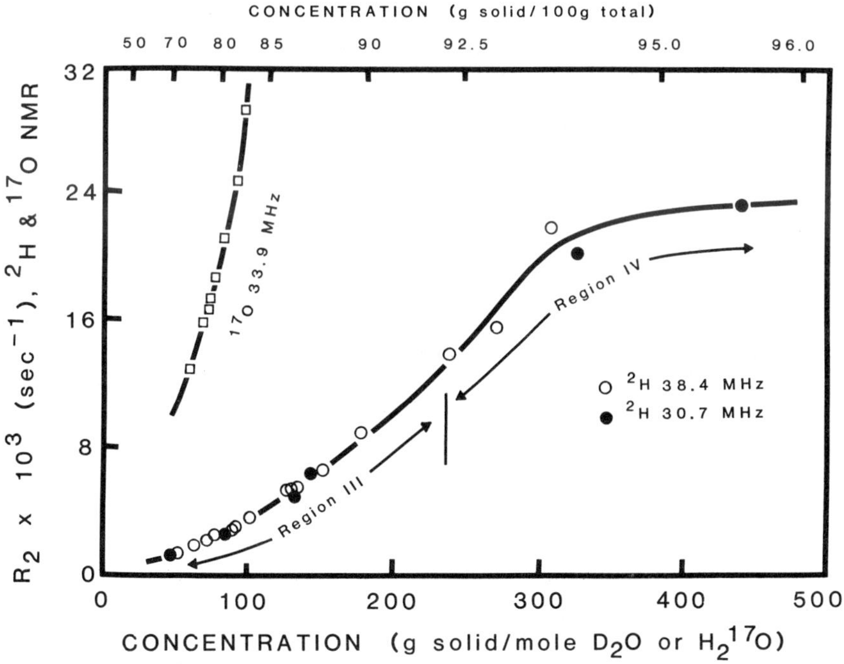

FIG. 11.8 Dependence of NMR transverse relaxation rate (R_2) on corn starch concentration in D_2O or ^{17}O-enriched water for ^{2}H and ^{17}O, respectively, in the high solids (powder) range. (*From Richardson et al., 1986c.*)

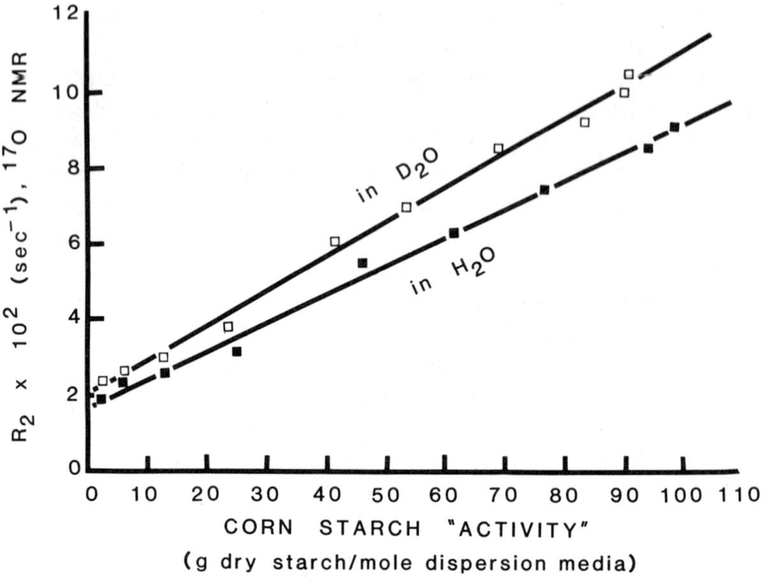

FIG. 11.9 Linear dependence of the ^{17}O NMR transverse relaxation rate (R_2) on the chemical "activity" of a corn starch suspension in H_2O and in D_2O. (*From Richardson et al., 1986b.*)

relaxation rate of ^{17}O-enriched water was enhanced in monosaccharide solutions, and was greatest in hexose solutions as compared to solutions of pentose and ribose. The transverse relaxation rates of glucose (R_2, sec^{-1}) increased with concentration in a nonlinear manner. Nonlinearity, however, decreased as temperature increased. The viscosity dependence of the relaxation rate for the glucose solutions was linear as plotted against $\eta/\rho T$, where η is viscosity and ρ is the density. This linear dependence was shown to hold for 2.8 molar glucose over a very wide range of $\eta/\rho T$, corresponding to a temperature interval of 2–63°C. Lastly, the NMR data for glucose, in combination with the dielectric relaxation data, showed that glucose increases the hydrogen bonding in aqueous solution. This conclusion was based on the significant decrease of the quadrupole coupling constant in glucose solution over that in pure water.

In a subsequent investigation by Suggett (1976), a strategy was outlined for the complimentary use of dielectric and NMR methods. In the NMR portion of the study (Suggett et al., 1976), a range of aqueous mono- and di-saccharide solutions were investigated. ^{17}O NMR relaxation was carried out for solvent and ^{1}H, ^{2}H, ^{13}C, and ^{17}O relaxation for the various solutes. The NMR data were used to test alternative models for the resolution of the dielectric spectra into their component relaxation processes. The most appropriate dielectric relaxation model based on the collateral NMR data was the two-Debye process model. Thus, sugar solutions were envisaged in terms of a major solvent process exhibiting a single correlation time somewhat longer than pure water (22 psec for 2.8 molar monosaccharide solution at 5°C, compared to 15 psec for water at the same temperature), together with a slower process due to the coupled motion of the hydrated side chains and the sugar ring.

Richardson et al. (1986d) studied the molecular mobility of solute bound water by ^{2}H and ^{17}O high-field NMR. The solute–water system was modeled by systems of sucrose and water or D_2O over the solids concentration range 5–80%. Supersaturation to 70% sucrose in D_2O was included. Figure 11.10 shows the dependence of the ^{17}O R_2 in D_2O on increasing sucrose concentration. Four regions of water mobility are observed. Region I, from 5–40% sucrose, shows a linear behavior of R_2 with concentration. This is described by the isotropic two-state model with fast exchange. Region II, from 40–60%

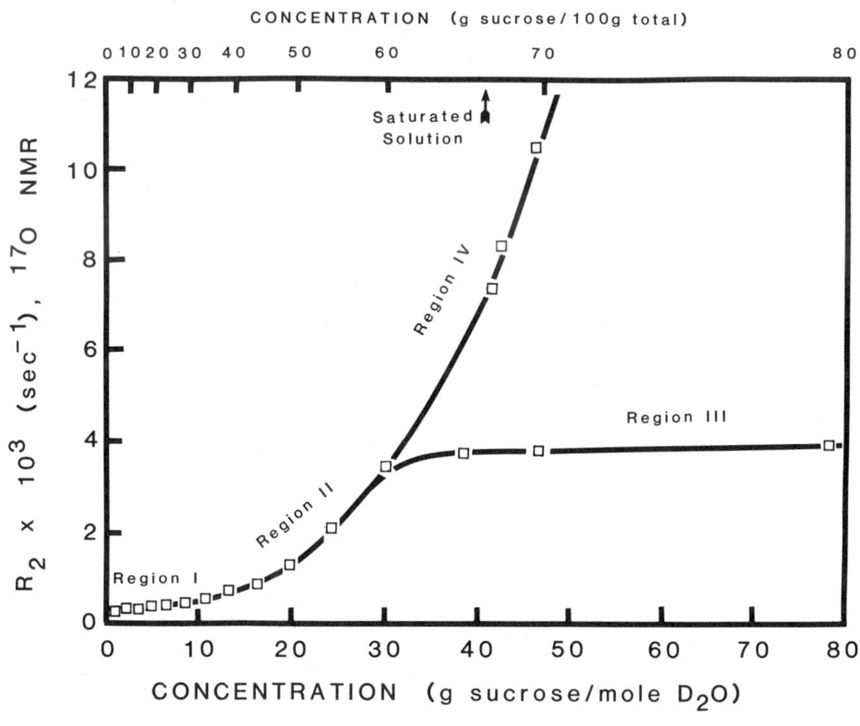

FIG. 11.10 Dependence of ^{17}O NMR transverse relaxation rate (R_2) in D_2O on sucrose concentration below and above saturation. (*From Richardson et al., 1986d.*)

sucrose, exhibits nonlinear behavior. The deviations from linearity are hypothesized to be due to formation of intermolecular hydrogen bonds between water and sucrose, hydrogen bond bridging of water between sucrose molecules, and sucrose–sucrose hydrogen bonding. This extensive development of hydrogen bonding is reflected as a significant increase in the R_2 of the water, which translates into a decreased water mobility. Region III, from 60–80% sucrose, shows a slight rise in R_2 between 60 and 67% sucrose; thereafter it shows a horizontal line that represents a mixture of sucrose crystals in a saturated sucrose solution. Region IV represents samples in the supersaturated condition. Region IV is a smooth continuation of Region II. This continous increase in the R_2 response for the supersaturated condition is attributed to the increased formation of hydrogen bonding structures as described for Region II.

Harvey and Symons (1978), using ^{1}H NMR (100 MHz), investigated the hydration of glucose, mannose, galactose, ribose, sorbose, and fructose. Based on chemical shifts as a function of temperature and the effect of added dimethyl sulfoxide, they concluded that each hydroxyl group of the monosaccharide bonds an average of two water molecules. Hydration studies of glucose with ^{1}H NMR were also carried out by Harvey and Symons (1976), Bociek and Franks (1979), and Mora-Gutierrez and Baianu (1986). Mora-Gutierrez and Baianu (1986) also reported low field (10 MHz) ^{1}H NMR measurements for sucrose, fructose, and corn syrup.

Other Polymers: Various polysaccharide gels and films have also been extensively studied by NMR. These include agar (Cope, 1969; Woessner et al., 1970; Labuza and Busk, 1979), agarose (Child and Pryce, 1972; Ablett et al., 1976; Lillford et al., 1980), and carrageenan (Ablett et al., 1976; Labuza and Busk, 1979). Labuza and Busk (1979) measured ^{1}H NMR T_1 as a function of gelatin concentration; plotting the inverse of T_1 against concentration yielded a linear relation followed by two departures from linearity, one at 30% and the other at 53% solids. These departures are explained by an increase in polymer–polymer interactions or increased helix formation with the exclusion of water.

Other single sorbent systems investigated are: soybean and ovalbumin protein (Hansen, 1976), cellulose (Hsi et al., 1979), and myofibrillar protein (Lioutas, 1984).

Two-Sorbent Systems: The number of two-sorbent systems which have been studied is considerably less than one-sorbent systems. The probable reason for this is that as sorbents are added the system becomes increasingly complex. For example, increasing the quantity of sorbents from one to two significantly increases the number and type of possible interactions. One must now consider not only water–sorbent but also sorbent–sorbent, sorbent–sorbent–water, and sorbent–water–sorbent interactions. Also, the rate of water exchange among these various combinations as well as the microhomogeneity of the system must be taken into account.

Collagen and salt systems have been studied by ^1H NMR (Berendsen and Migchelsen, 1965) and ^2H NMR (Fung and Trautmann, 1971). Fung and Trautmann (1971) reported that the deuterium quadrupole splitting of D_2O in oriented collagen fibers decreased in the presence of 0.2M $MgCl_2$ and NaCl. They postulated two physical models to explain this phenomenon: Either the presence of the ions causes a change in the structure of the collagen, making water binding more difficult, or the ions block some of the sites available for hydrogen bonding.

Lioutas (1984) studied changes in lysozyme hydration in the presence of electrolytes LiCl and NaCl by ^{17}O and ^7Li or ^{23}Na NMR. In the case of the lysozyme–LiCl system in D_2O, the NMR analyses revealed five water populations. It was observed that both the "monolayer" hydration of lysozyme as well as the total amount of water bound increased due to the presence of the LiCl. The correlation times of all water populations were increased threefold by the presence of Li^+, and the correlation times of Li^+ were increased tenfold in the presence of 10% lysozyme (w/w) in D_2O. In the case of the lysozyme–NaCl system in D_2O similar results were reported: The NMR analyses identified five water populations and both the "monolayer" hydration and the total amount of water bound increased due to the presence of the NaCl. In both cases, the ion NMR data (^7Li and ^{23}Na) proved to be very useful in the hydration analyses.

A similar study was carried out by Lioutas (1984) on the interaction of myofibrillar proteins with NaCl in D_2O by ^{17}O and ^{23}Na NMR. Incremental amounts of protein from 0 to 96% were added to D_2O containing 4% NaCl. The presence of NaCl caused a marked change in the dependence of the ^{17}O relaxation rate on concentration as compared to the protein alone; the rate decreased with increasing concentration to 1.5% protein and thereafter increased sharply, approaching an asymptote at very high concentration. From the ^{23}Na NMR data, correlation times for the Na^+ bound to the myofibrillar proteins were found to be about three times more mobile than the Na^+ bound to lysozyme (Lioutas, 1984). This indicates a significant difference in the interaction between these two two-sorbent systems.

Richardson et al. (1986e) expanded the work of Lang and Steinberg (1983) on NMR characterization by studying the mobility of water in polymer-solute systems with ^{17}O and ^2H high-field NMR. The polymer–solute system was modeled by freeze-dried starch:sucrose (FSS) (90:10) in D_2O or enriched oxygen-17 water ($H_2{}^{17}O$) over the solids concentration range 10–93% solids. The ^2H NMR R_2 data were significantly influenced by the chemical exchange process. Therefore, the ^{17}O data was the most useful for water mobility information. The ^{17}O NMR spectra showed a single Lorentzian peak for concentrations ranging from 10.0 to 72.4% solids (Fig. 11.11A). At concentrations ranging from 73.0 to 83.3% solids the ^{17}O NMR spectra showed two Lorentzian peaks, one broad and one sharp but both with the same central resonance

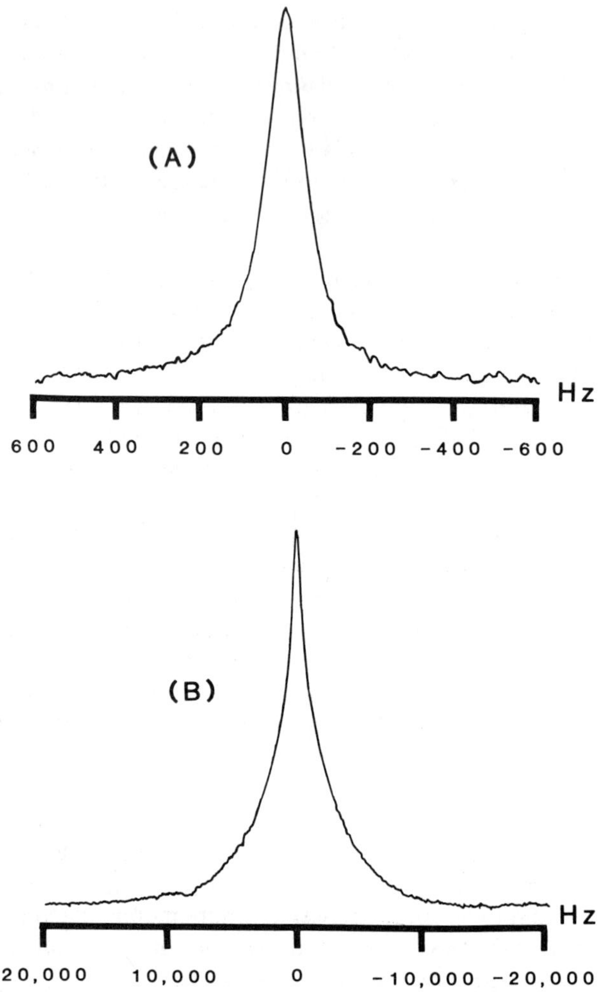

FIG. 11.11 Fourier transform ^{17}O NMR spectra for freeze-dried (90:10) starch:sucrose systems at (A) 40.0% solids in D_2O, (B) 80.3% solids in $H_2{}^{17}O$, and (C) 84.3% solids in $H_2{}^{17}O$. (*From Richardson et al., 1986e.*)

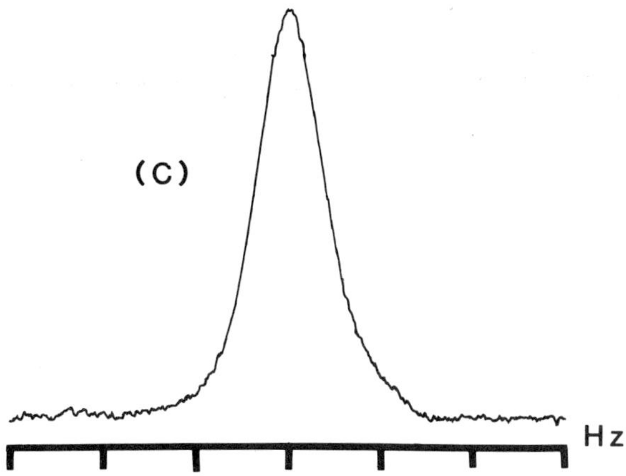

FIG. 11.11 Continued

frequency (Fig. 11.11B). However, at concentrations beyond 83.3% solids, the ^{17}O NMR spectra again showed only a single Lorentzian peak (Fig. 11.11C).

This peak behavior was explained in terms of FSS concentration and a_w. The presence of two peaks in the ^{17}O NMR spectra was hypothesized to be due to the development in the FSS-D_2O system of two distinct water phases that do not exchange at a rate greater than that of the NMR frequency. Phase I was attributed to the bulk sucrose–water population and Phase II was attributed to the water trapped between aggregated starch granules. This trapped water has the same sucrose concentration as the bulk sucrose–water in Phase I. The FSS concentration at which the system shifts from one to two peaks corresponds to the saturation a_w of the sucrose, 0.86. The ^{17}O NMR R_2 data expressed as a function of increasing FSS concentration did not significantly deviate from the R_2 data shown by starch as the only sorbent (Richardson et al., 1986b) until approximately 67% solids. At this concentration, the starch-only system showed the beginning of a sharp increase in R_2 versus concentration (Fig. 11.12). This increase was attributed to the introduction of water trapped between the starch granules (Richardson et al., 1986b). However, the two sorbent FSS system showed very little increase in R_2 as the concentration increased from 67 to 72.5% (Fig. 11.12). The sucrose-only system, included in Fig. 11.12 for comparison, has been previously discussed (Fig. 11.10).

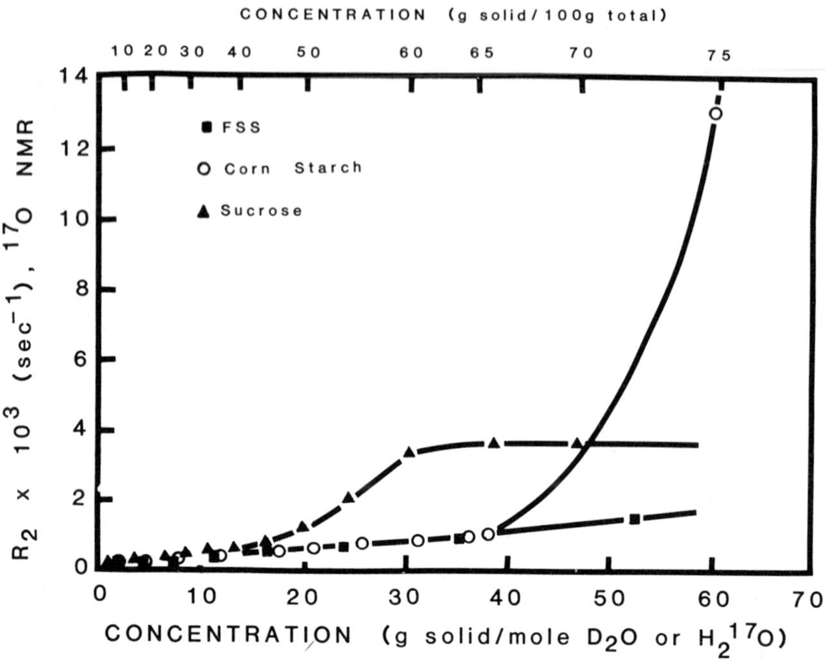

FIG. 11.12 Dependence of ^{17}O NMR transverse relaxation rate (R_2) on concentration of freeze-dried starch:sucrose (FSS), starch, and sucrose in D_2O. (*From Richardson et al., 1986e.*)

This difference between the FSS and starch-only systems is ascribed to the ability of the solvated sucrose in the FSS to create channels between the starch granules, allowing the water to continue to move about freely in the system. In other words, the trapped water that begins to develop in the starch-only system is able, through these sucrose-water channels, to continue to communicate with the other water populations at a higher solids content. When the sucrose–water is no longer able to provide this "communication network," the ^{17}O NMR spectrum begins to show two peaks (Fig. 11.11B).

Other studies dealing with two-sorbent systems are Duckworth and Kelly (1973), Duckworth et al. (1976) and Duckworth (1981), discussed under wide-line NMR, and Lang and Steinberg (1983) and Chinachoti and Steinberg (1986), discussed under water characterization.

Multi-Sorbent Systems: Wheat flour is one of the few multisorbent systems that has been investigated (Leung et al., 1979, 1983; Richardson et al., 1986a). Leung et al. (1979) determined the NMR T_2 of water protons in wheat flour doughs made with both hard and soft wheat flour. They reported nonexponential decay of the signal for each dough tested, indicating the existence of two or more water species which exchange slowly on the NMR time scale compared to their relaxation time. They resolved the relaxation curve into two components; a long and a short component. The long component represents the more mobile water fraction with a T_2 of about 60 nsec. The short component represents the less mobile water fraction with a T_2 of about 20 nsec and accounts for about 0.62 g water/g dry solid. They observed a rather small variation in the magnitude of the two components and concluded that, in the moisture range 0.64–0.93g water/g dry solid, the mobility of each water fraction as measured by pulsed NMR did not change appreciably with water content. The observed T_2 values seemed to be independent of both flour strength (hard vs. soft) and mixing time.

In a subsequent study, Leung et al. (1983) employed ^{2}H NMR to study water in doughs and breads. In this study they were unable to resolve the two fractions of water; they attributed this to the different relaxation mechanisms for protons (^{1}H) and deuterons (^{2}H). Hard and soft wheat flour doughs showed similar T_1 and T_2 values with increasing moisture content. They concluded that water mobility by either ^{1}H or ^{2}H relaxation does not reflect the different rheological properties of soft and hard wheat flour doughs. For the bread-staling portion of the study, there was a decrease in T_1 and T_2 with time, indicating an overall decrease in water mobility and an increase in water binding during the staling process.

Richardson et al. (1986a), realizing the potential problems of cross-relaxation with ^{1}H NMR and chemical exchange with ^{1}H and ^{2}H NMR, carried out both ^{17}O and ^{1}H high-field NMR studies of wheat flour suspensions and doughs (30–95% moisture). The isotropic two-state model with fast exchange was used to interpret these data by means of the Derbyshire (1982) and Kumosinski (Kumosinski and Pessen, 1982) models. Both the ^{17}O and ^{1}H NMR results showed the same trend in the dependence of R_2 on flour concentration in both water and deuterium oxide. From the ^{17}O NMR data, a correlation time of 16.7 ps was calculated for the water "bound" by the wheat flour. This value is in good agreement with ^{17}O NMR work on lysozyme (Lioutas et al., 1986) and other proteins (Halle et al., 1981).

Lillford et al. (1980) studied soya protein fibers (composed of water, protein, oil and salt) by measuring ^{1}H NMR, T_1 and T_2. Nonexponential relaxation was detected in both T_1 and T_2 measurements. The T_2 decay was fitted for both three and four processes; the four process fit was found superior. The theory developed to explain this complex relaxation behavior is given in detail by Lillford et al. (1980).

Other multi-sorbent systems which have been investigated are: collagen and muscle (Berendsen and Migchelsen, 1965; Civan and Shporer, 1972,

1975; Migchelsen and Berendsen, 1973; Duff and Derbyshire, 1974; Resing et al., 1976; Civan et al., 1978; Edzes and Samulski, 1978; Hoeve, 1980; Lillford et al., 1980; Renou et al., 1983) and powdered milk (Brosio et al., 1983).

Water diffusion. The final aspect of water probed by pulsed NMR is water diffusion. Lechert et al. (1980) carried out self-diffusion measurements on potato starch using the pulsed field gradient NMR technique developed by Steijskal and Tanner (1965). The self-diffusion coefficients for the mobile region of bound water showed values ranging between 8×10^{-7} cm²/sec for 20% water samples and 14×10^{-7} cm²/sec for 30% water samples at 30°C. The authors suggested that this type of water diffusion measurement may be important for a number of water diffusion dependent reactions, such as reactant mobility in the Maillard Browning reaction.

Callaghan and co-workers (1983a,b) also employed the pulsed field gradient technique of Stejskal and Tanner (1965) to investigate the diffusional mobility of water in cheese (Callaghan et al., 1983a) and wheat starch pastes (Callaghan et al., 1983b). In the cheese study, no significant difference in diffusion coefficients was observed between Swiss and Cheddar. The coefficients were about one-sixth of the value in bulk water at 30°C. They reported strong evidence that water diffusion is confined to the protein surface.

For the two wheat starch pastes studied, no significant difference in water diffusion coefficient was observed despite the differences in their rheological properties. Furthermore, the observed diffusion coefficients were only slightly less than the pure water values. This indicates that the majority of the water molecules are in the unassociated state and are not experiencing a major obstruction within the NMR time scale. Hence, they concluded that there was no restricted diffusion in the pastes. Basler and Lechert (1974) reported similar unrestricted diffusion of water in maize starch gels ranging in concentration and temperature between 50 and 95% water and 1 and 47°C, respectively. They concluded that the boundaries of the swollen starch grains were not a barrier to water diffusion.

Relation of NMR to physical properties. The relation between NMR and the physical properties of different systems has been a subject of more recent investigation. The relations covered in this review are swelling, gelatinization, retrogradation, rheology, and water activity.

Swelling. The swelling mechanism of starch was studied by Hennig et al. (1976). They postulated a new method for studying swelling. This method consists of extracting the starch with D_2O to replace all the exchangeable protons by deuterons, swelling of the starch in D_2O, and subsequent measurement of the ^1H resonance of the CH-protons of the starch molecules, which allows the determination of the mobility of these molecules.

Gelatinization. Gelatinization of starch was investigated by Lelievre and Mitchell (1975). NMR T_2 relaxation times of proton as a function of temperature were reported. Prior to gelatinization, T_2 values decreased with increasing temperature up to approximately 50°C; at this point, T_2 passes through a minimum and then rises to a T_2 value similar to that of an ungelatinized sample. They ascribed this decrease in T_2 to the onset of the melting of the crystalline polymer arrangements within the granules.

Goldsmith and Toledo (1985) used pulsed ^{1}H NMR to monitor the gelation of egg albumin. T_1 measurements were carried out as a function of heating time and temperature for a 10% egg albumin dispersion (Fig. 11.13). For each temperature, T_1 declined rapidly at first, then gradually leveled off asymptotically. As shown in Fig. 11.13, the asymptotic value of T_1, as well as the heating time required to reach it, decreased as the heating temperature increased. The effect of heating time and temperature on gel strength was also measured and gel strength was related to T_1; they reported a high negative correlation between these parameters.

Retrogradation. A study of retrogradation processes in gelatinized and rice starch, both glutinous and nonglutinous, was done by Nakazawa et al. (1983). They measured both T_1 and T_2 by pulsed ^{1}H NMR. Changes in the relaxation time and, in turn, changes in the correlation time and the amount of water bound during the retrogradation process were small for the nonglutinous rice starch but large and dependent on water content for the glutinous rice starch.

Rheology. The relation between NMR parameters and viscosity arise from the connection between equations that describe intermolecular motions and those used to describe intramolecular motions. The intramolecular motions of a molecule in solution are given by the Debye–Stokes–Einstein theory:

$$\tau_c = \frac{4\pi r^3 \eta}{3kT} \tag{6}$$

$$\tau_t = \frac{2\pi r^3 \eta}{kT} \tag{7}$$

where r is the Stokes radius of the molecule, η is the microviscosity of the solution, k is Boltzmann's constant, T is the temperature, τ_c is the rotational correlation time, and τ_t is the translational correlation time (Berliner and Reuben, 1980). Because of the relation between $T_1{}^{-1}$ and $T_2{}^{-1}$ with τ_t, these equations imply that $T_1{}^{-1}$ and $T_2{}^{-1}$ should vary linearly with T/η, but often they do not (Berliner and Reuben, 1980).

Several attempts have been made to account for the discrepancies that often arise between correlation times calculated from the Debye–Stokes–Einstein equations [Eq. (6) and (7)] and the correlation times observed by NMR spectroscopy (Berliner and Reuben, 1980). The main problem is associated with the meaning of η. The NMR relaxation process depends on rotational microviscosity, whereas the intramolecular process depends on macroscopic transla-

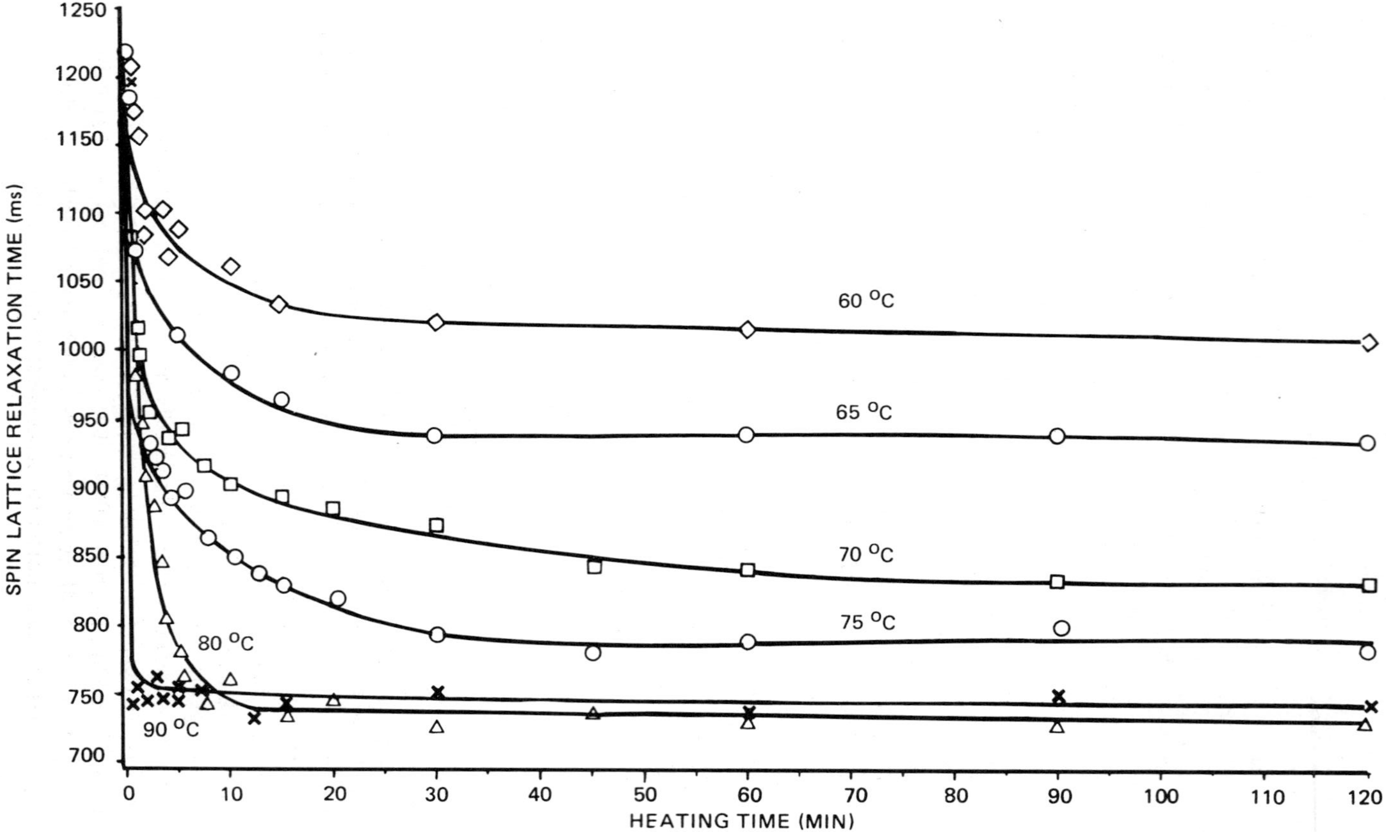

FIG. 11.13 Dependence of NMR longitudinal relaxation time (T_1) on gelatinization of a 10% suspension of egg albumin at various heating times and temperatures. (*From Goldsmith and Toledo, 1985.*)

tional viscosity (Berliner and Reuben, 1980). This microscopic vs. macroscopic issue is discussed on an applications level by Lechert et al. (1980) investigating starch and Richardson et al. (1985) studying wheat flour suspensions and doughs.

In a study by Deslauriers and Smith (1977), it was found that the dependence of T_1 on the macroscopic solution viscosity of proline in glycerol and water solutions was nonlinear. The explanation for this behavior was the differential solubility of the water and glycerol solvents used to manipulate the solution viscosities. The relationship between macroscopic viscosity and NMR relaxation in viscoelastic liquids was investigated by Barker et al. (1974). He reported a correlation between NMR linewidths and rheological properties of viscoelastic liquids, sodium dodecyl sulphate, and octyltrimethylammonium bromide. The specific nature of the correlation resulted in a cylindrical and spherical micelle model for the viscoelastic properties of the liquids under study.

Baianu and Forster (1980) discussed a possible relationship between dynamic ^{13}C NMR observations and gluten rheology. In another investigation, Baianu et al. (1982) proposed the use of NMR measurements for wheat variety identification and correlation with rheological testing as a practical application of NMR technology. This proposal was investigated in two recent studies. First, Leung et al. (1983), based on his previous work (Leung et al., 1979), reasoned that, due to a lack of correlation between water mobility and flour strength, the importance of water binding on the rheological properties of wheat flour doughs was of little significance. More specifically they concluded that water mobility as determined by their proton and deuteron relaxation did not reflect the different rheological properties of soft and hard wheat flour doughs.

Secondly, a study by Richardson et al. (1985a) tested the hypothesis that the rheological properties of a wheat flour–water system are related to the mobility of its water component as measured by ^{17}O NMR T_2 values. It was observed that there was an inverse relationship between the consistency coefficient of the power law equation and water mobility. These data indicated three distinct regions, each showing a linear relationship with water mobility decreasing as apparent viscosity increased. However, there was no overall relationship between NMR response and apparent viscosity. These results are consistent with the current view that the mobility and conformation of the macromolecules determine the different properties of doughs containing different types of wheat flours (Leung et al., 1983; Richardson et al., 1985a).

Water activity. Both NMR and a_w have been used to measure the degree of water binding, which leads to the concept that they should be related. However, as was the case with NMR and rheology, one must keep in mind the differences in physical properties being measured: a_w is a measure of water molecule volatility and NMR is a measure of water molecule mobility.

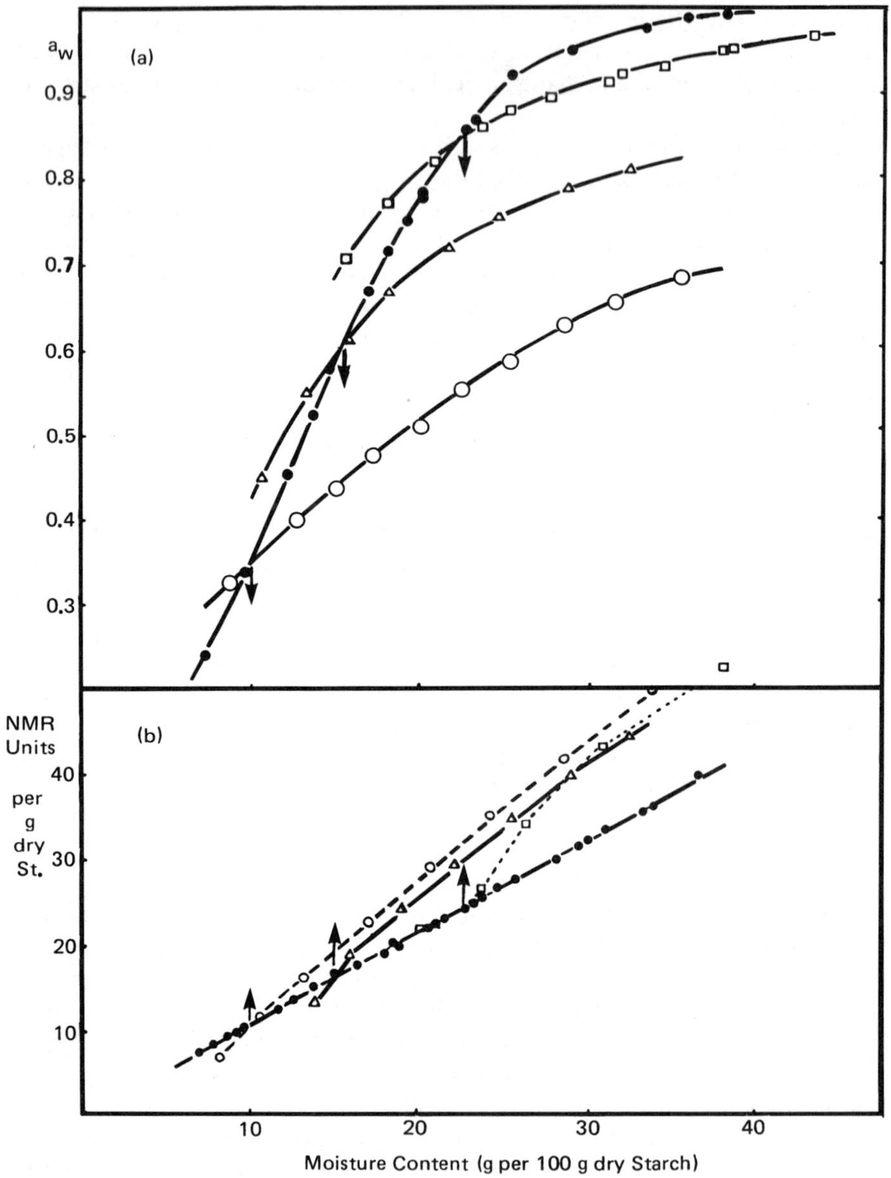

FIG. 11.14 Relationships between the intersection points of water sorption isotherms for gelatinized potato starch with or without added solutes and the mobilization points, determined by NMR, of the respective solutes in hydrated starch gel. Control starch, ●; starch + glucose (10%), □; starch + COONa (10%) and glucose (10%), △; starch + HCOONa (10%), CH_3COOK (10%), and C_2H_5COONa (10%), ○. (*From Duckworth, 1981.*)

Leung et al. (1976), studying water binding by several macromolecules with pulsed ^1H NMR, reported that T_2 increased exponentially with a_w. Richardson et al. (1986c) replotted their data as R_2 against a_w and obtained a curve showing rapidly increasing mobility from a_w 0.23 to 0.65. With further increase in a_w to 0.95, R_2 decreased linearly with a slope of -2.57×10^{-3} and correlation coefficient of 0.997.

Duckworth (1981), in studying the mobility of solutes by wide-line ^1H NMR, reported a close agreement between the mobilization point of the solutes as determined by NMR and the point of intersection of the water sorption isotherms for a macromolecule alone and that for the mixture of this macromolecule plus a solute (Fig. 11.14).

Lang and Steinberg (1981, 1983) reported that both isotherm and pulsed ^1H NMR data differentiated the same states of water in their model systems. In a subsequent study by Chinachoti and Steinberg (1986), this relationship was quantitatively expressed by equations relating T_1 and a_w.

The relation between ^2H and ^{17}O NMR mobility and a_w in starches was studied by Richardson et al. (1986c). Both nuclei showed R_2 linear against a_w, within their respective concentration ranges (Fig. 11.15). In case of ^{17}O data,

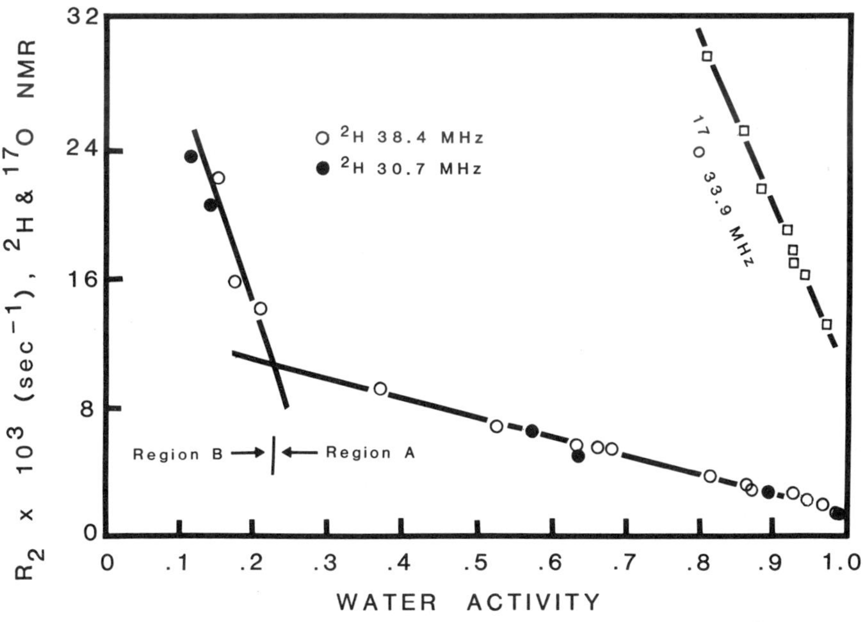

FIG. 11.15 Dependence of deuterium (at two frequencies) and ^{17}O NMR transverse relaxation rate (R_2) on water activity of corn starch in D_2O or $H_2{}^{17}O$, respectively. (*From Richardson et al., 1986c.*)

this linear behavior extended from a_w 0.97 to 0.80, below which the ^{17}O measurement could not be made because of bandwidth limitations on the spectrometer used. In case of the ^2H data, there were two linear regions: Region A, a_w 0.99 to 0.23, and Region B, a_w 0.23 to 0.11. The linear relation in Region A was hypothesized to be due to the hydrophobic nature of the starch granules, which results in diffusion limited water motion. In a similar study for hydrated lysozyme (Lioutas et al., 1986), the relationship between R_2 and a_w was found to be nonlinear; R_2 remained high as a_w was increased from 0.10 to 0.90 and decreased sharply as a_w increased above 0.90. This contrast in hydration and mobility behavior between starch and lysozyme was attributed to the presence of polar groups in lysozyme and nonpolar groups in starch (Richardson et al., 1986c).

Richardson et al. (1985b) compared water binding data, as measured by moisture sorption isotherms, to water mobility data, as measured by ^2H and ^{17}O NMR T_2 values, of amylopectin and corn starch. The isotherms for both starches showed similar water binding capacity. However, the NMR data showed differences in the molecular mobility of the water between the two starches. They concluded that these differences in water mobility in contrast to the similarity in water binding may serve, at least in part, to explain the observed differences in the physical properties exhibited by amylose and amylopectin starch fractions.

Quality Control Applications

The theory and applications of NMR for moisture determination have been reviewed extensively by Pande (1975). The application of NMR to the quantitative determination of moisture content is based on the absorption of radio frequency (RF) energy by hydrogen nuclei (Pande, 1975). Absorption of this RF energy gives rise to a free-induction decay (FID) signal. The amount of energy absorbed by the nuclei and, in turn, the intensity of resonance FID signal is directly proportional to the total number of hydrogen nuclei in the sample. Two basic methods are usually employed to convert this proportionality into sample moisture content.

In the first method, the resonance FID signal is resolved into two components corresponding to the signal from the hydrogen nuclei of the solid and those of the liquid. From the latter, the moisture content of the sample can be calculated. In some cases it is possible to further resolve the liquid signal and thus determine the quantity and state of water present, i.e., tightly bound, less tightly bound, and "bulk-like" water. Details of this method are given by Brosio et al. (1978) and Rollwitz (1985).

In the second method, a spin-echo pulse sequence (90°-τ-180°) is used which selectively measures the amplitude of the signal due only to the liquid.

Thus, the amplitude resulting from the spin echo is directly proportional to the quantity of liquid material in the sample. Details of this method are given by Pande (1975) and Jones (1984).

Two concerns must be noted with both of these methods. One is the problem of chemical exchange between protons of the solid and protons of the water. This may directly affect the resolution of the solid and liquid signals. The second concern is the contribution to the signal from hydrogen nuclei by other liquid components containing protons, such as oils and other organic molecules (Pande, 1975). Signal measuring delay techniques (Shih, 1983) and high resolution magic angle spinning techniques (Shoolery, 1983) have been investigated in attempts to quantify this contribution, specifically of the oil component.

The major advantages and disadvantages of using NMR as a means of measuring moisture content in industry are discussed by Pande (1975), Miller et al. (1980), and Jones (1984). The major advantages are that it is noninvasive, rapid, and not affected by particle size, optical properties, or sample homogeneity. The major disadvantages are its cost in comparison with other instruments and the chemical exchange phenomenon discussed above.

NMR has been applied to measurement of moisture in a variety of products. These include wheat, oats, rice, sugar, starch and its derivatives, candy, corn, cheese, skim milk powder, flours, and breads (Pande, 1975; Hester and Quine, 1976; Brosio et al., 1978; Miller et al., 1980; DeJovin et al., 1983; Jones, 1984; Rollwitz, 1985).

Hester and Quine (1976), employing a commercial pulsed ^{1}H NMR, quantitatively determined water content in the range 1–5% for skim milk powder and 77–81% for cottage cheese curds. They reported using several different modes of measurement where the optimum mode depended on the nature of foodstuff being analyzed. Brosio et al. (1978), in analyzing starch-rich food products, developed a liquid to solid ratio procedure for the determination of moisture content. This procedure is based on extrapolation of liquid and solid amplitude and time values from the FID. These extrapolated values are, in turn, related to the moisture content of the sample.

NMR is currently being used to study and determine water on both a quality control and on-line monitoring basis. These applications are reviewed by Pande (1975) and Rollwitz (1985).

ACKNOWLEDGMENTS

The scientific assistance of Dr. I. C. Baianu and the clerical assistance of Rebecca Jackson in the preparation of this manuscript are gratefully appreciated.

REFERENCES

Ablett, S., Lilliford, P. J., Baghdadi, S. M. A., and Derbyshire, W. 1976. NMR relaxation in polysaccharide gels and films. In *Magnetic Resonance in Colloid and Interface Science*. Resing, H. A. and Wade, C. G. (Ed.). ACS Symposium Series 34. American Chemical Society, Washington, DC.

Abragam, A. 1961. *The Principles of Nuclear Magnetism*. Clarendon Press, Oxford.

Allen, A. T. and Wood, R. W. 1974. Molecular association in the sucrose-water system. *Sugar Technol. Rev.* 2: 165.

Arakawa, T. and Timasheff, S. N. 1982a. Preferential interactions of proteins with salts in concentrated solutions. *Biochemistry* 21: 6545.

Arakawa, T. and Timasheff, S. N. 1982b. Stabilization of protein structure by sugars. *Biochemistry* 21: 6536.

Baianu, I. C. 1986. Personal communication. University of Illinois, Urbana, IL.

Baianu, I. C., Boden, N., Lightowlers, D., and Mortimer, M. 1978. A new approach to the structure of concentrated aqueous electrolyte solutions using pulsed NMR methods. *Chem. Phys. Letters* 54(1): 169.

Baianu, I. C. and Forster, H. 1980. Cross-polarization, high-field carbon-13 NMR techniques for studying physicochemical properties of wheat grain, flour, starch, gluten, and wheat protein powders. *J. Applied Biochem.* 2: 347.

Baianu, I. C., Johnson, L. F., and Waddell, D. K. 1982. High-resolution proton, carbon-13, and nitrogen-15 nuclear magnetic resonance studies of wheat proteins at high magnetic fields: Spectral assignments, changes with concentration, and heating treatments of flinor gliadins in solution—comparison with gluten spectra. *J. Sci. Food Agric.* 33: 373.

Barker, C. A., Saul, D., Tiddy, G. J. T., Wheeler, B. A., and Willis, E. 1974. Phase structure, nuclear magnetic resonance and rheological properties of viscoelastic sodium dodecyl sulphate and trimethylammonium bromide mixtures. *J. Chem. Soc., Faraday Trans.* 70: 154.

Basler, Von W. and Lechert, H. 1973. Wide-line NMR of water in starch gels at 295K. *Starch* 25(9): 289.

Basler, Von W. and Lechert, H. 1974. Diffusion of water in starch gels. *Starch* 26(2): 39.

Bates, R. G. 1982. The activity concept in analytical chemistry. In *Ions and Molecules in Solution*. Tanaka, N., Ohtaki, H., and Tamamushi, R. (Ed.). Elsevier Science Publishers B.V., Amsterdam.

Berendsen, H. J. C. 1975. Specific interactions of water with bipolymers. In *Water—A Comprehensive Treatise, Vol. 5. Water in Disperse Systems*. Franks, F. (Ed.). Plenum Press, New York.

Berendsen, H. J. C. and Migchelsen, C. 1965. Hydration structure of fibrous macromolecules. *Ann. N.Y. Acad. Sci.* 125: 365.

Berliner, L. J. and Reuben, J. 1980. *Biological Magnetic Resonance.* Plenum Press, New York.

Bociek, S. and Franks, F. 1979. Proton exchange in aqueous solutions of glucose. *J. Am. Chem. Soc. Farady Trans. I.* 2: 262.

Borah, B. and Bryant, R. G. 1982. Deuterium NMR of water in immobilized protein systems. *Biophys J.* 38: 47.

Brevard, C. and Granger, P. 1981. *Handbook of High Resolution Multinuclear NMR.* John Wiley & Sons, Inc., New York.

Brey, W. S., Evans, R. E., and Hitzrot, H. L. 1968. Nuclear magnetic resonance times of water sorbed by proteins, lysozyme and serum albumin. *J. Colloid Interface Sci.* 26: 306.

Bryant, R. G. 1978. NMR relaxation studies of solute-solvent interactions. *Ann. Rev. Phys. Chem.* 29: 167.

Bryant, R. G. and Halle, B. 1982. NMR relaxation of water in heterogeneous systems—consensus view? In *Biophysics of Water.* Franks, F. and Mathias, S. F. (Ed.). John Wiley & Sons Limited, New York.

Bryant, R. G. and Jarvis, M. 1984. Nuclear magnetic relaxation dispersion in protein solutions. A test of proton-exchange coupling. *J. Phys. Chem.* 88: 1323.

Bryant, R. G. and Shirley, W. M. 1980a. Dynamical deductions from nuclear magnetic resonance relaxation measurements at the water-protein interface. *Biophys. J.* 32: 3.

Bryant, R. G. and Shirley, W. M. 1980b. Water-protein interactions: Nuclear magnetic resonance results on hydrated lysozyme. In *Water in Polymers.* Rowland, S. (Ed.). ACS Symposium Series 127. American Chemical Society, Washington, DC.

Brosio, E., Altobelli, G., Yu, S., and Di Nola, A. 1983. A pulsed low resolution NMR study of water binding to powdered milk. *J. Fd. Technol.* 18: 219.

Brosio, E., Conti, F., Lintas, C., and Sykora, S. 1978. Moisture determination in starch-rich food products by pulsed nuclear magnetic resonance. *J. Food Technol.* 13: 107.

Bull, H. B. and Breese, K. 1970. Water and solute binding by proteins. 1. Electrolytes. *Arch. Biochem. Biophys.* 137: 299.

Bull, T. E., Forsen, S., and Turner, D. L. 1979. Nuclear magnetic relaxation of spin $5/2$ and spin $7/2$ nuclei including the effect of chemical exchange. *J. Chem. Phys.* 70(06): 3106.

Callaghan, P., Jolley, K. W., and Humphrey, R. S. 1983a. Diffusion of fat and water in cheese as studied by pulsed field gradient nuclear magnetic resonance. *J. Colloid Interface Sci.* 93(2): 521.

Callaghan, P., Jolley, K. W., Lelievre, J., and Wong, R. B. K. 1983b. Nuclear magnetic resonance studies of wheat starch pastes. *J. Colloid Interface Sci.* 92(2): 332.

Campbell, I. D. and Dwek, R. A. 1984. *Biological Spectroscopy.* Ch. 6. The Benjamin/Cummings Publishing Co., Inc., Menlo Park, CA.

Carr, H. Y. and Purcell, E. M. 1954. Effects of diffusion of free precession in nuclear magnetic resonance experiments. *Phys. Rev.* 94: 630.

Child, T. F. and Pryce, N. G. 1972. Steady-state and pulse NMR studies of gelatin in aqueous agarose. *Biopolymers* 11: 409.

Chinachoti, P. and Steinberg, M. P. 1984. Interaction of sucrose with starch during dehydration as shown by water sorption. *J. Food Sci.* 49: 1604.

Chinachoti, P. and Steinberg, M. P. 1985. Interactions of sodium chloride with raw starch in freeze-dried mixtures as shown by water sorption. *J. Food Sci.* 50: 825.

Chinachoti, P. and Steinberg, M. P. 1986. Quantitative determination of polymer and solute bound waters by pulsed NMR. Paper #311, 46th Annual Meeting of the Inst. of Food Technologists, Dallas, TX, June 15–18.

Civan, M., Achlama, A. M., and Shporer, M. 1978. The relationship between the transverse and longitudinal nuclear magnetic resonance relaxation rates of muscle water. *Biophys. J.* 21: 127.

Civan, M. and Shporer, M. 1972. ^{17}O nuclear magnetic resonance spectrum of $H_2{}^{17}O$ in frog striated muscle. *Biophys. J.* 12: 404.

Civan, M. and Shporer, M. 1975. Pulsed nuclear magnetic resonance study of ^{17}O, ^{2}D, and ^{1}H of water in frog striated muscle. *Biophys. J.* 15: 299.

Cooke, R. and Kuntz, I. D. 1974. The properties of water in biological systems. *Ann. Rev. Biophys. Bioeng.* Mullins, L. J. (Ed.). 9035: 95.

Cooke, R. and Wien, R. 1973. Nuclear magnetic resonance studies of intracellular water protons. *Ann. N.Y. Acad. Sci.* 204: 197.

Cope, F. W. 1969. Nuclear magnetic resonance evidence using D_2O for structured water in muscle and brain. *Biophys. J.* 9: 303.

De Jovin, J. A., Shih, J. M., and Wilson, J. 1983. Determination of the moisture content of brown sugar by pulsed NMR. IBM Instruments, Inc., Danbury, CT.

Derbyshire, W. 1982. The dynamics of water in heterogeneous systems with emphasis on subzero temperatures. In *Water—A Comprehensive Treatise. Vol. 7. Water and Aqueous Solutions at Subzero Temperatures.* Franks, F. (Ed.). Plenum Press, New York.

Derbyshire, W. and Duff, I. D. 1973. NMR of agarose gels. *Chem. Soc. Faraday Discuss.* 57: 243.

Deslauriers, R. and Smith, I. C. P. 1977. *Biopolymers* 16: 1245.

Duckworth, R. B. 1981. Solute mobility in relation to water content and water activity. In *Water Activity: Influences on Food Quality.* Rockland, L. B. and Stewart, G. F. (Ed.). Academic Press, New York.

Duckworth, R. B., Allison, J. Y., and Clappertion, A. A. 1976. The aqueous environment for chemical change in intermediate moisture foods. In *Intermediate Moisture Foods.* Davies, R., Birch, G. G., and Parker, K. J. (Ed.). Applied Science Publishers Ltd., London.

Duckworth R. B. and Kelly, C. E. 1973. Studies of solution processes in hydrated starch and agar at low moisture levels using wide-line nuclear magnetic resonance. *J. Food Technol.* 8: 105.

Duff, I. D. and Derbyshire, W. 1974. NMR investigation of frozen muscle. *J. Magn. Reson.* 15: 310.

Dwek, R. A. 1973. *Nuclear Magnetic Resonance (N.M.R.) in Biochemistry.* Clarendon Press, Oxford.

Earl, W. L. and Niederberger, W. 1977. Proton decoupling in ^{17}O nuclear magnetic resonance. *J. Magn. Reson.* 27: 351.

Edzes, H. and Samulski, E. 1978. The measurement of cross-relaxation effects in proton NMR spin lattice relaxation of water in biological systems: Hydrated collagen and muscle. *J. Magn. Reson.* 31: 207.

Finney, J. L., Goodfellow, J. M., and Poole, P. L. 1982. The structure and dynamics of water in globular proteins. In *Structural Molecular Biology.* Davies, D. B., Saenger, W., and Danyluk, S. S. (Ed.). Plenum Press, New York.

Frank, H. S. and Evans, M. W. 1945. Free volume and entropy in condensed systems. *J. Chem. Phys.* 13: 507.

Franks, F. 1982. Water activity as a measure of biological viability and quality control. *Cereal Foods World* 27(9): 403.

Fuller, M. E. and Brey, W. S. 1968. Nuclear magnetic resonance study of water sorbed on serum albumin. *J. Biol. Chem.* 243(2): 1968.

Fullerton, G. D., Ord, V. A., and Cameron, I. L. 1986. An evaluation of the hydration of lysozyme by a NMR titration method. *Biochimica et Biophysica Acta* 869: 230.

Fung, B. M. and Trautman, P. 1971. Deuterium NMR and EPR of hydrated collagen fibers in the presence of salts. *Biopolymers* 10: 391.

Gal, S. 1975. Solvent versus non-solvent water in casein-sodium chloride-water systems. In *Water Relations in Foods.* Duckworth, R. B. (Ed.). Academic Press, New York.

Gaura, R. M. 1984. Moisture determination in cornstarch. IBM Instruments, Inc., Danbury, CT.

Goldsmith, S. M. and Toledo, R. T. 1985. Studies on egg albumin gelation using nuclear magnetic resonance. *J. Food Sci.* 50: 59.

Gould, G. W. and Measures, J. C. 1977. Water relations in single cells. *Philos. Trans. R. Soc.* London Ser. B278: 151.

Grigera, J. R. and Mascarenhas, S. 1978. A model for NMR, dielectric relaxation and electric behavior of bound water in proteins. *Studia Biophysica,* Berlin, Band 73: 19.

Hahn, E. L. 1950. Spin echoes. *Physical Review* 80(4): 580.

Halle, B., Andersson, T., Forsen, S., and Lindman, B. 1981. Protein hydration from oxygen-17 magnetic relaxation. *J. Am. Chem. Soc.* 103: 500.

Halle, B. and Wennerstrom, H. 1981. Interpretation of magnetic resonance data from water nuclei in heterogeneous systems. *J. Chem. Phys.* 75(4): 1928.

Hallenga, K. and Koenig, S. H. 1976. Protein rotational relaxation as studied by solvent ^{1}H and ^{2}H magnetic relaxation. *Biochemistry* 15: 4355.

Hansen, J. R. 1976. Hydration of soybean protein. *J. Agric. Food Chem.* 24(6): 1136.

Harvey, J. M. and Symons, M. C. R. 1976. Proton magnetic resonance study of the hydration of glucose. *Nature* 261: 435.

Harvey, J. M. and Symons, M. C. R. 1978. The hydration of monosaccharides—An NMR study. *J. Soln. Chem.* 7(8): 571.

Hennig, Von H. J. 1977. NMR-investigations of the role of water for the structure of native starch granules. *Starch* 29:1.

Hennig, Von H. J. and Lechert, H. 1974. Measurements of the magnetic relaxation times of protons in native starches with different water contents. *Starch* 26(7): 232.

Hennig, Von H. J. and Lechert, H. 1977. DMR study of D_2O in native starches of different origins and amylose of type B. *J. Colloid & Interface Sci.* 62(2): 199.

Hennig, Von H. J., Lechert, H., and Goemann, W. 1976. Examination of the swelling mechanism of starch by pulsed NMR-methods. *Starch* 28: 10.

Hester, R. E. and Quine, D. E. C. 1976. Quantitative analysis of food products by pulsed NMR. I. Rapid determination of water in skim milk powder and cottage cheese curds. *J. Fd. Technol.* 11: 331.

Hester, R. E. and Quine, D. E. C. 1977. Quantitative analysis of food products by pulsed NMR. I. Rapid determination of oil and water in flour and feedstuffs. *J. Sci. Fd. Agric.* 28: 624.

Hilton, B. D., Hsi, E., and Bryant, R. G. 1977. ^1H nuclear magnetic resonance relaxation of water on lysozyme powders. *J. Am. Chem. Soc.* 99(26): 8483.

Hoeve, C. A. J. 1980. The structure of water in polymers. In *Water in Polymers*. Rowland, S. P. (Ed.). ACS Symposium Series 127. American Chemical Society, Washington, DC.

Hsi, E. and Bryant, R. G. 1977. Nuclear magnetic resonance relaxation in cross-linked lysozyme crystals: An isotope dilution experiment. *Arch. Biochem. Biophys.* 183: 588.

Hsi, E., Jentoft, J. E., and Bryant, R. G. 1976. Nuclear magnetic resonance relaxation in lysozyme crystals. *J. Phys. Chem.* 80(4): 412.

Hsi, E., Vogt, G. J., and Bryant, R. G. 1979. Nuclear magnetic resonance study of water adsorbed on cellulose. *J. Colloid & Interface Sci.* 70(2): 338.

Hubbard, P. S. 1970. Nonexponential nuclear magnetic relaxation by quadrupole interactions. *J. Chem. Phys.* 51(3): 985.

Jelinski, L. W. 1984. Modern NMR spectroscopy. *Chem. Eng. News* Nov. 5: 26.

Jones, S. A. 1984. Determination of the water content of glucose by pulsed NMR. IBM Instruments, Inc., Danbury, CT.

Kalk, A. and Berendsen, H. J. C. 1976. Proton magnetic relaxation and spin diffusion in proteins. *J. Magn. Reson*\ 24: 343.

Kashkina, L. V., Verkhovtseva, E. E., and Abramov, V. L. 1979. Proton magnetic resonance study of properties of water adsorbed on starch. *Colloid J. USSR* 41: 368.

Kirkwood, J. G. and Shumaker, J. B. 1952. Forces between protein molecules in solution rising from fluctuations in proton charge and configuration. *Proc. Nat. Acad. Sci. USA* 38: 863.

Klotz, I. M. 1958. Protein hydration and behavior. *Science* 128: 815.

Koenig, S. H. 1980. The dynamics of water-protein interactions: Results from measurements of nuclear magnetic relaxation dispersion. In *Water in Polymers*. Rowland, S. P. (Ed.). ACS Symposium Series 127. American Chemical Society, Washington, DC.

Koenig, S. H., Bryant, R. G., Hallenga, K., and Jacobs, G. S. 1978. Magnetic cross-relaxation among protons in protein solutions. *Biochemistry* 17: 4348.

Koenig, S. H., Hallenga, K., and Shporer, M. 1975. Protein-water interaction studied by solvent ^1H, ^2H and ^{17}O magnetic relaxation. *Proc. Nat. Acad. Sci. USA* 72(7): 2667.

Kubo, R. and Tomita, K. 1954. A general theory of magnetic resonance adsorption. *J. Phys. Soc.* Japan 9: 888.

Kumosinski, T. F. and Baianu, I. C. 1986. Personal communication. USDA-ARS, ERRC, Philadelphia, PA and Univ. of Illinois, Urbana, respectively.

Kumosinski, T. F. and Pessen, H. 1982. A deuteron and proton magnetic resonance relaxation study of β-lactoglobulin A association: Some approaches to the Scatchard hydration of globular proteins. *Arch. Biochem. Biophys.* 218(1): 286.

Kuntz, I. D. and Kauzmann, W. 1974. Hydration of proteins and polypeptides. *Adv. Protein Chem.* 28: 239.

Labuza, T. P. and Busk, G. C. 1979. An analysis of the water binding in gels. *J. Food Sci.* 44: 1379.

Lang, K. W. and Steinberg, M. P. 1981. Linearization of the water sorption isotherm for homogeneous ingredients over a_w 0.30–0.95. *J. Food Sci.* 46: 1450.

Lang, K. W. and Steinberg, M. P. 1983. Characterization of polymer and solute bound water by pulsed NMR. *J. Food Sci.* 48: 517.

Laszlo, P. (Ed.). 1983. *NMR of Newly Accessible Nuclei.* Vol. 1. Academic Press, New York.

Lechert, Von H. 1976. Possibilities and limits of pulsed NMR spectroscopy for the investigation of problems of starch-research and starch technology. *Starch* 28(11): 369.

Lechert, Von H. and Hennig, Von H. J. 1976. NMR investigation on the behavior of water in starches. In *Magnetic Resonance in Colloid and Interface Science.* Resing, H. A. and Wade, C. G. (Ed.). ACS Symposium Series 34. American Chemical Society, Washington, DC.

Lechert, Von H., Maiwald, W., Kothe, R., and Basler, W. D. 1980. NMR-study of water in some starches and vegetables. *J. Food Proc. Preserv.* 3: 275.

Lechert, Von H. and Schwier, I. 1982. Nuclear magnetic resonance investigations on the mechanisms of water mobility in different starches. *Starch* 34(1): 6.

Leistner, L. and Rodel, W. 1976. The stability of intermediate moisture foods with respect to microorganisms. In *Intermediate Moisture Foods.* Davies, R., Birch, G. G., and Parker, K. J. (Ed.). Applied Science Publishers, Ltd., London.

Lelievre, J. and Mitchell, J. 1975. A pulsed NMR study of some aspects of starch gelatinization. *Starch* 27(4): 113.

Leung, H. K., Magnuson, J. A., and Bruinsma, B. L. 1979. Pulsed nuclear magnetic resonance study of water mobility in flour doughs. *J. Food Sci.* 44: 1408.

Leung, H. K., Magnuson, J. A., and Bruinsma, B. L. 1983. Water binding of wheat flour doughs and breads as studied by deuteron relaxation. *J. Food Sci.* 48: 95.

Leung, H. K., Steinberg, M. P., Wei, L. S., and Nelson, A. I. 1976. Water binding of macromolecules determined by pulsed NMR. *J. Food Sci.* 41: 297.

Leung, H. K. and Steinberg, M. P. 1979. Water binding of food constituents as determined by NMR, freezing, sorption, and dehydration. *J. Food Sci.* 44: 1212.

Lillford, P. J., Clark, A. H., and Jones, D. V. 1980. Distribution of water in heterogeneous food and model systems. In *Water in Polymers*. Comstock, M. J. (Ed.). ACS Symposium Series 127. American Chemical Society, Washington, DC.

Lindman, B. 1983. Amphiphilic and polyelectrolyte systems. In *NMR of Newly Accessible Nuclei. Vol. 1. Chemical and Biochemical Applications*, p. 193. Academic Press, New York.

Lioutas, T. S. 1984. Interaction among protein, electrolytes and water determined by nuclear magnetic resonance and hydrodynamic equilibria. Ph.D. thesis, Univ. of Illinois, Urbana.

Lioutas, T. S., Baianu, I. C., and Steinberg, M. P. 1986. Oxygen-17 and deuterium nuclear magnetic resonance studies of lysozyme hydration. *Arch. Biochem. Biophys.* 247(1): 136.

Luallen, T. E. 1985. Starch as a functional ingredient. *Food Technol.* 39(1): 59.

Maciel, G. E. 1984. High-resolution nuclear magnetic resonance of solids. *Science* 226: 282.

Mantsch, H. H., Saito, H., and Smith, I. C. P. 1977. Deuterium magnetic resonance, applications in chemistry, physics, and biology. *Adv. NMR Spectra* 11(4): 211.

Markley, J. L. and Ulrich, E. L. 1984. Detailed analysis of protein structure and function by NMR spectroscopy: Survey of resonance assignments. *Ann. Rev. Biophys. Bioeng.* 13: 493.

Meiboom, S. and Gill, D. 1958. Modified spin-echo method for measuring nuclear relaxation times. *Rev. Sci. Instr.* 29: 688.

Migchelsen, C. and Berendsen, H. J. C. 1973. Proton exchange and molecular orientation of water in hydrated collagen fibers. An NMR study of H_2O and D_2O. *J. Chem. Phys.* 59(1): 296.

Miller, B. S., Lee, M. S., Hughes, J. W., and Pomeranz, Y. 1980. Measuring high moisture content of cereal grains by pulsed nuclear magnetic resonance. *Cereal Chem.* 57(2): 126.

Mora-Gutierrez, A. and Baianu, I. C. 1986. ^{1}H, ^{2}H and ^{17}O nuclear magnetic resonance studies of hydration and hydrogen bonding in polysaccharide solutions, suspensions of modified starch and model systems. In preparation, University of Illinois, Urbana.

Mousseri, J., Steinberg, M. P., Nelson, A. I., and Wei, L. S. 1974. Bound water capacity of corn starch and its derivatives by NMR. *J. Food Sci.* 39: 114.

Nagashima, N. and Suzuki, E. 1981. Pulsed NMR and state of water in foods. In *Water Activity: Influence on Food Quality*. Rockland, L. B. and Stewart, G. F. (Ed.). Academic Press, New York.

Nagashima, N. and Suzuki, E. 1984. Studies of hydration by broad-line pulsed NMR. *Appl. Spectroscopy Rev.* 20(1): 1.

Nakazawa, F., Takahashi, J., Noguchi, S., and Kato, M. 1980. Water binding in gelatinized nonglutinous and glutinous rice starch determined by pulsed NMR. *J. Home Econ. of Japan* 31(8): 541.

Nakazawa, F., Takahashi, J., Noguchi, S., and Takada, M. 1983. Pulsed NMR study of water behavior in retrogradation process of rice and rice starch. *J. Home Econ. of Japan* 34(9): 566.

Oakes, J. J. 1976. Protein hydration. *Chem. Soc., Faraday Trans.* 72: 216.

Okamura, T., Steinberg, M. P., Tojo, M., and Nelson, A. I. 1978. Water binding by soy flour as measured by wide line NMR. *J. Food Sci.* 43: 553.

Ostroff, E. D. and Waugh, J. S. 1966. Multiple spin echoes and spin locking in solids. *Phys. Rev. Letters* 16(24): 1097.

Pande, A. 1975. *Handbook of Moisture Determination and Control: Principles, Techniques and Applications,* Vol. 2, Ch. 7. Marcel Dekker, Inc., New York.

Peemoeller, H., Kydon, D. W., Sharp, A. R., and Schreiner, L. J. 1984. Cross relaxation at the lysozyme-water interface: An NMR line-shape-relaxation correlation study. *Can. J. Phys.* 62: 1002.

Peemoeller, H., Shenoy, R. K., and Pintar, M. M. 1981. Two-dimensional NMR time evolution correlation spectroscopy in wet lysozyme. *J. Magn. Reson.* 45: 193.

Peemoeller, H., Yeomans, F. G., Kydon, D. W., and Sharp, A. R. 1986. Water molecule dynamics in hydrated lysozyme. A deuteron magnetic resonance study. *Biophys. J.* 49: 943.

Potter, N. N. 1978. *Food Science,* 3rd ed. Avi Publishing Co, Inc., Westport, CT.

Rabideau, S. W. and Hecht, G. H. 1967. Oxygen-17 linewidths as influenced by proton exchange in water. *J. Chem. Phys.* 47(2): 544.

Renou, J. P., Alizon, J., Dohri, M., and Robert, H. 1983. Study of the water-collagen system by NMR cross-relaxation experiments. *J. Biochem. Biophys. Methods* 7: 91.

Resing, H. A., Garroway, A. N., and Foster, K. R. 1976. Bounds on bound water: Transverse and rotating frame NMR relaxation in muscle tissue. In *Magnetic Resonance in Colloid and Interface Science*. Resing, H. A. and Wade,

C. G. (Ed.). ACS Symposium Series 34. American Chemical Society, Washington, DC.

Richards, E. G. 1980. *An Introduction to the Physical Properties of Large Molecules in Solution.* Cambridge University Press, New York.

Richardson, S. J., Baianu, I. C., and Steinberg, M. P. 1985a. Relation between oxygen-17 NMR and rheological characteristics of wheat flour suspensions. *J. Food Sci.* 50: 1148.

Richardson, S. J., Chinachoti, P., and Steinberg, M. P. 1985b. Molecular mobility determinations of amylopectin and corn starch by ^{17}O nuclear magnetic resonance. Paper #479, 45th Annual Meeting of the Institute of Food Technologists, Atlanta, GA, June 9–12.

Richardson, S. J., Baianu, I. C., and Steinberg, M. P. 1986a. Mobility of water in wheat flour suspensions as studied by proton and oxygen-17 nuclear magnetic resonance. *J. Agric. Food Chem.* 34(1): 17.

Richardson, S. J., Baianu, I. C., and Steinberg, M. P. 1986b. Mobility of water in corn starch suspension determined by deuterium and oxygen-17 nuclear magnetic resonance. *Starch.* Accepted.

Richardson, S. J., Baianu, I. C., and Steinberg, M. P. 1986c. Mobility of water in starch powders by deuterium and oxygen-17 nuclear magnetic resonance in relation to its concentration and activity. *Starch.* Accepted.

Richardson, S. J., Baianu, I. C., and Steinberg, M. P. 1986d. Mobility of solute water determined by deuterium and oxygen-17 nuclear magnetic resonance measurements of sucrose systems. *J. Food Sci.* Accepted.

Richardson, S. J., Baianu, I. C., and Steinberg, M. P. 1986e. Mobility of water in polymer-solute systems determined by deuterium and oxygen-17 nuclear magnetic resonance. *Starch.* In submission.

Rollwitz, W. 1985. Using radiofrequency spectroscopy in agricultural applications. *Agric. Engr.* May: 12.

Schwier, I. and Lechert, H. 1982. X-ray and nuclear magnetic resonance investigations on some structure problems of starch. *Starch* 34: 11.

Shanbhag, S., Steinberg, M. P., and Nelson, A. I. 1970. Bound water defined and determined at constant temperature by wide-line NMR. *J. Food Sci.* 35: 612.

Shih, J. M. 1983. Determination of the oil and water content of rice by pulsed NMR. IBM Instruments, Inc., Danbury, CT.

Shirley, W. M. and Bryant, R. G. 1982. Proton-nuclear spin relaxation and molecular dynamics in the lysozyme-water system. *J. Am. Chem. Soc.* 104: 2910.

Shoolery, J. 1983. MAS of lipids in soybeans. *ACS Abstracts*. 186th American Chemical Society National Meeting, Washington, DC, Aug. 28.

Smith, S. H. 1947. The sorption of water vapor by high polymers. *J. Am. Chem. Soc.* 69: 646.

Solomon, I. 1955. Relaxation processes in a system of two spins. *Phys. Rev.* 99: 559.

Steijskal, E. O. and Tanner, J. E. 1965. Spin diffusion measurements: Spin-echoes in the presence of a time dependent field gradient. *J. Chem. Phys.* 42: 288.

Steinberg, M. P. and Leung, H. 1975. Some applications of wide-line and pulsed NMR in investigations of water in foods. In *Water Relations of Foods*. Duckworth, R. B. (Ed.). Academic Press, New York.

Suggett, A. 1976. Molecular motion and interactions in aqueous carbohydrate solutions. III. A combined nuclear magnetic and dielectric strategy. *J. Soln. Chem.* 5(1): 33.

Suggett, A., Ablett, S., and Lillford, P. J. 1976. Molecular motion and interactions in aqueous carbohydrate solutions. II. Nuclear-magnetic-relaxation studies. *J. Soln. Chem.* 5(1): 17.

Tait, M. J., Ablett, S., and Franks, F. 1972a. An NMR investigation of water in carbohydrate systems. In *Water Structure at the Water-Polymer Interface*. Jellinek, H. H. G. (Ed.). Plenum Press, New York.

Tait, M. J., Ablett, S., and Wood, F. W. 1972b. The binding of water on starch, an NMR investigation. *J. Colloid Interface Sci.* 41(3): 594.

Tait, M. J., Suggett, A., Franks, F., Ablett, S., and Quickenden, P. A. 1972c. Hydration of monosaccharides: A study by dielectric and nuclear magnetic relaxation. *J. Soln. Chem.* 1(2): 131.

Toledo, R., Steinberg, M. P., and Nelson, A. I. 1968. Quantitative determination of bound water by NMR. *J. Food Sci.* 33: 315.

Tricot, Y. and Niederberger, W. 1979. Water orientation and motion in phospholipid bilayers: A comparison between ^{17}O- and ^{2}H-NMR. *Biophysical Chem.* 9: 195.

Ulmius, J., Wennerstrom, H., Lindblom, G., and Arvidson, G. 1977. Deuteron nuclear magnetic resonance studies of phase equilibria in a lecithin-water system. *Biochemistry* 16(26): 5742.

van den Berg, C. and Bruin, S. 1981. Water activity and its estimation in systems: Theoretical aspects. In *Water Activity: Influences on Food Quality*. Rockland, L. B. and Stewart, G. F. (Ed.). Academic Press, New York.

Walmsley, R. H. and Shporer, M. 1978. Surface-induced NMR line splittings and augmented relaxation rates in water. *J. Chem. Phys.* 68(6): 2584.

Westlund, P. and Wennerstrom, H. 1982. NMR lineshapes of I=5/2 and I=7/2 nuclei. Chemical exchange effects and dynamic shifts. *J. Magn. Reson.* 50: 451.

Woessner, D. E. 1962. Spin relaxation processes in a two-proton system undergoing anisotropic reorientation. *J. Chem. Phys.* 36(1): 1.

Woessner, D. E. 1974. Proton exchange effects on pulsed NMR signals from preferentially oriented water molecules. *J. Magn. Reson.* 16: 483.

Woessner, D. E. 1977. Nuclear magnetic relaxation and structure in aqueous heterogeneous systems. *Mol. Phys.* 34(4): 899.

Woessner, D. E., Snowden, B. S., and Meyer, G. H. 1969. Calculation of NMR free induction signals for nuclei of molecules in a highly viscous medium or a solid-liquid system. *J. Chem. Phys.* 51(7): 2968.

Woodhouse, D. R. 1974. NMR in systems of biological significance. Ph.D. thesis, University of Nottingham.

Zimmerman, J. R. and Brittin, W. E. 1957. Nuclear magnetic resonance studies in multiple phase systems: Lifetime of a water molecule in an absorbing phase on silica gel. *J. Phys. Chem.* 61: 1328.

12

FDA Views on the Importance of a_w in Good Manufacturing Practice

Melvin R. Johnston* **and Rong C. Lin**

Food and Drug Administration
Washington, D.C.

INTRODUCTION

Food processing and preservation involves destruction or inactivation of microorganisms. The availability of free moisture can be a significant factor in the effectiveness of these processes and the potential for biochemical reactions and subsequent stability of a food.

Water in foods exerts a vapor pressure, the magnitude of which depends upon the amount free to vaporize. Water activity (a_w) is the quotient of the water vapor pressure of the substance (food) divided by the vapor pressure of pure water at the same temperature. The moisture content of most intermediate moisture foods (IMF) ranges from 15% to 40% resulting in an a_w greater than 0.85. When IMF, such as puddings, bread, processed cheese, anchovy, caviar, salted fish, and brined vegetables, are thermally processed in hermetically sealed containers, they are considered to be low-acid canned foods (LACF) and must comply with Title 21, Code of Federal Regulations (CFR), Parts 108.35 and 113 (FDA, 1985).

**Current affiliation:* Canned food consultant, New Braunfels, Texas

The primary public health concern associated with improperly processed low-acid foods is the possibility of foodborne intoxication due to botulinum toxin.

This is a time of great scientific and technological changes. The U. S. Food and Drug Administration (FDA) is in a unique position relative to emerging technologies. Available analytical instruments and methodologies are having a tremendous influence on regulation enforcement and assurance of compliance with the regulations.

GOOD MANUFACTURING PRACTICE REGULATIONS (GMPR) GOVERNING THE PROCESSING REQUIREMENTS

Regulations which detail the specific requirements are codified in the FDA Code of Federal Regulations (1985). The current GMPR for Thermally Processed Low-Acid Foods Packaged in Hermetically Sealed Containers, 21 CFR 113, establishes specific levels of a_w and pH to define low-acid foods. FDA's

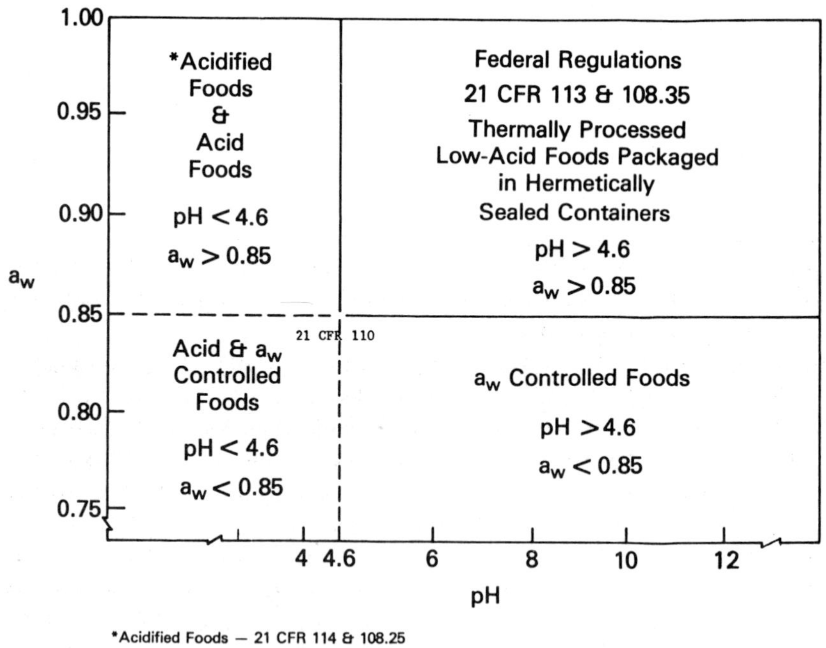

FIG. 12.1 FDA Good Manufacturing Practice Regulations governing processing requirements and classification of foods.

GMPRs governing the processing requirements and the classification of foods are shown in Fig. 12.1. Low-acid foods packaged in hermetically sealed containers must achieve commercially sterile conditions either by retorting or a combined treatment of pasteurization and a_w control or a combined treatment of pasteurization and acidification. However, it should be pointed out that microorganisms are not known to grow at the two extreme levels of pH, nor are there any thermally resistant pathogenic bacteria known to grow in foods with a_w levels below 0.85.

The FDA's compliance programs require that sample(s) be taken for analysis if a_w may be a critical factor of the scheduled process for the packaged food. The concern over a_w controlled LACF stems from their ability to support the outgrowth of pathogenic bacteria under normal nonrefrigerated conditions of storage and distribution. Figure 12.2 is a compilation of FDA data on a_w and pH of selected foods.

Establishing Scheduled Processes

Water activity is one of the critical factors governing thermal processing requirements of low-acid food packaged in hermetically sealed containers. Unlike many commonly recognized LACF protected by thermal destruction of

Foods	a_w		Sample Units	pH	
	Mean	Range	n	Mean	Range
Lupini Beans	0.94	.92—.96	18	4.8	4.2—5.1
Shrimp Curry	0.94	.93—.95	4	6.8	6.6—7.0
Cheese Spread	0.94	.93—.96	6	5.6	5.5—5.7
Imitation Truffles	0.95	.94—.96	4	5.9	5.8—6.0
Alimentary Paste	—	> .95	6	11.3	11.2—11.4
Capers	0.92	.91—.93	2	4.4	4.3—4.7
Red Bean Paste	0.93	.92—.94	6	5.5	5.4—5.6
Pickled Radish	0.93	.92—.95	100	4.8	4.6—5.3
Pickled Lettuce	0.93	.92—.94	12	5.9	5.8—5.9
Preserved Radish w/chili	0.90	.90—.91	6	4.9	4.8—5.0
Soy Sauce	0.84	.83—.85	6	4.8	4.7—5.0
Caviar	0.92	.91—.94	—	—	> 4.6
Black Bean Sauce	0.90	.89—.90	4	—	> 4.6
Sanbai-Zuke	0.91	.90—.92	12	4.8	4.6—5.3
Pound Cake	0.83	.82—.84	8	5.85	5.8—5.9
Chocolate Syrup	0.83	.83—.85	6	5.4	5.3—5.5

FIG. 12.2 Compilation of data on relationships of selected foods, a_w, water content, and their susceptibility to foodborne bacterial pathogens.

bacterial spores, some are protected by achieving a specific a_w level which is unfavorable for germination and outgrowth of bacterial spores having public health significance. The combination of pasteurization treatment and a_w control has been recognized as an acceptable method to render a food commercially sterile.

The traditional approaches in establishing a scheduled process for a low-acid food are by process calculations based on product heat penetration data, or the results of inoculated packs, or the results of bacterial population reduction tests. However, when the scheduled process sets forth critical factors, such as a_w control, to prevent the growth of microorganisms not destroyed by the thermal process, the factors shall be measured and controlled to ensure that the limits are met. Process establishment information should provide evidence to show that the a_w together with pasteurization treatment provide for the safety of the product. The process adequacy should not be determined based on pasteurization data alone. The thermal resistance, e.g., D, z, and F values, of spores of specific pathogenic bacteria in product(s) with a_w levels between 0.85 and 0.95 must be determined. Paragraph 21 CFR 113.8(f) states that when the scheduled process sets forth critical factors to prevent the growth of microorganisms not destroyed by the thermal process, the factors shall be carefully controlled to ensure that the limits established in the scheduled process are delivered and recorded.

Paragraph 21 CFR 113.83 states that critical factors, e.g., a_w, that may affect the scheduled process shall be specified in the scheduled process. The scheduled processes for low-acid foods shall be established by qualified persons having expert knowledge of thermal processing requirements for low-acid foods in hermetically sealed containers and having adequate facilities for making such determinations.

Production and Process Control

Paragraph 21 CFR 113.81(f) also requires when normally low-acid foods require sufficient solute (a_w control) to permit safe processing at low temperatures, such as in boiling water, there shall be careful supervision to ensure that the equilibrium a_w of the finished product meets that of the scheduled process. The scheduled thermal process for food having an a_w greater than 0.85 shall be sufficient to prevent the outgrowth of microorganisms capable of reproducing in the food under normal nonrefrigerated conditions of storage and distribution. Paragraph 21 CFR 113.40(i) requires that critical factors, e.g., a_w, specified in the scheduled process shall be measured with instruments having the accuracy and dependability adequate to ensure that the requirements of the scheduled process are met.

Paragraph 21 CFR 113.10 requires that operators of processing systems . . . and product formulating systems, including systems wherein a_w is used in conjunction with thermal processing . . . shall be under operating supervision of a person who has attended a Better Process Control School approved by the Commissioner of FDA.

All process deviations involving a failure to satisfy the minimum critical factors of the scheduled process, including a change in product formulation and its resultant a_w level, must be handled in accordance with 21 CFR 113.89.

LABORATORY ANALYTICAL METHODS FOR a_w DETERMINATIONS

Analyses of samples are necessary to determine if a product is in or out of compliance with the requirements of the regulation. The FDA is in the process of developing multi-instrument data and experience generated from field laboratories. Analytical methods for a_w value determinations are detailed in the 14th edition of *Official Methods of Analysis* of the AOAC (1984), 32.004 to 32.009. Analytical instruments or methods for a_w value determination are abundant. However, we shall limit our discussion to several instruments which are available in the FDA Districts and Headquarters.

Hair Hygrometry

Measurement is based on the magnitude of longitudinal change in length of water-sorbing fiber in the sample container at equilibrium. This measurement is based on the principle that the keratinaceous proteins in hair strands under tension tend to stretch when they absorb moisture. If the hair strand is fixed at one end and attached to an indicating lever arm at the other end, the relative humidity within an enclosure can be read directly.

The hair hygrometer is a dial-type polyamide thread hygrometer. This type of hygrometer is relatively inexpensive. Its accuracy is comparable to others. An accuracy of ± 0.02 up to ± 0.01 is the best that can be expected. Some selected and artificially aged plastic threads have resulted in improved accuracy over the hair type. This instrument provides a convenient screening test by the FDA field investigators to determine the a_w of foods and whether a follow-up confirmation test is needed. In addition, three reports from FDA's quality assurance samples showed that all test results, using the hair hygrometer, were within acceptable limits of $\pm$ 0.01.

It is very important that all FDA District Laboratories making a_w measurements have at least one instrument in common to enable the check sample system to monitor their effectiveness. By maintaining the hair hygrometer we can maintain a common instrument rather inexpensively.

Isopiestic Method

This method provides the measurement of equilibrium relative humidity of a_w of any aqueous systems in a closed container at a specified temperature.

The isopiestic sorption isotherm method was first developed by a physical chemist about 60 years ago. Modifications by many researchers have resulted in a great improvement in its value, particularly when high a_w measurements are encountered. Measurement of weight changes of certain anhydrous hydrophilic solids as well as evacuation significantly improve its applicability.

Most food substances are consistently adjusting their moisture content through adsorption or desorption processes depending upon the moisture condition and temperature of the environment until the food substances approach equilibrium. In other words, the equilibrium vapor pressure of the reference salt slush will be identical to the vapor pressure of the sample at equilibrium condition.

This technique enjoys the freedom from chemical contamination. However, multiple samples must be measured at different equilibrium conditions using different salt slushes in desiccators. The weight change must be carefully measured using an analytical balance. Since this is a method with discontinuous registration of weight changes, the frequency of weighing should be kept at a minimum (e.g., at 1 hr, 1½ hr, 2 hr) to avoid the environmental effect. Comparisons of a_w levels among various foods analyzed by this technique and by electronic hygrometers have given good correlations at $a_w > 0.90$, and superior precision has been claimed at a_w levels < 0.90.

Sample dishes, spatulas, and other tools that are in contact with foods should be of sanitary quality to prevent microbial contamination, since respiration due to microbial growth may prevent vapor pressure equilibrium. Antimicrobial agents, such as 3% potassium sorbate, have been used by some researchers. To prevent the microbial growth in salt slushes with high a_w, sterile distilled water and sterile jars and tools may be used.

Electronic Hygrometers

This type of instrument features the use of calibrated aluminum oxide or lithium chloride humidity sensor(s). Recalibration or standardization of sensor response is by a set of reference salt slushes. Water activity measurement is carried out by connecting the appropriate sensor to an air-tight (food) sample container and equilibrated at a specified temperature. Since the sensor is sealed in a small sample container, it usually takes less than 2 hr for a sample to approach equilibrium condition. There are three basic types of measuring systems:

1. The first system measurement is based on the ability of the hygroscopic film of lithium chloride to change its electrical resistance or conductance with microchange in relative humidity in an air-tight container. The resistance change is measured in terms of electric current flowing through the sensing element. The sensor is usually connected to a balancing potentiometer or a recorder with a scale calibrated in terms of a_w.
2. The second system is based on the measurement of electrical impedance of a liquid hygroscopic substance which absorbs or desorbs moisture or water vapor. The impedance value reflects the moisture content of the substance. The hygroscopic substance of the electrode is calibrated in terms of a_w.
3. The third system is based on the capacitance change of a thin polymer film capacitor. The sensor element of the probe is a small thin film chip capacitor composed of an upper and lower electrode with an organic polymer dielectric strip about one micron thick which absorbs water molecules through a thin metal electrode and causes a capacitance change proportional to the relative humidity.

These instruments provide a better and convenient means of a_w measurement with adequate accuracy and precision. However, they are susceptible to contaminants and poisons such as SO_2, H_2S, chlorine, and oil vapors. Temporary contaminants include ammonia, acetic acid, alcohols, glycol, and glycerols, depending upon the sensor material used. Food is a complex aqueous system. Water is likely the major constituent in food; however, most foods also contain a certain amount of nonsolute components (insoluble solids). In additon, other volatiles may also enter into the headspace of an enclosed test chamber, which may contaminate or create errors in a_w measurement. Contaminants may be removed or retained using a polypropylene or carbon filter, although their use may result in a longer equilibration time to approach equilibrium condition for a sample.

CONCLUSIONS

Ensuring safety in the food supply in the United States is the major function of FDA and the one that utilizes a substantial portion of total agency resources.

The purposes of GMP regulations are to detail the specific requirements and practices to be followed by industry to assure that foods are produced under sanitary conditions and are pure, wholesome, and safe to eat.

Specific parts and paragraphs of the GMP regulations delineate a_w in relation to process control and safety requirements. Production records and the results of a_w determinations shall also be maintained for compliance with the mandatory requirements of low-acid canned food regulations.

To determine the adequacy of a process, it may be necessary to demonstrate that the pathogenic bacteria in a_w-controlled products do not present any potential public health hazard. However, if a_w is not considered as a critical factor, then the traditional approaches in generating thermal processing parameters and all relevant details should be obtained to demonstrate the adequacy of the process. The combination treatments of a_w adjustment and pasteurization process have also been recognized as an acceptable method to render a food commercially sterile.

Water activity instrument standardization has been adopted by the AOAC (1984) as the Official Final Action. FDA generally uses indirect methods to measure the comparative a_w rather than the absolute a_w of a sample. There can be no substitute for a uniform method of a_w determination. Only through periodic checking and proper calibration of an instrument can an accurate a_w value be obtained. Judgment in the light of expert opinion should be exercised whenever a questionable condition exists. By and large, good laboratory practices as delineated in 21 CFR 58 should be followed and documented with complete records.

REFERENCES

AOAC. 1984. *Official Methods of Analysis.* 14th ed., Methods 32.004-32.009. Association of Official Analytical Chemists, Arlington, VA.

U.S. Food and Drug Administration. 1985. Title 21 Code of Federal Regulations Parts 58, 108, 113, and 114. U.S. Government Printing Office, Washington, DC.

13

Shelf-Stable Products and Intermediate Moisture Foods Based on Meat

Lothar Leistner

Federal Centre for Meat Research
Kulmbach, Federal Republic of Germany

INTRODUCTION

Since Scott (1957) introduced water activity (a_w) into food science, much has been learned about the basic aspects of a_w in relation to the physical, chemical, and microbiological properties of foods. Fewer published data are available on the application of a_w in the manufacturing of foods or feeds. However, it became apparent that in most foods for which a_w is important for quality and stability, other factors (hurdles) contribute to the desired product. This is particularly true for foods, which, in spite of a mild heat treatment, should be storable without refrigeration. Thus, the interest taken in a_w by food formulaters was extended to other factors (e.g., pH, Eh), and the goal became an intelligent product based on combination preservation technology (hurdle technology), which requires less energy input, minimum quality-damaging preservation methods, and is stable at ambient temperature. Furthermore, it became apparent that most traditional foods are based on combinations of preservation technologies and common sense. Therefore, by elucidating the principles of the stabilization of traditional foods, their processes could be improved, and ideas for new products could be derived.

In this contribution some recent developments in hurdle technology are discussed, and examples of foods based on combination preservation technology are given. These examples are shelf-stable products (SSP) as well as intermediate moisture foods (IMF); however, all are meat products, since this is my principal area of research.

HURDLE TECHNOLOGY

Numerous preservation methods (including heating, chilling, freezing, freeze-drying, drying, curing, salting, sugar addition, acidification, fermentation, smoking, oxygen removal, carbon dioxide addition, and irradiation) are used to make meat stable (inhibition of spoilage) and safe (avoidance of food poisoning). However, these preservation methods are based only on relatively few parameters (hurdles), i.e., F (high temperature), t (low temperature), a_w (less available water), pH (sufficient acidification), Eh (reduced oxygen supply), preservatives (e.g., nitrite, smoke, carbon dioxide), competitive microflora (e.g., lactic acid bacteria), and radiation (e.g., gamma rays). Thus, more processes are distinguishable than parameters which govern them, as has been pointed out by Leistner et al. (1981a) and Leistner (1985a). For the microbiological stability of almost all foods a combination of parameters (hurdles) is decisive.

Hurdle Effect

The so-called hurdle effect or hurdle concept was introduced by Leistner and Rödel (1976) and Leistner (1978), and it has since been modified and improved by our laboratory. The present version of the hurdle concept is given in Fig. 13.1, using nine examples.

Example 1 illustrates the principle of the hurdle effect and represents a food which contains six hurdles (i.e., F, t, a_w, pH, Eh, and preservatives). The microorganisms present cannot overcome ("overjump") these hurdles, thus the food is microbiologically stable and safe. However, Example 1 is only a theoretical case, because all hurdles are of the same height, i.e., have the same intensity. A more likely situation is presented in Example 2, since the microbial stability of this product is based on hurdles of different intensity. In this product the main hurdles are the a_w and preservatives (e.g., nitrite in meats), while additional and less important hurdles are storage temperature, pH, and redox potential. These five hurdles are sufficient to inhibit the usual types and

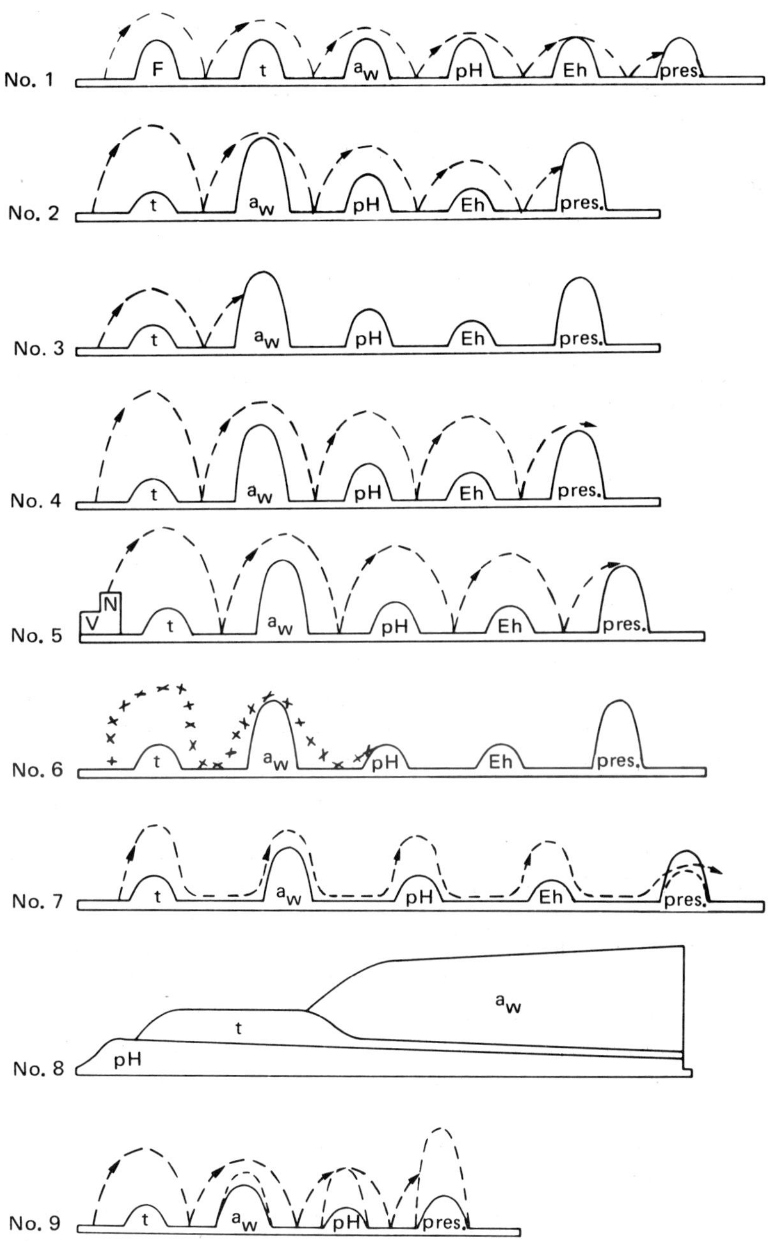

FIG. 13.1 Illustration of the hurdle effect, using nine examples. Symbols have the following meaning: F = heating, t = chilling, a_w = water activity, pH = acidification, Eh = redox potential, pres. = preservatives, V = vitamins, and N = nutrients.

numbers of organisms associated with such a product. If there are only a few microorganisms present at the start (Example 3), then a few or low hurdles are enough for the stability of the product. The aseptic packaging of perishable meat products is based on this principle, which has been developed in Japan to perfection. Due to the hygienic discipline in Japanese meat plants and the use of robots, an initial count of less than 100 organisms per gram is achieved in prepackaged meats, which allows a shelf life of 20 days at 10°C. Even Japanese meat products in general contain little salt, i.e., have a relatively low a_w hurdle. On the other hand, in Example 4, due to bad hygienic conditions, too many undesirable organisms are initially present. Therefore, the usual hurdles inherent in the product cannot prevent spoilage or food poisoning. This example emphasizes the importance of good hygiene during the slaughtering of animals and the processing of meat. Example 5 is a food superior in nutrients and vitamins, which foster the growth of microorganisms ("trampoline effect"), and thus the hurdles in such products must be enhanced, otherwise they will be overcome. Example 6 illustrates the behavior of sublethally damaged organisms in foods. If, for instance, bacterial spores in meat products are damaged sublethally by heat (as occurs in F-SSP, discussed later), then the vegetative cells derived from such spores lack vitality, and therefore are already inhibited by fewer or lower hurdles. However, as shown in Examples 7 and 8, the stability of foods is also related to the processing and storage time. On the one hand, if a food becomes more dry during processing or storage, then the a_w hurdle will increase with time, which consequently will improve the microbial stability of the product. On the other hand, in canned meat products the hurdles might decrease with time, and therefore could be overcome by microorganisms during storage of the product. The failing stability of canned cured meats in the course of storage could be due to a depletion of nitrite in the products (Example 7), which allows the growth of spoilage or food-poisoning organisms from dormant spores. In some foods, such as fermented sausages and raw hams, the microbial stability is achieved during processing by a sequence of hurdles. Example 8 illustrates the sequence of hurdles in a raw ham (the hurdles in fermented sausages will be discussed later.) For the stability of raw hams it is essential that the initial count of organisms in the interior of the product be low, ph below 6.0, and at the beginning of the curing process the temperature should be below 5°C. The low temperature should be maintained until sufficient salt (i.e., 4.5% NaCl, corresponds to an a_w below 0.96) has penetrated into all parts of the ham. After the a_w in the interior of the ham has been decreased to below 0.96, the product can be further ripened and smoked at room temperature to achieve the desired flavor by enzymatic action (Leistner, 1985d). Finally, there is some indication that the product of the hurdles, rather than their number, determines the microbial stability of a food. Example 9 illustrates the intensifying effect that hurdles in a food might have on each other. Hammer and Wirth (1984) reported that different additives which cause a depression of the a_w in meat products do not enhance each other; nevertheless, an intensifying effect of different hurdles, e.g., a_w and pH, on the microbial stability of a food is likely. However, this question needs further study.

Leistner et al. (1981a) pointed out that the hurdle effect is of fundamental importance for the preservation of foods, since the hurdles inherent in a stable product control microbial spoilage and food poisoning as well as desired fermentation of foods. Furthermore, these authors acknowledged that the hurdle concept illustrates only the well-known fact that complex interactions of a_w, pH, and temperature are significant in the microbial stability of foods.

In addition, Leistner et al. (1981a) and Leistner (1985a) stated, that the hurdle concept could be used in food control as well as in food design. In food control the objective is to evaluate the stability of a product quickly and to predict its shelf life. This could be achieved by the measurement of the physical and chemical hurdles in a particular food (which is less time-consuming than a microbiological investigation) and by computer evaluation of the results. The computer could indicate which organisms would be likely to grow, if a food has a certain pattern of hurdles, and what shelf life could be expected.

If new food products are formulated according to needs, the hurdle concept could be helpful (Leistner et al., 1981a; Leistner, 1985a). For instance, if energy conservation during storage of meat products is desired, then the energy consuming hurdle t (chilling) could be replaced by less energy expensive hurdles, such as a_w and/or pH, because the hurdles that enhance microbial stability of foods are to a certain extent interchangeable. Also the reduction of the input of nitrite and nitrate into meat products, which has been achieved in several countries in recent years (Leistner, 1985b, c), shifted attention from these preservatives to other hurdles in meat products. A sensible and intelligent combination of hurdles could secure the microbial stability of a food, and at the same time should, if possible, improve the sensory, nutritive, toxicological, and economic properties of the product.

These aims could be achieved by an extension of the hurdle concept to hurdle technology (Brimelow, 1985; Leistner, 1985a; Leistner, 1986a). Hurdle technology foods (HTF), were defined by Brimelow (1985) as a food whose shelf life with respect to microorganisms is extended by the use of two or more factors, none of these factors being individually sufficient to inhibit spoilage or food-poisoning organisms. Brimelow (1985) suggested that the hurdle technology (HT) approach to food development should be investigated in more detail, and all the currently available microbiological data should be "pulled together to provide a coherent, albeit complex, framework on which the technology may be securely based." One method of achieving this might be, according to Brimelow (1985), to assemble an interactive computer program which incorporates a data bank of the microbial limiting combinations. Brimelow (1985) stated, "The program would be freely available and the aim would be that it would be possible to come to the computer with a proposal for a recipe/process method/packaging concept and then be able to ascertain, in theoretical terms at least, the microbiological stability of the product concept. The program would hopefully provide prompts and explanations and advice on how to bring an unstable product back into line." This suggestion of Brimelow (1985) for using a computer program in food design coincides with our suggestion (Leistner et al., 1981a; Leistner, 1985a), to use the hurdle concept in food control.

Magic Square

Optimal adjustment of hurdles F, a_w, pH, and Eh is of prominent interest for heated meat products, based on hurdle technology, from both the theoretical and practical point of view. These factors can bring about a stable product, which can be stored without refrigeration, in spite of a mild heat treatment. This concept was suggested by Fox and Loncin (1982) and could be called the "magic square" (Leistner, 1986a), as illustrated by Fig. 13.2.

The stability of autoclaved sausages in casings (classified as F-SSP), which are sold at discount chains in West Germany, is based on this principle. These products (discussed later in detail) are given only a relatively mild heat treat-

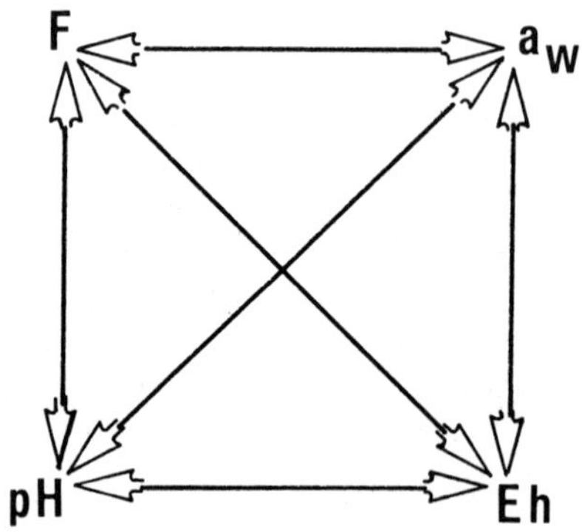

FIG. 13.2 The "magic square," which refers to the interaction of the hurdles F, a_w, Eh, and pH in mild-heated meat products (e.g., F-SSP) which are storable without refrigeration.

ment (F value about 0.4), which inactivates all vegetative microorganisms and sublethally damages bacterial spores. Bacteria deriving from such damaged spores having a diminished vitality, and therefore are already inhibited by a_w and pH values that are not detrimental to the sensory properties of the meat products. Undoubtedly also the Eh value, i.e., the amount of available oxygen, contributes to the microbial stability of meat products, including F-SSP, since not only aerobic but also many facultative anaerobic bacteria do not grow well at a low Eh. Therefore, a_w-tolerant bacilli, which grow in laboratory media aerobically at a_w 0.86, are inhibited in sausage items at a_w 0.97–0.96, if the redox potential is low (Leistner et al., 1980; Leistner, 1985a). Because the redox potential (Eh value) apparently is of major importance for the inhibition of aerobic spore formers (bacilli) in meat products, reliable methods for measuring the Eh in meats are essential, and should be developed to the same accuracy level as pH and a_w measurements. Some time ago Leistner and Mirna (1959) as well as Leistner and Wirth (1965) suggested methods for the determination of the Eh in cover brines and meat products, respectively. However, these methods call for improvements, which are feasible by using up-to-date electrodes and electronic devices. The fourth factor in the "magic square" is the pH, which interacts with F, a_w, and Eh directly and indirectly with respect to the microbial stability of foods and often is an essential hurdle. Reliable quantitative data on F, a_w, pH, and Eh are required for the optimal adjustment of hurdles in meat products, designed according to the "magic square."

Balance

Experimental work of recent years suggests that even small enhancements of the individual hurdles in a food in summation have a definite effect on the microbial stability of a product (Hechelmann and Leistner, 1984; Lücke, 1984; Lücke and Hechelmann, 1986). Figure 13.3 illustrates this phenomenon. For instance, for the stability of meat products it would be dependent upon whether the F value is 0.3 or 0.4, the a_w value is 0.975 or 0.970, the pH value is 6.4 or 6.2, and the Eh value is somewhat higher or lower. Every small improvement or enhancement of a hurdle brings some weight on a balance, and the sum of these weights determines whether a food is microbiologically unstable, uncertain, or stable (Fig. 13.3). The quantification of these influences on the microbial stability of foods will probably become a challenging research area, and work to fine tune the hurdles F, a_w, pH, and Eh in foods should produce results of practical importance.

In this endeavor technologists and microbiologists must work together. The technologists must determine which additives are suitable for the enhancement of hurdles in foods by taking toxicological, sensory, nutritive, and legal

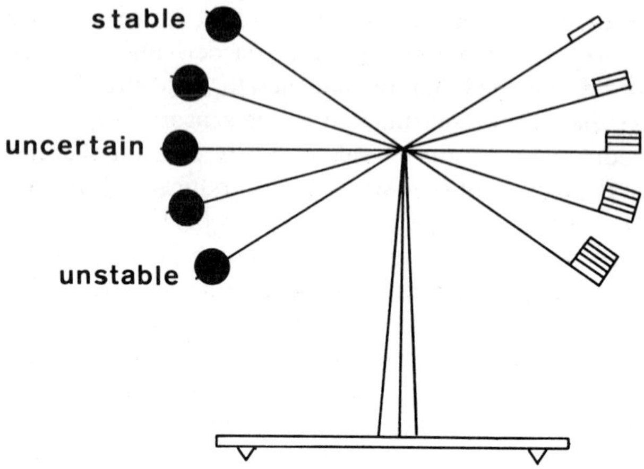

FIG. 13.3 The balance should illustrate that even small enhancements of different hurdles could bring about in summation a substantial improvement of the microbial stability of a food.

limitations into consideration. For instance, Hammer and Wirth (1984) investigated the influence of eleven different additives on the a_w of a meat product. They concluded that the a_w of liver sausage could be adjusted below 0.96 with acceptable additions of fat (about 36%) and sodium chloride (1.6–2.0%). The microbiologists must determine which intensity of factors or hurdles in a particular food are needed for desired microbiological stability. This was done, for instance, by Hechelmann and Leistner (1984) for industrially produced autoclaved sausages in casings (F-SSP) and by Lücke (1984) for canned meats produced in butcher shops. Both groups of meat products are quite similar from the microbiological point of view, and are stable at ambient temperature with the following hurdle combination: $F > 0.4$, $a_w < 0.97$ (if nitrite is effective in the product) or < 0.96 (if nitrite is not effective), $pH < 6.5$, and the Eh should be low (has not yet been quantified).

Sequence of Hurdles

In Examples 1 to 7 in Fig. 13.1 the given sequence of hurdles is arbitrary. However, in certain foods, such as raw hams and fermented sausages, the sequence of the hurdles is fixed, because during the processing and storage of these foods particular hurdles develop or fade out again. The sequence of hurdles characteristic for raw hams have been described by Leistner (1986c), and is illustrated in Example 8 in Fig. 13.1. The sequence of hurdles is more intricate in fermented sausages (salami), as illustrated by Fig. 13.4, and has been discussed by Leistner (1985a, 1986b).

In salami the hurdles occur in a sequence and are particularly important in certain ripening stages of the product to effectively inhibit food-poisoning organisms (*Salmonella* spp., *Clostridium botulinum, Staphylococcus aureus*) as well as other bacteria, yeasts, and molds which might cause spoilage. On the other hand, this sequence of hurdles also favors the selection of the desired competitive flora (lactic acid bacteria, nonpathogenic staphylococci), which contribute to the flavor and stability of fermented sausages.

An important hurdle in the early stage of the ripening process of salami is nitrite (pres. in Fig. 13.4), added with curing salt, because an addition of at least 125 mg/kg sodium nitrite inhibits the growth of salmonellae. The nitrite hurdle diminishes during the ripening process, since the nitrite depletes. Due to the multiplication of bacteria in salami the redox potential of the product decreases, and this in turn enhances the Eh hurdle, which inhibits the growth of aerobic organisms and favors the selection of the competitive flora (cf. Fig. 13.4). The growth and metabolic activity of lactic acid bacteria, which then flourish, cause acidification of the product and thus an increase of the pH hurdle. This is of particular importance in the microbial stability of quick-ripened fermented sausages, which are not completely dried. The preservative (nitrite), Eh, competitive flora, and pH diminish with time, because in long-ripened salami the nitrite level and the count of lactic acid bacteria decrease, while the Eh and pH increase somewhat. Therefore, only the a_w hurdle is strengthened with time, and this hurdle, therefore, is mainly responsible for the stability of long-ripened fermented sausages (Leistner, 1986b).

Certainly also in the processing and storage of some other foods a sequence of hurdles is responsible for the microbial stability, and it would be challenging to investigate this phenomenon in various food items.

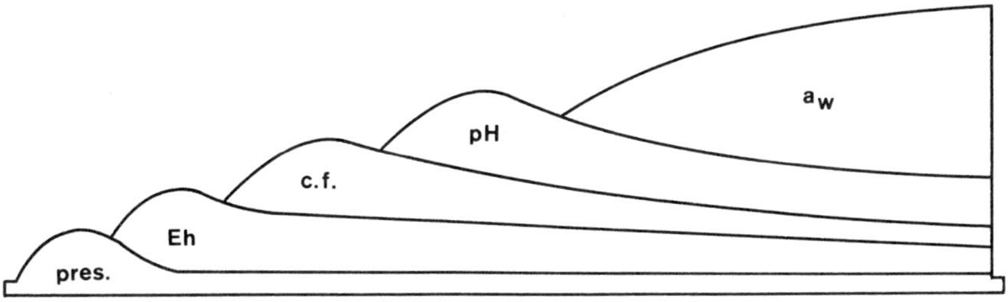

FIG. 13.4 Sequence of hurdles occuring during ripening and storage of fermented sausages (salami). pres. = nitrite, Eh = decrease of redox potential, c.f. = growth of competitive flora, pH = acidification, and a_w = the drying process.

SHELF-STABLE PRODUCTS

By an intelligent use of hurdle technology new products could be developed and traditional foods improved. This applies to SSP as well as IMF. In our laboratories we have studied SSP and IMF based on meat for several years and some of the results obtained are summarized here.

Both product groups are storable without refrigeration, a feature important for developing countries as well as for industrialized countries. In developing countries refrigeration is not readily available. In industrialized countries foods which need no refrigeration save costs by saving energy.

The term SSP was introduced by our laboratories (Leistner, 1977; Leistner and Rödel, 1979) for high moisture meats ($a_w > 0.90$), which may be stored without refrigeration after a mild heat treatment. The definition for SSP (Leistner et al., 1979; Leistner, 1985a) is given in Table 13.1.

SSP meats offer the following advantages:

1. The mild heat treatment improves the sensory and nutritional properties of the product.
2. The lack of refrigeration simplifies distribution and saves energy during storage.
3. For industrialized countries production of SSP is more attractive than IMF, because the required a_w for SSP is not as low, and thus less humectants and/or less drying of the product are necessary.

The SSP concept was experimentally explored first by Leistner and Karan-Djurdjić (1970) with canned liver sausage. Reichert and Stiebing (1977) confirmed that liver sausage processed according to the SSP concept is shelf stable. Today we distinguish between F-SSP, a_w-SSP, and pH-SSP, depending on the hurdle which is most important for the particular product group. Additional hurdles contribute to the microbiological stability in each group

TABLE 13.1 Definition of Shelf-Stable Products (SSP)

Heating:	Mildly heated (70–110°C core temperature) in sealed containers, but sufficient to inactivate all but sporulated bacteria. Recontamination after heat processing is avoided, and therefore only spores of bacilli and clostridia are of concern.
Stabilization:	Growth of surviving spores of bacilli and clostridia is inhibited by sufficient decrease of a_w, pH, and Eh.
Storage:	No refrigeration required.

(Leistner, 1985a). The primary reason for stability of F-SSP is the inactivation or sublethal damage of bacterial spores, for a_w-SSP the reduction of a_w, and for pH-SSP an increased acidity.

Traditional SSP meats (both a_w-SSP and pH-SSP) have been on the market for many years, and newly developed SSP, especially F-SSP, are a major recent achievement of empiric meat technology. All these products are readily accepted by the consumer and many have an unusual stability.

F-SSP

For decades in West Germany, canned meat products have been on the market and enjoyed a good safety record even though they are not fully processed, i.e., they are not heated to F_c 4.0–5.5, but only to F_c 0.6–0.8. These products are called three-quarter-preserved meats (*Dreiviertelkonserven*) and should be stored ≤ 10°C (Leistner, 1979). During the past five years, several West German meat processors have introduced a new line of mildly heated meat products, which are sold in huge quantities by large discount chains without refrigeration. These products are autoclaved sausages in casings, called F-SSP (Leistner, 1985a), since their stability rests primarily on sublethal damage of surviving spores of bacteria.

F-SSP consist of liver, blood, and Bologna-type (*Brühwurst*) sausages (100–500g), filled in artificial PVDC-casings (30–45 mm diameter), impermeable to water vapor and to air, and closed by clips. These sausages are autoclaved for 20–40 min at 103–108°C under stringently controlled counter pressure (1.8–2.0 bar during heating, 2.0–2.2 bar during chilling). Supposedly such autoclaved sausages have a shelf life of 6–8 weeks without refrigeration. However, in big discount chains a turnover within two weeks is usual, and the housewife most likely keeps these products in the refrigerator, as they look like sausages. Therefore, the storage time without refrigeration is in reality much shorter. The autoclaved sausages have been developed by meat processors without much scientific support, to supply discount chains that would like to sell foods without refrigeration, to save energy costs. These F-SSP have not caused botulism or severe spoilage problems during the approximately five years they have been on the market. However, from the scientific viewpoint they may involve risk, and therefore their stability was studied and recommendations were made to prevent spoilage and food-poisoning organisms.

Hechelmann and Leistner (1984) investigated 208 batches of autoclaved sausages in casings, immediately after purchase and after 60 days of storage at 25°C. Of the liver and blood sausages examined 19% proved unstable, and of the Bologna-type sausage (*Brühwurst*) 68% were unstable because bacterial counts increased to > 10^5/g during storage (Leistner, 1985a). The organisms detected were bacilli and clostridia; other bacteria were recovered only in small numbers and close to the clips of the sausages. These clips are not tight against microorganisms, but at low relative humidity (ambient temperature) are more reliable than under refrigeration. About 4.5% of the autoclaved sausages spoiled due to growth of clostridia during storage of the products. If bacilli caused the instability, then multiplication of these bacteria in the sausages quite often ceased after reaching 10^5–10^6/g, without leading to obvious spoilage of the product. The consumption of oxygen by the developing culture of bacilli most likely inhibited further growth.

It was observed by Hechelmann and Leistner (1984) that the a_w of autoclaved sausages definitely influences their stability. If the a_w of the F-SSP is < 0.97 (or preferably 0.96), these products are much more stable. Most manufacturers add slightly more sodium chloride (2.0–2.3%) and/or fat (30–36%) to F-SSP in order to adjust the a_w. Nitrite probably contributes relatively little to the stability of the F-SSP, since not much residual nitrite is present in such products (Hechelmann and Leistner, 1984; Leistner, 1985a). However, nitrite is somewhat more effective in Bologna-type sausage (Leistner et al., 1981b) than in liver and blood sausage (Lücke et al., 1981), and therefore for F-SSP the a_w essential for stability in Bologna-type sausage (*Brühwurst*) can be higher (0.97) than in liver and blood sausage (0.96).

Hechelmann and Leistner (1985) reported that autoclaved sausages are apparently more stable in casings than in cans. The reason for this surprising observation is the nonequilibrium state of a_w, which is likely in cans with headspace. During chilling of the cans after autoclaving some water condensation may occur inside the can, and if drops of water fall back on the surface of the sausage mix, locally the a_w increases and thus growth of clostridia can start in this portion of the product. This is more probable if the a_w of the F-SSP has been adjusted close to the limit. If autoclaved sausages fill casings tightly, a water condensation inside the container cannot occur, and therefore F-SSP in casings are more stable than in cans with a headspace.

Hechelmann and Leistner (1984, 1985) concluded from their investigations that stable and safe autoclaved sausages (F-SSP) can be produced under the following conditions:

1. The count of viable *Bacillaceae* spores in the product should be as low as possible, because if few spores are present, growth of organisms is easily inhibited by hurdles.
2. The sausage mix must be heated to F > 0.4, since this results in sublethal damage of surviving spores.
3. The a_w of autoclaved sausages must be adjusted to < 0.97 (Bologna-type sausage) or < 0.96 (blood and liver sausage) to inhibit the growth of clostridia and bacilli. In Bologna-type sausage (*Brühwurst*) the re-

sidual nitrite contributes somewhat to the inhibition of these organisms, but not in blood and liver sausages.

4. The Eh in the product should be low (use of airtight casings), since a reduced redox potential inhibits the growth of low a_w-tolerant bacilli.
5. The pH should be adjusted < 6.5 in blood sausage; other meat products have pH values below 6.5.
6. Autoclaved sausages should be packed in casings. If cans are used, a headspace should be avoided.

Autoclaved sausages (F-SSP) are a good example of how food processers are often ahead of food science and are more prepared to take risks. However, in the interest of consumer safety, food scientists should understand the principles behind empirical processes, delineate the risks associated with new products, and help to minimize these risks.

a_w-SSP

The term a_w-SSP was chosen for shelf-stable products stabilized mainly by a_w, although other hurdles are also important for their microbial stability (Leistner, 1985a). Traditional meat products of the a_w-SSP type are those common in Italy, are preferred by the consumer, and have been available for decades. Two groups of a_w-SSP meats are distinguishable. One group is represented by Italian Mortadella and the other by German Brühdauerwurst. In Italian Mortadella the reduction of a_w is achieved mainly by the formulation of the sausage and some drying during heating of the product. On the other hand, German Brühdauerwurst acquires a desired a_w primarily by drying of the finished product. The processing and stability of both groups of a_w-SSP meats have been studied in our laboratories.

The formulas and processing condition of Italian Mortadella and German Brühdauerwurst have been described earlier (Leistner et al., 1979; Wirth and Leistner, 1982), and were recently summarized by Leistner (1985a). In both product groups a_w is adjusted empirically. However, in samples taken from the market the a_w values were close to 0.95, the desired a_w for these meat products (Leistner et al., 1979). This was surprising, since none of the manufacturers measured the a_w of their products or recognized the significance of an a_w of 0.95 for optimal stability of Italian Mortadella and German Brühdauerwurst. Due to the a_w adjustment, both product groups may be stored without refrigeration. Since the lipases are inactivated by heat in a_w-SSP meats, they are even more stable than fermented sausages. According to Wirth (1979), fermented sausages can be stored 15 months and German Brühdauerwurst 18 months without sensory deterioration.

Leistner et al. (1980) challenged the microbial stability of Italian Mortadella and of German Brühdauerwurst by inoculating batches with spore-pools of clostridia (including *Clostridium botulinum* type A and B) and bacilli (including a_w-tolerant strains of *Bacillus licheniformis* and *B. subtilis*). The results were summarized by Leistner (1985a). It was concluded that clostridia and bacilli are not of concern in a_w-SSP meats if the a_w is close to or < 0.95. Bacilli, including a_w-tolerant strains, did not multiply in such products during several months of storage at 25°C, in Italian Mortadella at a_w 0.97. Probably the relatively low redox potential in a_w-SSP meats prevents the multiplication of bacilli at a_w at which they grow readily under aerobic conditions in culture media. However, to inhibit clostridia, including *C. botulinum*, in a_w-SSP meats the a_w has to be adjusted to ≤ 0.95.

There is general agreement that spores of bacilli and clostridia may initiate germination at a_w levels appreciably lower than those which will permit vegetative cell growth to occur. Therefore, it was not surprising that the number of bacterial spores decrease in a_w-SSP meats during storage, probably because many spores germinate but the vegetative cells are not able to multiply. Thus, the stability of a_w-SSP can improve during storage. This phenomenon was observed by Leistner and Karan-Djurdjić (1970) and Leistner et al. (1980), and discussed by Leistner et al. (1981a).

Molds may be troublesome for a_w-SSP meats. The surface a_w of Mortadella or Brühdauerwurst corresponds to the a_w of the interior, if the sausages are filled into casings penetrated by water vapor. Therefore, a_w-SSP meats support mold growth on the surface. These undesirable molds can be inhibited by smoke treatment, vacuum packaging, or by dipping the finished product in a 20% potassium sorbate solution (Leistner et al., 1975).

It is concluded that stable and safe meat products of the a_w-SSP type can be produced under following conditions:

1. a_w-SSP meats should be heated to an internal temperature ≥ 75°C, to inactivate vegetative microorganisms.
2. The a_w-SSP meats should be heated in sealed containers (preferably casings), which avoid recontamination after processing.
3. The a_w of a_w-SSP meats must be adjusted to ≤ 0.95. Thus, a lower a_w is more essential than in F-SSP. Due to the milder heat treatment of a_w-SSP, bacterial spores are damaged less than in F-SSP.
4. The Eh of the product should be low, because a reduced redox potential contributes to the growth inhibition of a_w-tolerant bacilli.
5. The growth of molds on the surface of a_w-SSP meats must be avoided by smoke or potassium sorbate treatment, or by vacuum packaging of the product.

pH-SSP

It is well known that pasteurized fruit and vegetable preserves with pH < 4.5 are bacteriologically stable, in spite of only a mild heat treatment. In such products vegetative microorganisms are inactivated by heat, and the multiplication of surviving bacilli and clostridia is inhibited by the low pH. Such products could be called pH-SSP, because their stability is due mainly to the low pH (Leistner, 1985a). However, the a_w hurdle must contribute to the stability of these preserves if pH-tolerant clostridia (*C. pasteurianum, C. butyricum,* etc.) are of concern. For instance, Jakobsen and Jensen (1975) observed that pasteurized pears proved stable at pH 4.5 and $a_w < 0.97$ or at pH 4.0 and a_w 0.97–0.98. However, if the a_w of this product was in the range 0.98–0.99 even pH 3.8 was not low enough for microbial stability.

Bacterial spores are able to germinate at lower pH levels than vegetative bacilli and clostridia are able to multiply (Gould, 1969). Therefore, also in pH-SSP, as in a_w-SSP, the number of spores tends to decrease during storage. On the other hand, while the heat resistance of bacteria and their spores is enhanced with decreasing a_w, it is diminished with decreasing pH. Thus pH-SSP need less heat treatment for the inactivation of microorganisms than do a_w-SSP.

Meat products with pH < 4.5 are unpalatable and hence are not marketed. Nevertheless, the stability of some meats is based primarily on the pH-hurdle. This is true for short-ripened fermented sausage, in which lactic acid bacteria or the addition of glucono-delta-lactone lower the pH to 5.2–4.8. The stability of long-ripened Italian salami, with a final pH of 5.9–6.0, is due mainly to the a_w hurdle (Leistner, 1986b). However, even quick-ripened fermented sausage does not belong to the group of pH-SSP, since this product is not heated.

True pH-SSP meats, in which the vegetative microorganisms are inactivated by heat and the stability is primarily based on the pH hurdle, are brawns and Gelderse Rookworst. Brawns are jelly sausages with pH 5.0–4.5 adjusted by the addition of acetic acid. Such products can be stored without refrigeration, if recontamination after heat processing is avoided (Rödel et al., 1976). Gelderse Rookworst is a Bologna-type sausage (*Brühwurst*) in which the pH is adjusted to 5.4–5.6 by the addition of 0.5% glucono-delta-lactone. This product is microbiologically stable for several weeks without refrigeration, if vacuum packaged in a pouch and pasteurized at 80°C for about 1 hr. This treatment inactivates vegetative organisms in and on these sausages. Bacterial spores are apparently not of much concern in this product, since their population decreases during the heating process and the surviving spores are inhibited by the low pH and other hurdles. Gelderse Rookworst is common in the Netherlands and is exported in large quantities to Britain. Rookworst (pH 5.4–5.6) is acceptable from the sensory point of view. However, the pH of this product should not be <

5.4, because then the sour taste becomes irritating, especially if the product is eaten warm (Leistner, 1985a).

Apparently stable and safe Gelderse Rookworst as pH-SSP can be produced under the following conditions:

1. The pH of the product should be adjusted to 5.4 by the addition of 0.5% glucono-delta-lactone. This reduced pH should inhibit the multiplication of clostridia and bacilli, if their spores are present only in small numbers.
2. The product should be vacuum packaged and heated to an internal temperature of 80°C to inactivate vegetative organisms in the interior and on the surface of the sausages and to avoid recontamination. Furthermore, vacuum packaging will secure a relatively low redox potential and contribute to the inhibition of pH-tolerant bacilli.

The microbiological stability of pH-SSP meats stored without refrigeration could probably be improved if, besides the pH-hurdle, an a_w-hurdle is introduced. The sensory quality of pH-SSP meats could benefit from an a_w- hurdle since the pH must only be decreased to a moderate level.

INTERMEDIATE MOISTURE FOODS

Intermediate moisture foods are in the a_w-range 0.90–0.60, and are often stabilized by additional hurdles, such as heating, preservatives, pH, and Eh (Leistner and Rödel, 1976; Leistner et al., 1981a). These foods are storable without refrigeration.

Today there are newly developed and traditional IMF on the market. Only a few newly developed IMF are based on meat (Brimelow, 1985). The only novel intermediate moisture meats that have been very successful are pet foods. There are several reasons why novel IM meats for humans did not make the expected breakthrough: newly developed intermediate moisture meats are often not sufficiently palatable, are expensive, contain too many additives (chemical overloading of the food), and pose legal problems with respect to the need to obtain approval of new additives (Leistner and Rödel, 1976; Leistner et al., 1981a).

However, there are many traditional IMF meat products, which are highly acceptable in different parts of the world (Table 13.2). In Europe meat products in the a_w-range 0.90–0.60 are not very common. However, if traditional meats such as raw ham, fermented sausage, Brühdauerwurst, Speckwurst, and dried beef (*Bündnerfleisch*) are dried sufficiently, they acquire an $a_w < 0.90$

TABLE 13.2 Some Traditional Intermediate Moisture Foods (IMF) Based on Meat, Which Are Stable Without Refrigeration

Raw ham (Europe), fermented sausage (Europe), Brühdauerwurst and Speckwurst (Germany), Bündnerfleisch (Switzerland).
Pastirma (Turkey, Egypt), K'lich (North Africa), Khundi (West Africa), Quanta (East Africa), Biltong (South Africa).
Beef jerky (North America), Pemmican (North America), Carne de Sol (South America), Charque (Brasil).
Lup Cheong (China), Tsusou Gan (China), Njorsou Gan (China), Sou Song (China), Dendeng Giling (Indonesia).

(Leistner et al., 1981a). Traditional meat products in the intermediate moisture range are frequently found in countries where the climate is hot and refrigeration is expensive or unavailable. In developing countries, the shortage of meat is due not only to a scarcity of animals, but also often to spoilage of this precious food. Suitable technologies for the processing of meat are therefore needed in such countries (Leistner, 1983). The introduction of European-style meat products to developing countries presents difficulties if they require elaborate equipment for production and costly refrigeration for storage (Savić, 1981a, b; 1985). Therefore, IM meats, which are easy to prepare, storable without refrigeration, and do not need expensive packaging, may benefit developing countries.

For industrialized countries traditional IMF found in developing countries are of interest. If in cooperation with scientists from developing countries the principles behind these meats are studied, their processing and shelf life could be improved without impairment of their sensory and nutritive properties. The improved formulas should be made widely available, because they could be of benefit in many parts of the world, if they can be adapted to local tastes. Furthermore, from traditional meat-based IMF, new and promising ideas for product development in industrialized countries could emerge, since such meats are based upon centuries-old trial-and-error processes.

Our laboratory studied the formulation, processing, and stability of several traditional meat-based IMF. Results obtained with Chinese sausages, Chinese dried meats, South African biltong, and Turkish pastirma will be summarized here.

Chinese Sausage

The earliest known reference to Chinese sausage was made at the time of the Southern and Northern Dynasties (A.D. 420–589), and at that time, minced goat or lamb meat was the main ingredient, mixed with spring onion, salt, bean sauce, ginger, and pepper (Ho and Koh, 1984). Today various formulas for traditional Chinese sausage (called *Lup Cheong* in Cantonese or *La Zang* in Mandarin) are found in different provinces of China (Lo et al., 1980); the Cantonese variety is the more famous. The technology employed in the processing of traditional Chinese sausage is similar throughout all of China (Lo et al., 1980) and is also used by other Chinese communities of Asia, e.g., in Singapore (Ho and Koh, 1984) and Malaysia (Savić, 1985). Traditional Chinese sausage is an IM meat product, and therefore may be stored without refrigeration. There are also emulsion-type sausages (*Quan Zang*) known in China (Lo et al., 1980), which are perishable high-moisture meats, and thus will not be discussed here.

Traditional Chinese sausage (Lup Cheong) is a raw nonfermented product, which most Chinese know and cherish during Chinese New Year celebrations. Lup Cheong is made from coarsely ground pork (preferably ham) and pork fat, mixed with sugar, salt, soy sauce, Chinese wine, saltpeter (potassium nitrate), 5-spice-powder (anise, cinnamon, clove, fennel, watchau), and monosodium glutamate. Sometimes up to 25% water is added, so that after drying the sausage will have the desired wrinkled appearance (Ho and Koh, 1984). The batter is filled into small intestine hog casings, which are tied at about 15-cm intervals. The filled casings are punctured thoroughly to enable the escape of entrapped air and water vapor during the drying process. Lup Cheong is dried for 1–2 days at 45–50°C over charcoal, and thereafter kept for 2–3 days at room temperature for equilibration of moisture. Traditional Chinese sausage can be stored without refrigeration for 1–3 months, if mold growth is avoided. Lup Cheong should look reddish-brown and fat-speckled. Although processed and stored as a raw sausage, the product is always warmed before consumption and is eaten hot. It is often sliced and steamed with rice, noodles or various vegetables. Due to the intense aroma of Lup Cheong, a few slices are sufficient to flavor an entire dish.

The microbiological stability of traditional Chinese sausage is due mainly to the rapid reduction of a_w (Leistner and Dresel, 1986), and this is aided by the addition of salt (2.8–3.5%) and sugar (1–10%), the thin caliber of the casing (26–28 mm), and a high ripening temperature (45–50°C) at low relative humidity (65–75%). On the other hand, the pH hurdle is not important for stability, because the pH of the raw sausage is relatively high (5.7–5.9) and the number of lactic acid bacteria low ($< 10^6$/g).

We have investigated the traditional Chinese sausage by visiting production plants in Taiwan and Singapore, and by importing 24 samples (each consisting of several sausages) of Lup Cheong from different countries, including the Peoples Republic of China and Taiwan. Furthermore, we reproduced this IMF in our laboratories and challenged the stability of this product with food-poisoning and spoilage organisms (Leistner et al., 1984; Leistner and Dresel, 1986).

In our 24 imported samples the usual a_w and pH were 0.75 and 5.9, respectively. However, we observed considerable variations in physical properties among the samples. Variation was even greater for the chemical characteristics. On average the content of sodium chloride, sodium nitrite, and potassium nitrate in the investigated products were 4.5%, 30 mg/kg, and 500 mg/kg, respectively. Apparently the composition of Lup Cheong made by different manufacturers varies widely, especially with respect to an unnecessary overdosing of salpeter (nitrate), which occurs frequently. Ho and Koh (1984) observed similar pH values and levels of salt, nitrite, and nitrate in Lup Cheong samples obtained in Singapore; however, they reported much lower a_w-values (0.6–0.7) than we found. This is probably due to the characteristic high sugar content (15–20%) in Chinese sausages from Singapore.

In spite of the wide variation in the physical and chemical characteristics, the microbiological properties of our 24 Lup Cheong samples were favorable. In general, the total bacterial count (mainly *Micrococcaceae*) and the number of lactic acid bacteria were in the range 10^5–10^6 and 10^4–10^5 per gram, respectively. *Enterobacteriaceae* and *Staphylococcus aureus* were virtually absent in these samples. Therefore, the bacteriological status of Lup Cheong is good and provides no risks. If traditional Chinese sausage spoils, it is due to an insufficient drying and is caused by a vigorous growth of Gram-positive bacteria, especially lactic acid bacteria, which lead to a sour product. Even though Lup Cheong is a raw sausage, a fermentation is undesirable, because Chinese consumers object very much to a sour taste in sausages. Molds, which can grow on the surface, are another spoilage problem if Lup Cheong is not vacuum packaged. Of the food-poisoning bacteria only *S. aureus* is of concern for traditional Chinese sausage (Leistner et al., 1984; Leistner and Dresel, 1986). However, this risk can be avoided by properly drying the product.

We concluded from our investigations that under the following conditions traditional Chinese sausage (Lup Cheong) is stable and safe (Leistner and Dresel, 1986):

1. The number of Gram-positive bacteria in the raw material should be moderate.
2. The a_w of the sausage must be decreased within 12 hr to < 0.92, and within 36 hr to < 0.90. This is achieved by drying the product for 36 hr at 48°C and 65% relative humidity.
3. If the drying is not done over charcoal, then the product should be lightly smoked for several hours at 48°C and 65% relative humidity.
4. Subsequently the product should go through an equalization time of 3 days at 20–25°C and 75% relative humidity, until the a_w is < 0.80; the sausage is then ready for packaging.

5. Vacuum packaging of the final product is recommended since it improves the flavor of the sausage during storage and inhibits mold growth.

Leistner and Dresel (1986) developed two standardized formulas for Chinese sausage by using the same technology but different ingredients for the products. Ho and Koh (1984) suggested a standardized formula for Lup Cheong consumed in Singapore. Traditional Chinese sausage can generally be recommended for the preservation of meat in developing countries, because it is simple to prepare, stable, and safe. The principle used in the preservation of Lup Cheong, i.e., the quick decrease of a_w, is also of interest to industrialized countries, since this product demonstrates that raw sausage may also be successfully processed at 48°C and 65–75% relative humidity.

Chinese Dried Meats

In China, dried meat products (*Sou Gan*) have been known from time immemorial, and they are highly esteemed by Chinese communities throughout Asia for their flavor and nutritive value. Sou Gan is relatively simple to prepare, easy to store (no refrigeration) and to transport (light weight due to reduced water content). The total consumption of these foods is very large and their popularity is still increasing (Chen, 1983). Most Chinese dried meats are IMF (a_w 0.90–0.60), and some are in the low-moisture range ($a_w < 0.6$). At least 30 different products are known, depending on the species of meat, the type of technology and kind of spices used.

Our laboratories studied the physicochemical and microbiological properties of Chinese dried meats (Shin et al., 1982; Shin and Leistner, 1983; Shin, 1984). Furthermore, we attempted to reproduce and standardize Sou Gan (Shin, 1984) after visiting several manufacturers in Taiwan and Singapore.

In China empirical technologies have been employed for making dried meats for decades or even centuries. Three different processes are distinguishable, which have been described briefly by Huang (1974) as well as Ho and Koh (1984) and were outlined by Leistner et al. (1984) and Leistner (1985a):

Process I. Foods made according to this process are dried pork slices (also called dried sweet meat, barbequed dried pork) or dried beef slices (also called dried beef squares). In this process, lean meat (preferably from hams or loins) is cut along the grain into paper-thin (0.2 cm) slices, which are mixed with sugar, salt, soy sauce, monosodium glutamate, and spices (anise, cinnamon, clove, fennel, watchau). The pickle is held for 24 hr at room-temperature or preferably, for 36 hr at 4°C. Afterwards the meat slices are placed side by side and slightly overlapping on oiled bamboo baskets or wire racks and dried for several hours at 50–60°C, until they reach approximately 50% of their original weight or 35 ± 5% moisture. The meat slices are removed from the trays and cut into squares, which are grilled over charcoal for a few minutes at 130–180°C, and should be finally dried at room temperature to $a_w < 0.69$. According to Ho and Koh (1984) about 5% maltose can keep the dried pork slices wet and bright in color. Since the production of dried pork slices is time-consuming and labor intensive, modified processes were suggested by Lin et al. (1981) as well as Ockerman and Kuo (1982). Lin et al. (1981) recommended the injection of pickle and tumbling of the meat to replace the soaking and hand-spreading of the slices. Ockerman and Kuo (1982) observed that an addition of nitrate/nitrite combined with vacuum packaging could be used to retard oxidative rancidity of dried pork slices. The vacuum packaging of products also improves tenderness, since it retards moisture loss during storage (Ockerman and Kuo, 1982). A product containing 30% sugar and 2.5% salt, compared with dried pork slices made with less sugar and salt, had the highest panel scores when evaluated by a group of Oriental panelists (Kuo and Ockerman, 1985). In a textbook from mainland China in which the other processes for dried meats are described (Lo et al., 1980), Process I is not mentioned, probably because the health risks of this process are higher than in Processes II and III.

Process II. Meats made according to this process are beef, pork, or chicken in pieces, cubes, or strips. For flavoring, besides 5-spice-powder, other spices are sometimes added, such as curry, chilies, cayenne pepper, ginger, fruit juice, and wine. Many varieties are found in meats made with Process II; however, Lo et al. (1980) pointed out that these products are mainly based on beef. In Process II the meat, after removal of the fat, is cut into fairly large chunks and is cooked with 10% water over medium heat until the meat is tender. Then the meat is cooled, drained (liquid retained) and cut into pieces, cubes or strips. To the liquid, sugar, salt, soy sauce, monosodium glutamate, and spices are added, and the mixture is again heated. The meat is placed in a pan with the liquid and stirred over low heat until the mixture is almost dry. Finally, the meat is spread flat on wire racks or plates and dried for several hours at 50–60°C or until it has lost about 50% of the original weight. The a_w achieved should be < 0.69. According to Lo et al. (1980) such dried meats can be kept in glass jars or metal boxes (to exclude oxygen) for 3–5 months. These authors recommend that the meat be wrapped in paper and heated (inactivation of microorganisms) before storing the product in clean containers.

Process III. Pork processed in this way is called pork floss, shredded pork, meat flakes, or Sou Song in Chinese. In this process lean pork is cut along the grain in pieces and cooked with equal amounts of water until very soft. The meat is drained and the liquid is evaporated to 10% of its volume. To this broth, sugar, salt, soy sauce, wine, monosodium glutamate, fennel, ginger, or other spices are added. The cooked meat is mashed, i.e., separated into fibers, and added to the liquid. The mixture is held at low heat until all the liquid has evaporated. Finally, the flakes are a cottonlike mass and are stirred for several hours at 80–90°C until very dry ($a_w < 0.6$). To make the flakes crispy, about 20% of hot vegetable oil is added, and the product is further stirred over low heat until dry and golden brown ($a_w < 0.4$). Similar products are made from beef, chicken, or fish using the same technology. Singapore imports increasing amounts of pork floss from China, and the imported product is roasted again to make it crispy (Ho and Koh, 1984). Pork floss tends to absorb moisture and, therefore, must be stored dry. According to Lo et al. (1980) pork floss, while it is still warm, should be placed into clean glass containers, and thus can be stored for six months without refrigeration.

Chinese dried meats are prepared preferably from hot-boned meat (but chilled meat is suitable too), requiring little energy and only simple equipment for processing. Such meats are storable without refrigeration. However, if not sufficiently dried, the products are spoiled by molds. For dry products, rancidity is the limiting shelf life factor which can be prolonged by vacuum packaging of the products. Our laboratory has published the formulas and processing techniques for six traditional dried meats of China (Shin et al., 1984). Improvements in processing and packaging of dried shredded and sliced pork were suggested by Lin et al. (1980).

To study the physicochemical and microbiological properties of Chinese dried meats we imported 42 commercial samples from Taiwan, Singapore, and Hong Kong. At the time of arrival, the a_w of these samples was in the range of 0.78–0.20, and the pH-range was 6.2–5.3. We challenged the stability by inoculating portions of the samples with pools of xerotolerant molds of the *Aspergillus glaucus* group, and stored them for 3 months at 25°C. Of the 42 samples tested, 35 (83%) proved stable. These stable meats had an $a_w \leq 0.69$, which therefore could be regarded as the critical a_w for Chinese dried meats, which are stored unpackaged and without refrigeration (Shin et al., 1982; Shin, 1984). However, Ho and Koh (1984) are of the opinion that the a_w of these products should be decreased to < 0.61 in order to avoid mold growth.

From our investigation of the 35 stable samples, it was concluded that such meats, if prepared using Processes I and II, range in a_w from 0.55–0.69 in pH from 5.8–6.0, and contain 20–35% sugar, 3–5% NaCl, and 10–15% moisture. If Process III was used, we observed an a_w-range of 0.20–0.59 and a moisture content of 2–12%. Therefore, these foods are in the low-moisture range. The same was true for some products prepared with Process I and II, if they were dried more than required for microbial stability (Leistner et al., 1984; Leistner, 1985a).

The microbial stability of Chinese dried meats depends primarily on the a_w and F (heat-treatment) hurdles, while the pH hurdle is not important. Contrary to African biltong (discussed below), which depends on a_w and pH hurdles only, in Chinese dried meats only few microorganisms are present. This is due to the heating step. From stable imported products we rarely recovered more than 10^4 microorganisms/g. Most samples were in the range 10^2–10^3/g, which is impressive for uncanned meat products. Shin (1984) conducted inoculation studies using Chinese dried meats prepared in the laboratory and observed that salmonellae, pathogenic staphylococci, yeasts, and molds are eliminated during processing by the heat applied. Enterococci may survive, but die during storage of the products. Spores of bacilli and clostridia also decrease during processing and storage. However, most organisms encountered in imported Chinese meats were bacilli.

Recontamination of Chinese dried meats could easily occur after processing. Therefore Shin (1984) studied the survival of microorganisms inoculated onto imported and reproduced meats. He observed that during storage of stable products the number of organisms decreased, especially in meats close to the critical a_w 0.69. Staphylococci and yeasts decreased rapidly, salmonellae more slowly, and enterococci and bacilli survived best (Shin et al., 1982). Thus, Chinese dried meats are indeed safe products, because the heat treatment eliminates most organisms present in the raw material, and survivors as well as organisms which recontaminate the product are inhibited and are inactivated by the a_w control.

From our experimental work it may be concluded that stable and safe Chinese dried meats can be produced under the following conditions:

1. The microbial load of the raw material should be low. This is possible in quickly processed hot-boned meat or in properly chilled meat stored for a short time.
2. The quality of the meat required depends on the food product. Best cuts are necessary for Process I, and lowest quality cuts for Process III. Meat used in every process should contain very little fat.
3. For the stability of Chinese dried meats it is essential to reduce the a_w quickly and to heat the meat rapidly and completely. Process I is more hazardous than Processes II and III, because in Process I the meat is kept longer before heating.
4. For unpackaged, nonrefrigerated products an $a_w \leq 0.69$ is required for inhibition of microorganisms, including molds. For packaged products stored under refrigeration, the a_w may be higher.
5. Finished products should not be exposed to air and light. In closed jars or boxes as well as in vacuum packages the dried meats are protected from microbial recontamination, invasion of insects, absorption of water, and rapid onset of rancidity.

Chinese dried meats are easy to prepare and can be processed and stored without refrigeration. Therefore, they could be produced readily in developing countries, especially if Processes II and III are used. The traditional humectants employed, i.e., sugar and salt, are cheap and generally available. In addition, sugar has beneficial effects on the texture and plasticity of dried meats. However, in the Western world the sweet taste of Chinese dried meats, which is an advantage in China, is not readily accepted by many consumers. Therefore, alternative humectants should be tested. From the microbiological point of view Chinese dried meats can be recommended without reservations.

South African Biltong

Biltong is a well-known dried meat product originating from beef or game meat, and is regarded as a delicacy in South Africa. Its use dates back to the early settlers of the Cape.

Most muscles in the carcass may be used for Biltong, but the largest muscles are most suitable. The meat is cut with the grain into long strips and placed in brine, or frequently salted in the dry state. Common salt is the principal curing agent used, although other ingredients such as sugar, vinegar, pepper, coriander, or other spices are included in some pickling mixtures. Nitrate and nitrite as well as other preservatives (boric acid, Pimaricin, or potassium sorbate) are sometimes added. The addition of 0.1% potassium sorbate to the raw meat has been recommended and permitted in South Africa. Biltong is left in the cure for several hours, then dipped into hot water with vinegar and hung for 1–2 wk in the air to dry. It is sold in sticks, slices, and in the ground or pulverized form. These meats may be stored for months without refrigeration. In general, Biltong is not heated during processing or before consumption; thus it is eaten raw (Leistner, 1985a).

Van den Heever (1970) investigated 60 commercial Biltong samples and observed a mean of a_w 0.74, 25% moisture, 6.6% NaCl, and pH 5.9. The average data reported by van der Riet (1976a) of 20 Biltong samples were a_w 0.70, 23% moisture, 5.6% NaCl, and pH 5.7. In both studies a wide variation of these parameters was encountered, since specific standards for the processing of Biltong do not exist. Our laboratories studied 25 Biltong samples imported from South Africa (Shin and Leistner, 1983; Shin, 1984). We determined a_w of 0.36–0.93 (most samples were in the range 0.65–0.85) and a pH range 4.8–5.8 (the pH of most samples was about 5.5). Eight (32%) of our 25 samples were microbiologically unstable, since they were spoiled by molds, yeasts, and *Micrococcaceae* during transport or storage.

In our study, stable Biltong contained 5–10% NaCl (average about 7%), little sugar and nitrite, but 10–860 mg/kg nitrate. However, the addition of nitrate does not ensure stability, since spoiled samples often contained much residual nitrate. In our investigation, Biltong was stable with the hurdle combination of $a_w \leq 0.77$ and $pH \leq 5.5$, because such samples did not become moldy spontaneously. According to van der Riet (1976b, 1982) spoilage of Biltong is caused predominantly by molds of the xerotolerant *Aspergillus glaucus* group, and less frequently by yeasts or bacteria. Van der Riet (1976b, 1982) pointed out that Biltong with ≤ 24% moisture (presumed equivalent to a_w 0.68) is microbiologically stable. Rancidity is the limiting factor for shelf life of microbiologically stable Biltong. However, some consumers prefer Biltong with a slightly rancid flavor (van Wyk, 1980).

Since Biltong is an unheated meat product, it may harbor many microorganisms. In products taken from the market, total counts of $> 10^7$/g and $> 10^6$/g molds and yeasts have been reported (van den Heever, 1970). According to Taylor (1976) during the processing of Biltong the predominantly Gram-negative, salt-sensitive flora initially present is replaced by Gram-positive, salt-tolerant apathogenic staphylococci and micrococci, which are the dominant microflora of the finished product. Taylor (1976) attributed these changes in the microflora to the increasing NaCl concentration of the product, which reduces a_w. In our study (Shin, 1984) stable Biltong samples had total counts of 10^3–10^6/g and contained predominantly lactobacilli and *Micrococcaceae,* with less yeasts and molds. A few samples had total counts $< 10^2$/g, probably because of the use of preservatives.

Aflatoxins are unlikely to be found in Biltong with $a_w < 0.80$, even though *Aspergillus flavus* quite frequently occurs on this product (van der Riet, 1976a, 1982). *Staphylococcus aureus* and its enterotoxins are apparently not of much concern in Biltong with low a_w (Bokkenheuser, 1963; van den Heever, 1970; Shin, 1984). However, salmonellae are troublesome. The recovery of *Salmonella* spp. from Biltong has been reported by several authors (Bokkenheuser, 1963; van den Heever, 1970; Prior and Badenhorst, 1974), and salmonellae were detected in 7 (16%) of 45 samples investigated by Prior and Badenhorst (1974). It was observed by Bokkenheuser (1963) as well as by van den Heever (1965) that salmonellae survive for a long time in Biltong, especially if they were present in large numbers in the muscle of diseased animals. Biltong with such endogenous infection has caused salmonellosis in humans (Bokkenheuser, 1963; Botes, 1966). Therefore, care in the selection of meat as well as good hygiene is required in the processing of Biltong (van den Heever, 1965, 1970; Prior and Badenhorst, 1974; van der Riet, 1982). The introduction of Biltong for general use in developing countries is obviously hampered by these requirements (Leistner, 1985a). However, Biltong with minimized risk can be produced using the following guidelines:

1. The meat used must be free of salmonellae and other organisms which survive the process.
2. The addition of 0.1% potassium sorbate to the raw meat should reduce the hygienic risks by inhibition of *Enterobacteriaceae* and molds.
3. The a_w of the product should be decreased as rapidly as possible to < 0.80 by salt and drying.
4. Good hygiene during processing is essential, since the product is eaten in the raw state.

Turkish Pastirma

Pastirma is a meat product made of salted and dried beef, which is highly esteemed in Turkey and Egypt as well as in other Moslem countries, and even in some parts of the Soviet Union (Armenia).

In Turkey Pastirma is preferably produced from September to November, since during this season flies are not prevalent, the air temperature is not so high (as in the summer), and the relative humidity is moderate due to scanty rainfall. Most famous for Pastirma production in Turkey is the city of Kayseri in Anatolia, where every year about 15,000 beef cattle are slaughtered, and more than 500,000 kg Pastirma are produced by 40 manufacturers. From 80 kg of lean beef about 50 kg of Pastirma are obtained. The finished product contains about 30–35% moisture and 5% salt and can be stored at room temperature for 9 months (Berkmen, 1960).

The meat used for Pastirma originates from 5–6 year-old beef cattle and is taken from the hindquarters (e.g., *Musculus semitendinosus*) 6–12 hr after slaughtering. The meat is cut into 50–60-cm long strips, which should have a diameter of not more than 5 cm. These strips are rubbed and covered with salt which contains 0.02% potassium nitrate. Several incisions are made in the meat to facilitate salt penetration. The salted meat strips are arranged in piles about 1 m high and kept for one day at room temperature. They are turned over, salted again, and stored in piles for another day. Thereafter, the meat strips are washed and dried in the air, for 2–3 days in summer and for 15–20 days in the winter. After drying the strips are piled up again to a height of 30 cm and pressed with heavy weights for 12 hr. They are dried again for 2–3 days and once again pressed for 12 hr. Finally the meat is dried for 5–10 days in air. Thus, the production of Pastirma requires several weeks. However, not much energy is required since most of the salting and drying is done at room temperature.

After the salting and drying process, called ripening (*yetirme*) in Turkey, the entire surface of the meat is covered with a 3–5-mm thick layer of a paste called cemen. This cemen consists of 35% freshly ground garlic, 20% helba (ground *Trigonella foenum graecum*), 6% hot red paprika, 1% kammon, 1% mustard, and 37% water. In this paste the helba is used as a binder; the other ingredients are spices, but garlic is the most important. The meat strips covered with cemen are stored in piles for 1 day, and thereafter are dried for 5–12 days in a room with good air ventilation before the Pastirma is ready for distribution.

Berkmen (1960) studied the survival of salmonellae, anthrax bacilli, pathogenic clostridia, rinderpest virus, and tapeworm larvae in Turkish Pastirma, and concluded that this process is an outstanding preservation method for meat, because the organisms investigated hardly survived the manufacturing process and were virtually absent in the finished product. Krause et al. (1972) investigated 19 samples of Turkish Pastirma and recovered predominantly micrococci and lactobacilli, but rarely *Enterobacteriaceae*. On the other hand, Genigeorgis and Lindroth (1984) recovered salmonellae from Basturma, a dried beef manufactured in the U.S. with a process similar to that used in Armenia. The reasons for the survival of salmonellae in this product are probably due to modifications of the recipe (less salt and garlic), in order to meet consumer preference in the U.S. In view of the *Salmonella* risk, Genigeorgis and Lindroth (1984) suggested that the modified Basturma should be heated to 52°C for 6 hr. However, this process might change the typical sensory properties of Pastirma.

Our laboratories investigated 16 samples of Pastirma produced in Turkey or by Turkish manufacturers in West Germany (El-Khateib et al., 1986a). As survey averages we observed 6.5% NaCl, 12 mg/kg sodium nitrite, 400 mg/kg potassium nitrate, pH 5.5, and a_w 0.88 in these samples. There were considerable variations in these parameters among the samples. The nitrate content was quite often unnecessarily high. The total microbial counts in the paste and the meat were in general 10^7/g and 10^6/g, respectively. Predominantly lactic acid bacteria were recovered, whereas *Enterobacteriaceae* and molds were absent in the paste as well as in the meat. Lactic acid bacteria in such high numbers probably decreased the pH, and thus contributed to the preservation of Pastirma.

Surprisingly, mold growth on the surface of Pastirma is seldom found. El-Khateib et al. (1984; 1986b) studied this phenomenon by challenging Pastirma, produced with paste containing different ingredients and garlic in various amounts, with Aspergilli and Penicillia frequently occurring on meats. Since about 75% of the Penicillia found on meats are potentially toxigenic (Leistner, 1984), it is essential that mold growth be inhibited on Pastirma. We observed that of the usual ingredients in the paste of Pastirma, only garlic distinctively inhibits molds. This inhibition diminishes with storage time, because the fungistatic substances in garlic are volatile. However, since the surface a_w of Pastirma usually decreases during storage, this loss of inhibition is compensated. In general, a paste prepared with 35% fresh garlic will insure a mold-free product for months, even if the Pastirma is stored in the summer at ambient temperature. Thus, the garlic in the paste of Pastirma improves the hygienic properties of this product, since other raw hams easily become moldy and may contain mycotoxins (Leistner, 1984).

The microbiological stability of Turkish Pastirma is superior to Biltong. The stability of Biltong is apparently based only on a_w, with little contribution of the pH hurdle, whereas in traditional Pastirma several hurdles are inherent. In the meat portion of Pastirma the a_w and pH hurdles are effective, and the competitive flora (lactic acid bacteria) probably contribute to the inhibition of *Enterobacteriaceae,* including *Salmonella* spp. In the paste of Pastirma, besides a_w, pH, and competitive flora, preservatives (garlic) are also an effective hurdle, resulting in an inhibition of undesirable microorganisms, including toxigenic molds. Since garlic penetrates into the meat portion of Pastirma it also might inactivate *Enterobacteriaceae.* El-Khateib et al. (1984) demonstrated that garlic effectively inhibits salmonellae. Thus, Turkish Pastirma is a good example of a traditional food based on hurdle technology.

From the work of Berkmen (1960) and our investigations it may be concluded that safe and stable Turkish Pastirma can be produced under the following conditions:

1. Lean meat from the hindquarter of healthy beef cattle should be used.
2. The meat should be cut into strips about 5 cm in diameter.
3. The a_w of the meat must be decreased quickly by addition of sufficient salt and by pressing the meat to reduce moisture. The salt content of the finished product should be 4.5–6.0% and the a_w 0.85–0.90.
4. Growth of lactic acid bacteria (up to 10^7/g) in the paste and the meat is desirable for pH reduction, but for sensory reasons the pH should not be < 5.5.
5. After drying, the meat must be covered with a paste containing 35% garlic.

CONCLUSION

This contribution describes some examples of SSP and IMF meat products that are more or less effectively stabilized by hurdle technology. There are additional traditional foods that should be studied in order to improve their microbial stability, and at the same time suggest ideas for the development of new foods. New products should be developed with due respect for the past.

REFERENCES

Berkmen, L. 1960. Über die Haltbarkeit von Krankheitserregern in einem spezifisch türkischen Fleischerzeugnis. *Fleischwirtschaft* 40: 926.

Bokkenheuser, V. 1963. Hygienic evaluation of biltong. *S. A. Med. J.* 37: 619.

Botes, H. J. W. 1966. Biltong-induced, Salm. enteritidis var. typhimurium food poisoning—a case report. *J. S. Afr. Vet. Med. Assoc.* 37: 173.

Brimelow, C. J. B. 1985. A pragmatic approach to the development of new Intermediate Moisture Foods. In *Properties of Water in Foods*. Simatos, D. and Multon, J. L. (Ed.), p. 405. Martinus Nijhoff Publishers, Dordrecht, The Netherlands.

Chen, M. T. 1983. *Meat Science and Technology*. Yae-Sham Book Publication Co., Taipei, Taiwan, R.O.C.

El-Khateib, T., Schmidt, U., and Leistner, L. 1984. Hemmung von Salmonellen und unerwünschten Schimmelpilzen durch Knoblauch bei ägyptischen Fleischerzeugnissen. *Jahresbericht Bundesanstalt Fleischforschung, Kulmbach* C 26.

El-Khateib, T., Schmidt, U., and Leistner, L. 1986a. Mikrobiologische Stabilität von türkischer Pastirma. *Mitteilungsblatt Bundesanst. Fleischforsch. Kulmbach,* No. 94: 7198.

El-Khateib, T., Schmidt, U., and Leistner, L. 1986b. Hemmung von Schimmelpilzen auf Pastirma. *Mitteilungsblatt Bundesanst. Fleischforsch. Kulmbach* No. 94: 7205.

Fox, M. and Loncin, M. 1982. Investigations into the microbiological stability of water-rich foods processed by a combination of methods. *Lebensm.-Wiss. u.-Technol.* 15: 321.

Genigeorgis, C. and Lindroth, S. 1984. The safety of Basturma, an armenian-type dried beef with respect to Salmonella. *Proceedings* 30th European Meeting of Meat Research Worker, Sept. 9–14, 1984, Bristol, United Kingdom, p. 217.

Gould, G. W. 1969. Germination. In *The Bacterial Spore*. Gould, G. W. and Hurst, A. (Ed.), p. 397. Academic Press, London.

Hammer, G. F. and Wirth, F. 1984. Wasseraktivitäts-(a_w-)Verminderung bei Leberwurst. *Mitteilungsblatt Bundesanst. Fleischforsch. Kulmbach* No. 84: 5890.

Hechelmann, H. and Leistner, L. 1984. Mikrobiologische Stabilität autoklavierter Darmware. *Mitteilungsblatt Bundesanst. Fleischforsch. Kulmbach* No. 84: 5894.

Hechelmann, H. and Leistner, L. 1985. Ungleichmäßiger a_w-Wert als Ursache für mangelhafte Stabilität von F-SSP. *Jahresbericht Bundesanstalt für Fleischforschung, Kulmbach.* C 27.

Ho, H. F. and Koh, B. L. 1984. Processing of some Chinese meat products in Singapore. *Proceedings 4th SIFST Symposium Advances in Food Processing,* June 14–15, 1984, Singapore, p. 94.

Huang, S. H. 1974. *Chinese Snacks.* Dept. of Home Economics. Wei-Chuan Food Corp., Taipei, Taiwan, R.O.C. (in English).

Jakobsen, M. and Jensen, H. C. 1975. Combined effect of water activity and pH on the growth of butyric anaerobes in canned pears. *Lebensm.-Wiss.-Technol.* 8: 158.

Krause, P., Schmoldt, R., Tolgay, Z., and Yurtyeri, A. 1972. Mikrobiologische und serologische Untersuchungen an Lebensmitteln in der Türkei. *Fleischwirtschaft* 52: 83.

Kuo, J. C. and Ockerman, H. W. 1985. Effect of salt, sugar and storage time on microbiological, chemical and sensory properties of Chinese style dried pork. *J. Food Sci.* 50: 1384.

Leistner, L. 1977. Hürden-Effekt und mikrobiologische Stabilität von Lebensmitteln. *Jahresbericht Bundesanstalt Fleischforschung, Kulmbach* C 39.

Leistner, L. 1978. Hurdle effect and energy saving. In *Food Quality and Nutrition.* Downey, W. K. (Ed.), p. 553. Applied Science Publishers, London.

Leistner, L. 1979. Mikrobiologische Einteilung von Fleischkonserven. *Fleischwirtschaft* 59: 1452.

Leistner, L. 1983. Prospects of the preservation and processing of meat. *Proceedings Vth World Conference on Animal Production,* Aug. 14–19, 1983, Tokyo, Japan, Vol. I, p. 255.

Leistner, L. 1984. Toxigenic penicillia occurring in feeds and foods: a review. *Food Technol. in Australia* 36: 404.

Leistner, L. 1985a. Hurdle Technology applied to meat products of the Shelf Stable Product and Intermediate Moisture Food types. In *Properties of Water in Foods.* Simatos, D. and Multon, J. L. (Ed.), p. 309. Martinus Nijhoff Publishers, Dordrecht, The Netherlands.

Leistner, L. 1985b. Nitrate (NO_3^-) and meat products—situation in West Germany. In *Nitrites and the Quality of Meat Products.* Tsvetkov, Ts. (Ed.), p. 38. Institute of Meat Industry, Sofia, Bulgaria.

Leistner, L. 1985c. Nitrite (NO_2^-) and meat products—situation in West Germany. In *Nitrites and the Quality of Meat Products.* Tsvetkov, Ts. (Ed.), p. 49. Institute of Meat Industry, Sofia, Bulgaria.

Leistner, L. 1985d. Empfehlungen für sichere Produkte. In *Mikrobiologie und Qualität von Rohwurst und Rohschinken,* p. 219. Institute for Microbiology, Toxicology and Histology of the Federal Centre for Meat Research, Kulmbach, West Germany.

Leistner, L. 1986a. Hürden-Technologie für die Herstellung stabiler Fleischerzeugnisse. *Fleischwirtschaft* 66: 10.

Leistner, L. 1986b. Allgemeines über Rohwurst. *Fleischwirtschaft* 66: 290.

Leistner, L. 1986c. Allgemeines über Rohschinken. *Fleischwirtschaft* 66: 496.

Leistner, L. and Dresel, J. 1986. Die chinesische Rohwurst—eine andere Technologie. *Mitteilungsblatt Bundesanst. Fleischforsch. Kulmbach* No. 92: 6919.

Leistner, L., Hechelmann, H., and Lücke, F.-K. 1981b. Clostridium botulinum in Brühwurst. *Mitteilungsblatt Bundesanst. Fleischforsch. Kulmbach* No. 72: 4591.

Leistner, L. and Karan-Djurdjić, S. 1970. *Beeinflussung der Stabilität von Fleischkonserven durch Steuerung der Wasseraktivität. Fleischwirtschaft* 50: 1547.

Leistner, L., Maing, I. Y., and Bergmann, E. 1975. Verhinderung von unerwünschtem Schimmelpilzwachstum auf Rohwurst durch Kaliumsorbat. *Fleischwirtschaft* 55: 559.

Leistner, L. and Mirna, A. 1959. Das Redoxpotential von Pökellaken. *Fleischwirtschaft* 11: 659.

Leistner, L. and Rödel, W. 1976. The stability of Intermediate Moisture Foods with respect to micro-organisms. In *Intermediate Moisture Foods*. Davies, R., Birch, G. G., and Parker, K. J. (Ed.), p. 120. Applied Science Publishers, London.

Leistner, L. and Rödel, W. 1979. Microbiology of Intermediate Moisture Foods. In *Food Microbiology and Technology*. Jarvis, B., Christian, J. H. B., and Michener, H. D. (Ed.), p. 35. Medicina Viva Servizio Congressi, Parma, Italy.

Leistner, L., Rödel, W., and Krispien, K. 1981a. Microbiology of meat and meat products in high- and intermediate-moisture ranges. In *Water Activity: Influences on Food Quality*. Rockland, L. B. and Stewart, G. F. (Ed.), p. 855. Academic Press, New York and London.

Leistner, L., Shin, H. K., Hechelmann, H., and Lin, S. Y. 1984. Microbiology and technology of Chinese meat products. *Proceedings* 30th European Meeting of Meat Research Workers, Sept. 9–14, 1984, Bristol, United Kingdom, p. 280.

Leistner, L., Vuković, I., and Dresel, J. 1980. SSP: Meat products with minimal nitrite addition, storable without refrigeration. *Proceedings* 26th European Meeting of Meat Research Workers, Aug. 31–Sept. 5, 1980, Colorado Springs, USA, Vol. II, p. 230.

Leistner, L. and Wirth, F. 1965. Methoden zur Bestimmung des Redoxpotentials in Fleischkonserven. *Fleischwirtschaft* 45: 803.

Leistner, L., Wirth, F., and Vuković, I. 1979. SSP (Shelf Stable Products)—Fleischerzeugnisse mit Zukunft. *Fleischwirtschaft* 59: 1313.

Lin, S. Y., Chang, P. Y., Lai, C. S., and Li, C. F. 1980. Studies on improvement of processing and packaging for dried shredded and sliced pork. Research Report No. 149. Food Industry Research and Development Institute, Hsinchu, Taiwan, R.O.C. (in Chinese).

Lin, S. Y., Chang, P. Y., Lai, C. S., and Li, C. F. 1981. The new process for preparing dried pork slices. Research Report No. E-41. Food Industry Research and Development Institute, Hsinchu, Taiwan, R.O.C. (in English).

Lo, C. X. et al. 1980. *Processing of Foods of Animal Origin.* Provisional textbook for the Agricultural Universities of China, published by the Eastern-Northern Agriculture University, Peking, p. 157, 165, (in Chinese).

Lücke, F.-K. 1984. Mikrobiologische Stabilität im offenen Kessel erhitzter Wurstkonserven. *Mitteilungsblatt Bundesanst. Fleischforsch. Kulmbach* No. 84: 5900.

Lücke, F.-K. and Hechelmann, H. 1986. Assessment of botulism hazards from German-type shelf-stable pasteurized meat products. *Proceedings 2nd World Congress Foodborne Infections and Intoxications,* May 25–30, 1986, Berlin (West), p. 578.

Lücke, F.-K, Hechelmann, H., and Leistner, L. 1981. Clostridium botulinum in Rohwurst und Kochwurst. *Mitteilungsblatt Bundesanst. Fleischforsch. Kulmbach* No. 72: 4597.

Ockerman, H. W. and Kuo, J. C. 1982. Manufacture and acceptability of an oriental dried pork product. *Proceedings* 28th European Meeting of Meat Research Workers, Sept. 5–10, 1982. Madrid, Spain, Vol. I, p. 230.

Prior, B. A. and Badenhorst, L. 1974. Incidence of salmonellae in some meat products. *S. Afr. Med. J.* 48: 2532.

Reichert, J. E. and Stiebing, A. 1977. Herstellung von längerfristig haltbaren Leberwurstkonserven durch Pasteurisieren infolge a_w-Wertsenkung. *Fleischwirtschaft* 57: 910.

Rödel, W., Ponert, H. and Leistner, L. 1976. Einstufung von Fleischerzeugnissen in leicht verderbliche, verderbliche und lagerfähige Produkte. *Fleischwirtschaft* 56: 417.

Savić, I. 1981a. Stand der Fleischwirtschaft in den tropischen und subtropischen Ländern—Afrika. *Fleischwirtschaft* 61: 984.

Savić, I. 1981b. Stand der Fleischwirtschaft in den tropischen und subtropischen Ländern—Asien, Lateinamerika. *Fleischwirtschaft* 61: 1339.

Savić, I. 1985. Small-scale sausage production. FAO Animal Production and Health Paper No. 52, p. 83, Food and Agriculture Organization of the United Nations, Rome, Italy.

Scott, W. J. 1957. Water relations of food spoilage microorganisms. In *Advances in Food Research.* Vol. 7, Mrak, E. M. and Stewart, G. F. (Ed.), p. 83. Academic Press, New York and London.

Shin, H. K. 1984. Energiesparende Konservierungsmethoden für Fleischerzeugnisse, abgeleitet von traditionellen Intermediate Moisture Meats. PhD thesis, Universität Hohenheim, Stuttgart-Hohenheim, West Germany.

Shin, H. K., Hechelmann, H., and Leistner, L. 1982. Mikrobiologische Stabilität traditioneller IMF-Fleischerzeugnisse. *Jahresbericht Bundesanstalt Fleischforschung,* Kulmbach C 22.

Shin, H. K. and Leistner, L. 1983. Mikrobiologische Stabilität traditioneller IM-Meats, importiert aus Afrika und Asien. *Jahresbericht Bundesanstalt Fleischforschung* Kulmbach C 21.

Shin, H. K., Lin, S. Y., and Leistner, L. 1984. Rezepturen und Technologie einiger Chinesische Fleischerzeugnisse. *Mitteilungsblatt Bundesanst. Fleischforsch.* Kulmbach No. 84: 5965.

Taylor, M. B. 1976. Changes in microbial flora during biltong production. *S. Afr. Fd. Review* 3/2: 120.

Van den Heever, L. W. 1965. The variability of salmonellae and bovine cysticerci in biltong. *S. Afr. Med. J.* 39: 14.

Van den Heever, L. W. 1970. Some public health aspects of biltong. *J. S. Afr. Vet. Med. Assoc.* 41: 263.

Van der Riet, W. B. 1976a. Studies on the mycoflora of biltong. *S. Afr. Fd. Review* 3/1: 105.

Van der Riet, W. B. 1976b. Water sorption isotherms of beef biltong and their use in predicting critical moisture contents for biltong storage. *S. Afr. Fd. Review* 3/6: 93.

Van der Riet, W. B. 1982. Biltong ein südafrikanisches Trockenfleischprodukt. *Fleischwirtschaft* 62: 970.

Van Wyk, P. J. 1980. Bilton manufacture. In *Rural Food Processing Seminar,* p. 72. Nat. Fd. Res. Inst. C.S.I.R., Pretoria, R.S.A.

Wirth, F. 1979. Vergleich roher und erhitzter Fleischerzeugnisse bei langer Lagerung. *Proceedings* 25th European Meeting of Meat Research Workers, Aug. 27–31, 1979, Budapest, Hungary, Vol. II, p. 587.

Wirth, F. and Leistner, L. 1982. Informationsreise zum Studium der Herstellung und Stabilität traditioneller italienischer Fleischerzeugnisse. *Mitteilungsblatt Bundesanst. Fleischforsch.* Kulmbach No. 78: 5306.

14

Microbial Stabilization of Intermediate Moisture Food Surfaces

J. Antonio Torres

Oregon State University
Corvallis, Oregon

INTRODUCTION

The application of intermediate moisture food (IMF) technology to the development of new products and the improvement of traditional ones has been an unquestionable success (Bone, 1987; Leistner, 1987; Flink, 1977). However, IMF technology has not and will not reach its full industrial potential unless major research problems are solved. An example is the sensory acceptability of the solute concentrations required to reduce a_w to a safe level. Undoubtedly, sensory considerations will always limit the application of this technology to the development of products stable at room temperature. An interesting alternative that is being systematically investigated in our laboratories is the production of refrigerated IMF products. Such products require lower solute concentrations and are therefore less affected by sensory limitations. Due to their expected long-term microbial stability and lower energy consumption, these products can be viewed as a more economic alternative to frozen products.

Another area of needed research concerns the microbial stability of IMF. These products are generally packaged in moisture-proof materials and are not affected by the frequently changing relative humidity conditions of the external environment. However, changing temperatures to which food products are exposed during production, storage, distribution, and use should be recognized as an important microbial stability factor (Torres et al., 1985a, b, c; Torres and Karel, 1985; Torres, 1984). This possibility has been largely ignored by researchers working in this field, who have determined product shelf life under constant temperature and humidity conditions (e.g., Bhatia and Mudahar, 1982; Erickson, 1982; Theron and Prior, 1980; Hanseman et al., 1980; Flora et al., 1979; Pavey, 1972; Anonymous, 1972). For example, refrigerated IMF products are very sensitive to microbial spoilage when exposed to temperature-abuse situations (Hsu et al., 1983). For this reason the microbial stability of traditional and reduced a_w refrigerated products is the subject of current research efforts in our laboratories.

Temperature changes also cause surface condensation problems. These localized increases in a_w lead to growth of microorganisms on the surface of foods. Due to product development limitations inherent to intermediate moisture technology, these surface stability problems have been difficult to solve. This paper includes specific discussions concerning IMF product development limitations, some of the mechanisms leading to surface condensation, and recent research work aimed at improving surface microbial stability.

IMF Product Development Limitations

The development of commercial IMF products is severely limited by difficulties in achieving the a_w necessary for microbial growth inhibition and, at the same time, satisfying numerous sensory and safety restrictions (Torres et al., 1985a; Motoki et al., 1982; Troller and Christian, 1978). The product development process has been schematized in Fig. 14.1, where shelf life and organoleptic quality are represented as a function of various fabrication parameters, such as water content, use of solutes, pH, concentration of preservatives, heat processing conditions, and/or many of the other hurdles used to obtain desired microbial growth inhibition. The selection of a minimum quality and a minimum shelf life has been used to define the points F_q and F_s, respectively. Any product with F (fabrication parameters combination) larger than F_s will have a better than required shelf life; any product with an F smaller than F_q will have a quality better than the minimum acceptable.

According to the relative positions of F_q and F_s one can define three possible situations:

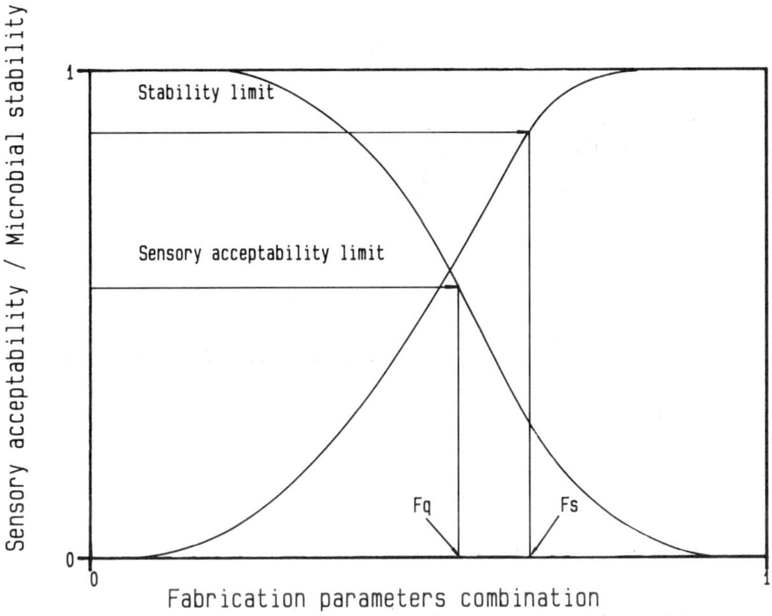

FIG. 14.1 Theoretical development of an IMF. See text for detailed description.

$F_q < Fs$: there is no combination of fabrication parameters that will satisfy the minimum requirements

$F_q = F_s$: there is only one possible combination of fabrication parameters

$F_q > F_s$: there is a range (F_s to F_q) of acceptable combinations that will satisfy the minimum requirements

An example of this approach, shown in Fig. 14.2 and corresponding to the work done to prepare an intermediate moisture cheese analog (Motoki et al., 1982), illustrates very dramatically the product development difficulties surrounding IMF technology. The identification of potentially acceptable product formulations was guided by three controlling parameters: (1) an acceptable texture, as measured mechanically with an Instron; (2) an acceptable taste, using sensory analysis tests; and (3) a minimum a_w, as measured with an electric hygrometer. As shown in Fig. 14.2, the region of acceptable combinations, represented by the shaded area, is quite limited. This points out why once a formulation is found to satisfy a design criteria, it is not always possible to include safety margins to provide protection against product abuse.

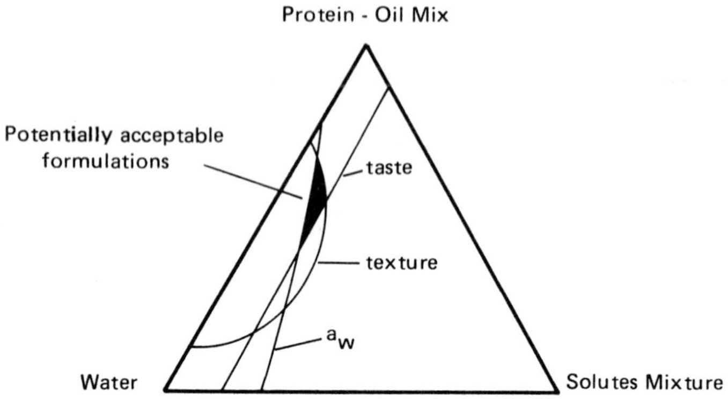

FIG. 14.2 Formulation of an intermediate moisture cheese analog.

Microbial Stability of Surfaces

Due to their effect on surface microbial stability, considerations of the effect of changing environmental conditions should occupy a central role in IMF product development. Examples of unsteady state temperature situations abound in the commercial production and distribution of IMF. A typical situation is the temperature pattern inside a food warehouse (Fig. 14.3). Grundke and Kuklov (1980) reported fluctuations that were especially severe during the winter season when energy was saved by turning the heating system on and off. A similar situation can be expected in a consumer's kitchen, where an air conditioning or heating system is similarly turned on and off. The constant repetition of these temperature cycles can result in severe microbial stability problems due to the repetitive creation of surface conditions with high a_w.

Another common situation is the transfer of products between environments with very different temperatures, e.g., a heated warehouse to a rail car, which in winter could be 30°C colder. Condensation caused by evaporative cooling can raise local surface a_w above the design safety value. A similar situation can be expected when products are packaged while still warm. These moisture gains disrupt the delicate moisture content balance provided to maintain microbial stability. Surface microbial growth is then often observed.

A different situation, but with similar consequences, is the lack of temperature uniformity within an IMF product. A typical example of this kind of situation is a refrigerated IMF sitting on a display shelf (Reid, 1976), where the product is heated by a light source while it is cooled down by a cold source. Let us assume a product with a_w 0.90 is under the influence of a 2°C tempera-

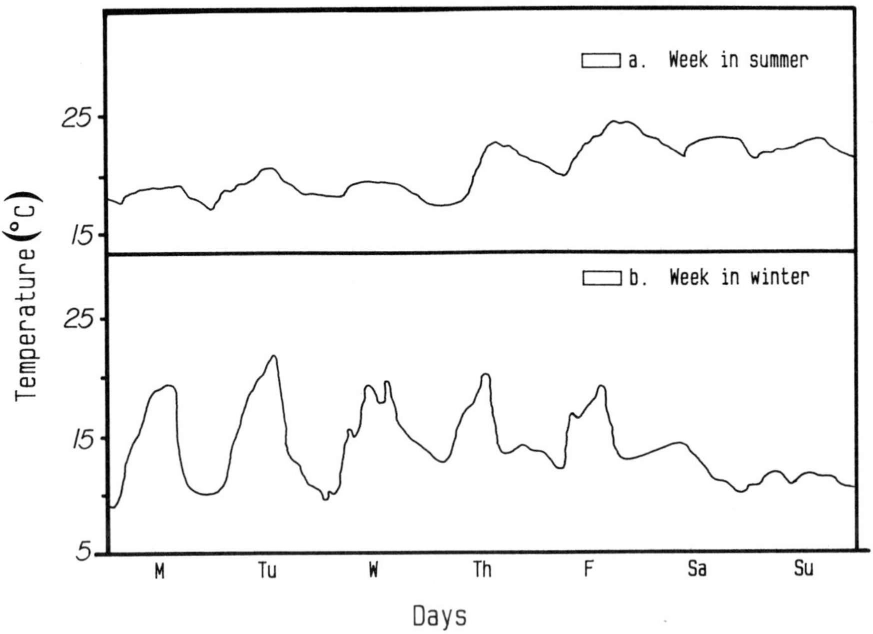

FIG. 14.3 Typical food warehouse temperature fluctuations during 1 wk in summer and winter.

ture difference. As shown in Fig. 14.4 and Table 14.1, the initial vapor pressure on the warm side will be even higher than the saturation vapor pressure on the cold side. The vapor pressure differential will result in water being transferred from the warm to the cool side until equilibrium conditions are established.

Product handling is another source of surface microbial stability problems. Even though some products are cooked or pasteurized, slicing and packaging give ample opportunity for surface recontamination (Stiles and Ng, 1979). A problem associated with surface contamination is that viable counts can be highly variable (Anderson et al., 1980; Gill, 1979). This is particularly important in IMF since bacteriostatic barriers can be overcome by a large number of cells (Anonymous, 1978a). Furthermore, for some microorganisms, in particular for *Staphylococcus aureus,* the minimal a_w requirement depends on oxygen concentration. Under anaerobic conditions the minimal a_w for growth of *S. aureus* is 0.91, while under aerobic conditions it is 0.86 (Scott, 1953). Therefore, the surface of foods, where oxygen could be more readily available, is the region where kne should be more concerned with potential outgrowth of this ubiquitous organism.

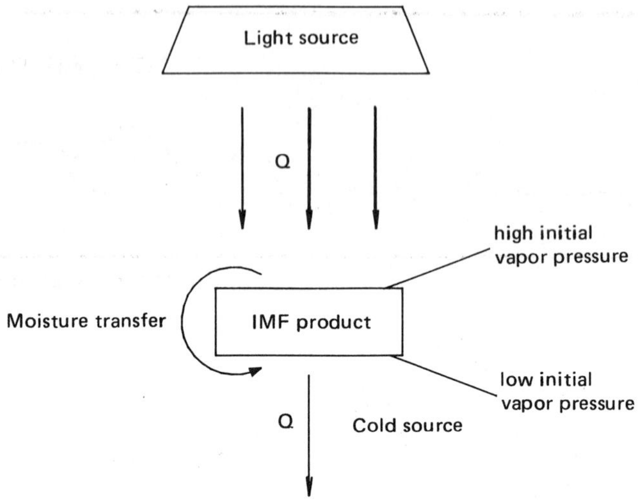

FIG. 14.4 Influence of a 2°C temperature difference within a refrigerated IMF product on vapor pressure and moisture transfer.

TABLE 14.1 Effect of Temperature Differences Within a Product on Surface Vapor Pressure[a]

Temp (°C)	Cold side (T) (mm Hg)		Warm side (T + 2°C) (mm Hg)	
	p_v°	p_v	p_v°	p_v
0	4.6	4.1	5.3	4.8
5	6.5	5.9	7.5	6.8

[a]Temperature difference caused by an illumination source illustrated in Fig. 14.4. $p_v = (a_w)\ p_v^\circ$; $a_w = 0.90$. p_v = surface vapor pressure. p_v° = pure water vapor pressure.
Source: Torres et al. (1985a).

MICROBIAL STABILIZATION OF IMF SURFACES

Based on the above considerations there is a need to consider the surface of an IMF as a separate region where stability has to be considered independently from the bulk of the product, and to develop processes that specifically enhance its microbial stability. These treatments should be acceptable by the U.S. Food and Drug Administration, i.e., safe and effective. Treatments should include food grade materials and, if surface condensation occurs, the modified properties should inhibit or at least reduce microbial growth rate to such an extent that a_w reequilibration will occur before high cell populations are achieved.

We have recently explored two approaches to enhance the surface microbial stability of IMF (Torres et al., 1985a, b; Torres and Karel, 1985; Torres, 1984):

1. Maintain an unequal distribution of preservative, i.e., start with a higher (initial) concentration of preservative(s) on the surface, and use a coating to maintain the concentration differential for as long as possible. This has been called the "reduced preservative diffusion" approach. The preservative selected for our model studies has been sorbic acid and its potassium salt.
2. Create a surface microenvironment where pH is lower than that of the food bulk. Most food preservatives, particularly sorbic acid, are lipophilic acids whose effectiveness is pH-dependent (Eklund, 1983; Freese et al., 1973) and growth itself is strongly affected by pH (Leistner and Rodel, 1976). This approach has been called the "reduced surface pH" approach.

REDUCED PRESERVATIVE DIFFUSION

The improvement of surface microbial stability by the reduced preservative diffusion approach requires the selection of a coating capable of reducing preservative diffusion from the food surface into food bulk (Fig. 14.5). Acceptance by the U.S. Food and Drug Administration of this approach is predicted since the high concentration of preservative is located just on the surface (Miller, 1979), and because surface spraying of preservatives is a common industrial practice (Bolin et al., 1980; Robach et al., 1980; To and Robach, 1980; Cunningham, 1979; Robach, 1979; Robach and Ivey, 1978; Anonymous, 1978b, 1977a, b, 1976b; Nury et al., 1960). The overall concentration is maintained within legal limits by lowering the concentration in the bulk, e.g., from

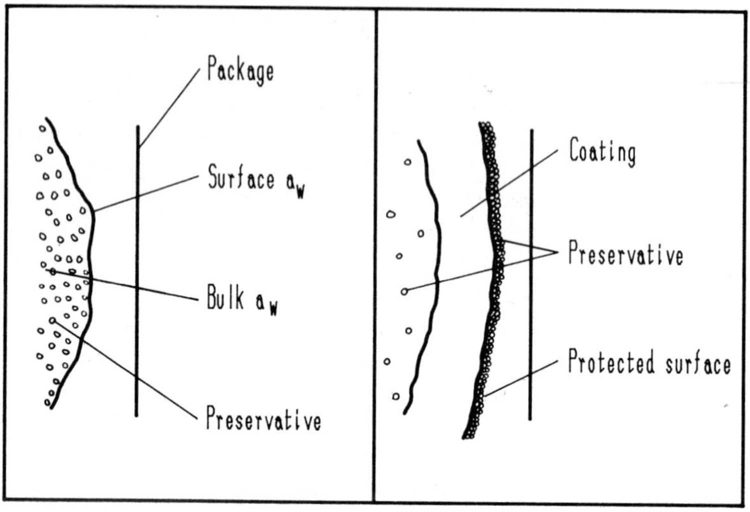

FIG. 14.5 Stability enhancement by application of an impermeable coating (right) and a high preservative concentration on the surface.

0.3% potassium sorbate to 0.1%. The amount corresponding to this reduction is the amount used on the surface. In our intended application, the selection of these values should be based on an acceptable coating thickness, the preservative coating diffusion rate, the food piece geometry, and preliminary microbial stability tests at high preservative concentrations. The latter tests are needed since there is a lack of information on the effectiveness of high preservative concentrations.

Model Development

Mass transfer estimations. Order of magnitude estimates can be made of the mass transfer rate required to achieve significant microbial stability improvements. Approximations for the required diffusion values (D) are based on the following unidimensional diffusion equation (Crank, 1975, Eq. 2–7):

$$C(x) = [M/(\pi Dt)^{1/2}] \exp(-x^2/4Dt) \tag{1}$$

where

$C(x)$ = preservative concentration at location x, g/cm^3
x = diffusion distance, cm
t = time, sec
M = preservative amount deposited on the surface, g/cm^2

For small x values, i.e., surface conditions, this expression can be simplified to:

$$C(x) = [M/(\pi Dt)^{1/2}]\,(1 - x^2/4Dt) \tag{2}$$

It is difficult to define surface accurately to determine what reduction in surface concentration will result in microbial stabilization loss. For estimation purposes, an average surface concentration can be defined as follows:

$$\begin{aligned} C'(O,\Delta x) &= \text{average between } x = 0 \text{ and } x = \Delta x \\ &= M/(\pi Dt)^{1/2} \end{aligned} \tag{3}$$

The average value at time 0 can be defined as:

$$C'_o(O,\Delta x) = M/\Delta x \tag{4}$$

An average reduction in surface concentration can be defined as:

$$\begin{aligned} f &= C'(O,\Delta x)/C'_o(O,\Delta x) \\ &= x/(\pi Dt)^{1/2} \end{aligned} \tag{5}$$

This equation can be modified to yield:

$$D = \Delta x^2/f^2\pi t \tag{6}$$

Assuming that Δx is the coating thickness h, Eq. (6) can be used to estimate the D (diffusion) values required for a given desired stability period and assumed f value (= 0.05, Table 14.2). To visualize how difficult it is to achieve such low diffusion rates, these values should be compared with the one obtained by Guilbert et al. (1983). Their value, determined in an IM agar model, was 2×10^{-6} cm^2/sec, i.e., to be effective, a coating will have to allow potassium sorbate diffusion about 1,000 times slower than IM food-like matrices.

TABLE 14.2 Estimated Sorbic Acid Coating Diffusion Coefficients (D) Required to Achieve Significant Microbial Stability Improvements[a]

	D, cm^2/sec			
	Assumed coating thickness, h			
Time (days)	0.01 mm $\times 10^{-10}$	0.03 mm $\times 10^{-9}$	0.05 mm $\times 10^{-9}$	0.1 mm $\times 10^{-8}$
1	14.7	13.3	36.8	14.5
5	2.9	2.7	7.4	3.0
10	1.5	1.3	3.7	1.5
30	0.5	0.4	1.2	0.5

[a]Eq. (6).
Source: Torres et al. (1985a).

Coating film permeability measurements: The permeability cell illustrated in Fig. 14.6 can be used to determine the permeability constants of coating films supported by regenerated cellulose. Under well-stirred conditions, the resistance to mass transfer for an effective coating, low D value, can be assumed to be due only to film properties. The use of large concentration differences and large reservoirs compared to the area of mass transfer results in a constant concentration difference between reservoirs and facilitates the analysis of experimental data. Under these experimental conditions, the permeability rate is constant and equal to

$$N = DA\,(c_2' - c_1')/x \tag{7}$$

where

N = flux of potassium sorbate
D = apparent diffusion constant
A = area of mass transfer
c_i' = concentration in the film which is in equilibrium with bulk solution concentration c_i; equilibrium expressed as $c_i' = kc_i$
x = a diffusion distance

Assuming constant k we obtain:

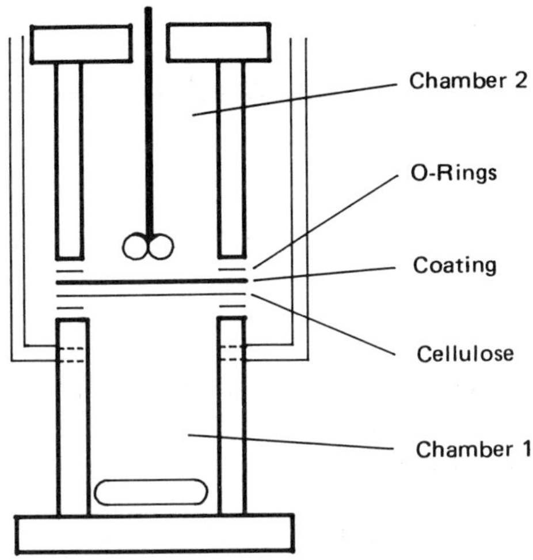

FIG. 14.6 Permeability cell used to determine permeability constants of coating films.

$$N = KA(c_2 - c_1) \tag{8}$$

where

$$K = kD/x \tag{9}$$

The use of regenerated cellulose as a reference permeability value provides an opportunity to obtain a rapid even though approximate measurement of coating effectiveness. The advantages of using regenerated cellulose are its homogeneity, inertness, mechanical strength, and low cost.

Assuming no interface resistance and k values independent of testing material, the following expressions are valid (r = cellulose, rc = cellulose-coating combination, and c = coating film):

$$K = k_i D_i / x_i \text{ with } i = r, rc, c \tag{10}$$

$$k_r = k_{rc} = k_c \tag{11}$$

$$x_{rc}/D_{rc} = x_r/D_r + x_c/D_c \tag{12}$$

where

$$x_{rc} = x_r + x_c \tag{13}$$

We can express x_c as:

$$x_c = f'x_r \tag{14}$$

Thus,

$$x_{rc} = (1 + f')x_r \tag{15}$$

and

$$(1 + f')/D_{rc} = (f'D_r + D_c)/D_rD_c$$

We are interested in $D_c \ll f' D_r$, thus,

$$1 + f' = f'D_{rc}/D_c \tag{16}$$

From Eqs. (10), (11), and (15) we obtain

$$\begin{aligned} K_{rc}/K_r &= (x_r/x_{rc})(D_{rc}/D_r) \\ &= D_{rc}/(1 + f')D_r \end{aligned} \tag{17}$$

Finally from Eqs. (14), (16), and (17) we obtain

$$D_c/D_r = (x_c/x_r)(K_{rc}/K_r) \tag{18}$$

This simple expression allows us to use permeability values for cellulose-supported films to predict if a coating film has the desired effect on sorbic acid diffusion rate.

As shown in Fig. 14.6 the permeability cell should be provided with mechanical and magnetic stirrers to reduce resistance at the interfaces. The side tubes are used to load the cell with a highly concentrated solution, eliminate air bubbles, and adjust levels to eliminate the influence of hydrostatic pressure on permeability. Small samples are taken from the upper chamber, and after appropriate dilution, the preservative concentration is determined spectrophotometrically with no need for an extraction procedure. Glycerol solutions can be used to conduct studies at the intermediate moisture a_w range. This is important since it is desirable to establish a coating film sorption status similar to that occurring under conditions of actual use.

Uncoated cellulose sorbic acid permeability measurements, at an initial concentration difference of 10 mg sorbic acid/ml solution, give an average value $K_r = (4.7 \pm 0.1) \times 10^{-2}$ (mg/hr cm^2)(mg/mL), $n = 5$ (Torres et al., 1985a). Determinations on zein–cellulose combinations (Fig. 14.7) show a

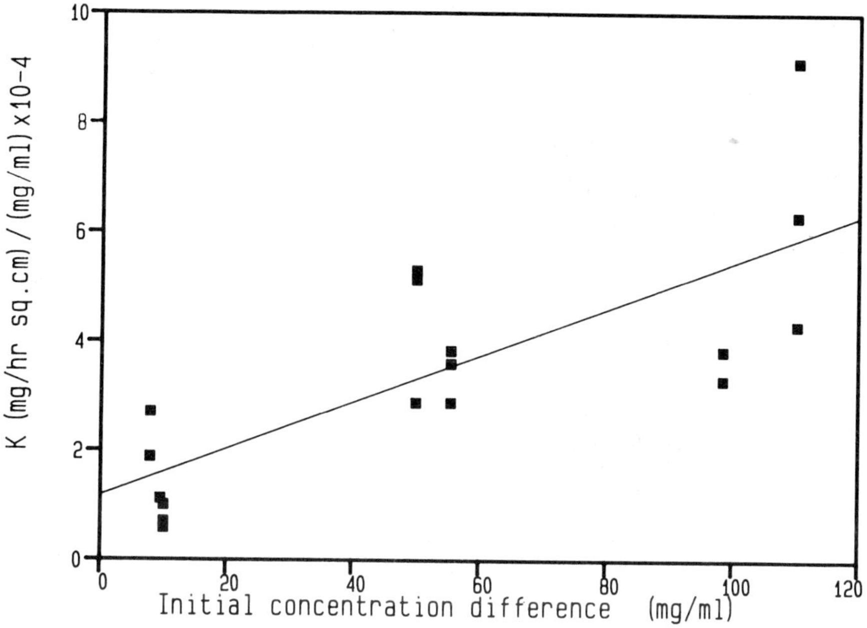

FIG. 14.7 Permeability of zein coating films.

concentration dependence (Torres et al., 1985a). The effectiveness of zein coatings can be evaluated by comparing experimental K-values obtained at the same initial concentration difference (10 mg sorbic acid/mL solution). These values are:

$$K_r = 4.7 \times 10^{-2} \text{ (mg/hr cm}^2\text{)(mg/mL)}$$

$$K_{rc} = 1.5 \times 10^{-4} \text{ (mg/hr cm}^2\text{)(mg/mL)}$$

$$x_r = 0.003 \text{ cm (product specification)}$$

$$x_c = 0.0008 \pm 0.00012 \text{ cm (SEM determination)}$$

Substituting these values in Eq. (18) shows that zein films transport sorbic acid at a rate 1,000 times lower than the cellulose film.

IMF Model Studies

A cheese analog has been used to test the effectiveness of the reduced diffusion approach (Torres and Karel, 1985; Motoki et al., 1982). An emulsion with the composition given in Table 14.3 was stuffed into cellulose casings and heated for 2 hr in a water bath held at 85°C. After cooling, pH was determined with a surface electrode probe. Finally a_w was measured using an electric hygrometer (SINA Equihygroscope, Beckman Industrial, Cedar Grove, NJ). Casings were then carefully removed under a laminar air flow hood. The cylinder thus obtained was cut into disks (r = 1.3 cm, h = 1 cm) and placed on sterile dissecting needles. After 12 hr storage they were coated with zein. The zein solution was prepared by dissolving *X*g zein (Colorcon, West Point, PA), (*X*/4)g glycerol, and 1g Myvacet (Kodak Chemical Reagents, Rochester, NY) in 95% ethanol (balance to 100g). The solution was kept warm in a 50°C water bath.

To show that zein coatings reduced preservative diffusion from the coated food surface into the food bulk to an extent consistent with the measured steady state permeability cell experiments, coated and uncoated IM model samples sprayed with sorbic acid were separated into a core and a surface fraction. From each disk-shaped sample (r = 1.3 cm, h = 1 cm) a smaller disk (r = 0.8 cm, h = 0.5 cm) was obtained; the rest was the surface fraction. Sorbic acid concentration in each fraction was then determined by HPLC (Torres et al., 1985a; Park and Nelson, 1981).

As shown in Table 14.4, sorbic acid determinations showed variation in the amount of sorbic acid deposited on each individual sample. Controls indicate

TABLE 14.3 Formulas for IMF Model

Ingredient	Amount (g)
Isolated soy protein	26.1
Sodium caseinate	5.9
Calcium caseinate	2.0
Hydrogenated vegetable oil	34.0
Decaglycerol monooleate	0.4
Salt	4.8
Glycerol	5.9
Sorbitol	19.3
Potassium sorbate	—[a]
Water	—[b]

[a]Variable, 0.1 or 0.3% (w/w).
[b]The initial water content is 100 mL/100g solids; this amount is reduced by drying the mixture with warm air so as to achieve the desired final a_w 0.85 or 0.88 (Motoki et al., 1982).
Source: Torres et al. (1985a).

TABLE 14.4 Total Sorbic Acid Sprayed on Each Individual Sample

Time[a] (hr)	Core and surface fraction[b] (mg)	Recovery controls[c] (mg)
2	18.9 ± 6.8 (n = 5)	—
18	22.6 ± 5.3 (n = 6)	—
44	17.7 ± 2.1 (n = 6)	13.7
68	15.4 ± 6.1 (n = 6)	14.9; 16.4
118	20.0 ± 3.9 (n = 5)	12.9; 15.0
168	21.5 ± 10.2 (n = 5)	13.6
228	14.8 ± 4.0 (n = 6)	12.3; 9.1
	18.6 ± 6.1 (n = 39)	

[a]Elapsed time after spraying sorbic acid.
[b]Corresponds to uncoated and coated samples; n = number of samples.
[c]These samples were not sliced to determine if losses had occurred during fractionation.
Source: Torres et al. (1985a).

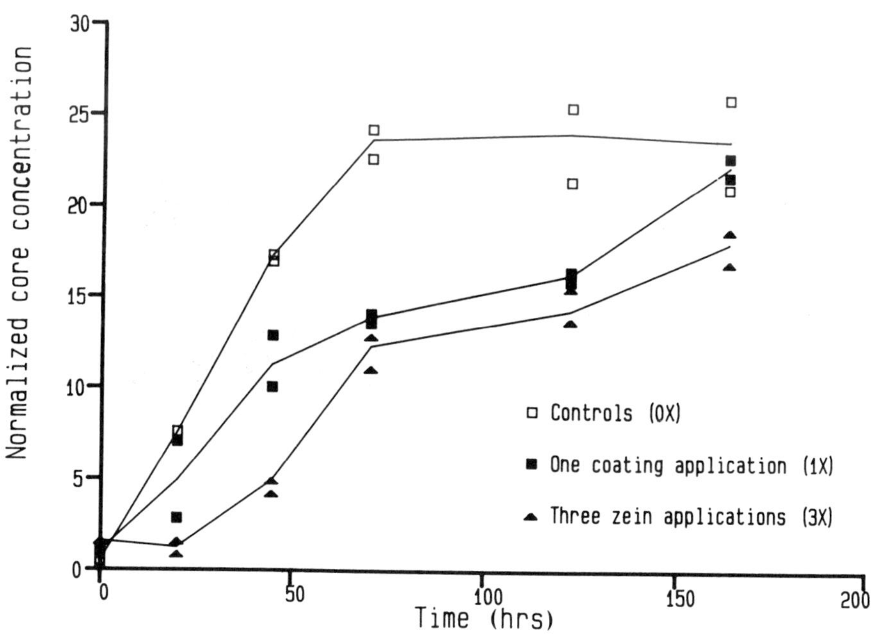

FIG. 14.8 Normalized concentrations of sorbic acid in core.

that this variation was not due to sample handling during fractionation into a surface and a core fraction. Data were normalized to solve this variation by dividing sorbic acid core concentrations by the total amount deposited on each individual piece. Normalized values were then plotted as a function of sampling time and surface treatment. As shown in Fig. 14.8, zein coatings reduced sorbic acid core concentrations significantly. This reduction shows the lower

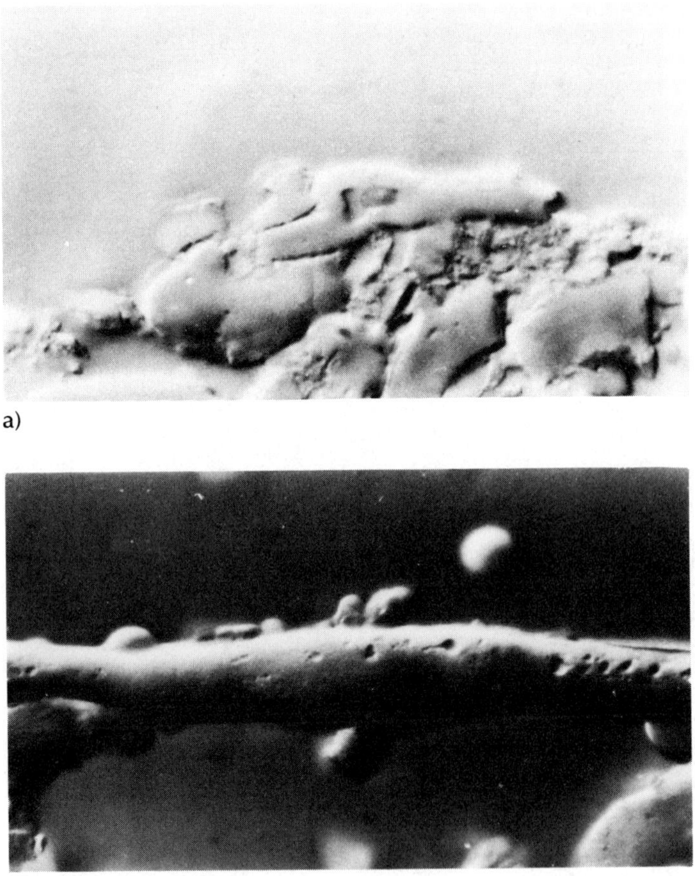

a)

b)

FIG. 14.9 Nomarsky photomicrographs of zein-coated IMF model. (a) Uncoated control (0×); (b) Sample with one zein spray-coated layer (1×), coating thickness = 12.0 ± 2.3 microns (n = 10); (c) Sample with two zein spray-coated layers (2×), coating thickness = 26.5 ± 0.9 microns (n = 10); (d) Sample with three zein spray-coated layers (3×), coating thickness = 38.0 ± 1.8 microns (n = 10).

diffusion rate of sorbic acid deposited on the surface into the food due to the barrier properties of zein films. As expected, the thickness, i.e., the number of zein spray applications, also had an effect.

Nomarsky microscopy (Peil, 1982) was used to observe the general appearance of coated samples as well as to estimate coating thickness. Photomicrographs in Fig. 14.9 show the IMF model with a continuous coating having a minimal thickness variation. Thickness measurements ($n = 10$) showed that each successive application resulted in similar thickness increments (12.0, 14.5, 11.5; average = 12.7 microns).

Apparent diffusion coefficients for sorbic acid in the food model and zein

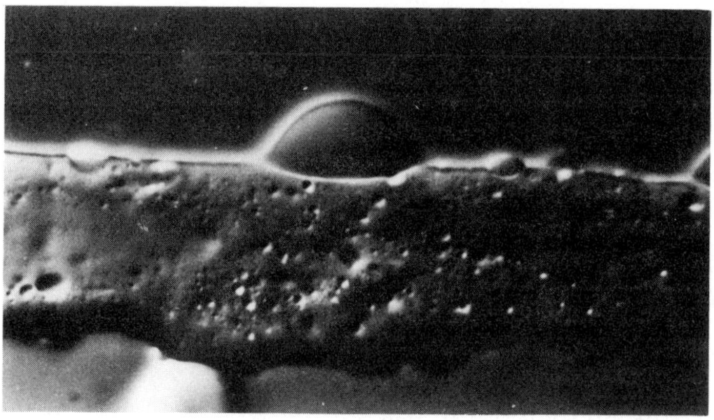

c)

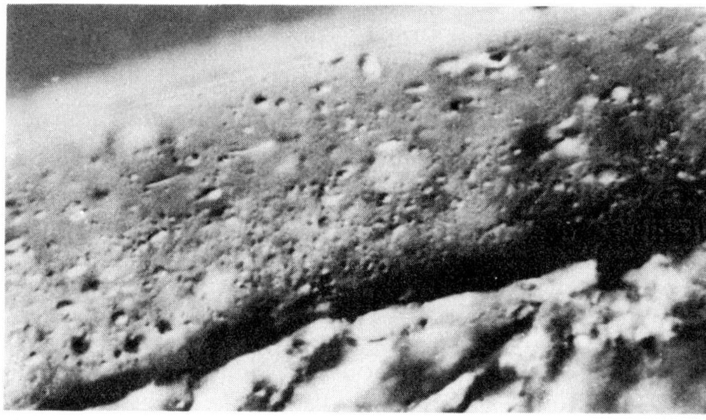

d)

coating were evaluated using data from Fig. 14.8. The following assumptions were necessary:

1. Experimentally determined average core concentrations of sorbic acid were assumed to represent the concentration in the center of the food piece. This is valid when the core sample is small. In our case, it was ⅕ of the food piece, or about 1 g.
2. Solutions for unidimensional diffusion in an infinite slab were assumed valid for the analysis of our disk-shaped samples ($r = 1.3$ cm, $h = 1$ cm) with diffusion occurring from every surface. This simplification was possible because edge effects were eliminated during sample fractionation, i.e., by obtaining a core piece shaped as a smaller disk ($r = 0.8$ cm, $h = 0.5$ cm).

Center conditions can be evaluated using Gurney–Lurie graphs (Adams, 1954) using the following definitions:

a. $Y = (C' - C)/C'$ (19)

Y = an unaccomplished core concentration change

C = core concentration

C' = average overall concentration

b. $X = D_f t / r_f^2$ (20)

X = relative time

r_f = half thickness = 0.5 cm

t = time, sec

D_f = food apparent sorbic acid diffusion constant, cm^2/sec

c. $m = (D_f r_f)/(D_c r_c)$ (21)

m = a resistance ratio

r_c = coating thickness = 0.0012 cm (for 1× samples)
= 0.0038 cm (for 3× samples)

D_c = coating apparent sorbic acid diffusion constant, cm^2/sec

Determinations were done at various time intervals, with individual values summarized in Table 14.5. Calculated average values for the apparent diffusion coefficients for sorbic acid in the food model were ($1.0 \pm 0.1 \times 10^{-6}$, $n = 6$), and in the coating ($3.3 \pm 0.7 \times 10^{-9}$, $n = 5$) and $6.8 \pm 0.9 \times 10^{-9}$, $n = 6$) cm^2/sec for samples coated one and three times, respectively. The value obtained for the uncoated IMF model, 1×10^{-6} cm^2/sec, agrees well with the one obtained by Guilbert et al. (1983). Their value, determined in an IM agar model at the same a_w, 0.88, was 2.0×10^{-6} cm^2/sec.

TABLE 14.5 Calculated Apparent Sorbic Acid Diffusion Constants

a. **Uncoated samples [0×]**

Time (hr)	Y[a]	X[b]		D_f, 10^6 cm²/sec
10	0.84	0.15		0.9
20	0.61	0.26		0.9
40	0.37	0.50		1.1
60	0.12	0.95		1.1
				1.0 ± 0.1

b. **Coated samples [1×]**

Time (hr)	Y	X	m[c]	D_c, 10^9 cm²/sec
60	0.46	0.85	0.55	4.4
80	0.39	1.14	0.70	3.4
100	0.35	1.42	0.75	3.2
120	0.31	1.71	1.00	2.4
140	0.18	1.99	0.75	3.2
				3.3 ± 0.7

c. **Coated samples [3×]**

Time (hr)	Y	X	m	D_c, 10^9 cm²/sec
60	0.60	0.85	1.00	7.6
80	0.47	1.14	0.95	8.0
100	0.43	1.42	1.10	6.9
120	0.40	1.71	1.35	5.6
140	0.32	1.99	1.30	5.8
160	0.23	2.28	1.10	5.9
				6.8 ± 0.9

[a]From Fig. 14.8, using Eq. (19).
[b]Eq. (20).
[c]Eq. (21).
Source: Torres et al. (1985a).

The calculated D_f/D_c ratios were 300 (1×) and 150 (3×). A ratio of 1,000 was obtained in the permeability cell experiments when compared to cellulose film. The difference may be due to several factors including the fact that permeability experiments compared zein coatings with a cellulose film. Secondly, the surface of the IMF model is rich in topographical features such as pores, valleys, etc., which makes it more difficult to cover than cellulose. Thirdly, the assumptions made to analyze Fig. 14.8—diffusion is unidimensional and measured core concentration represents the situation in the piece center—introduce errors leading to lower D values. The values are, however, quite comparable given the orders of magnitude of diffusion reduction that were obtained.

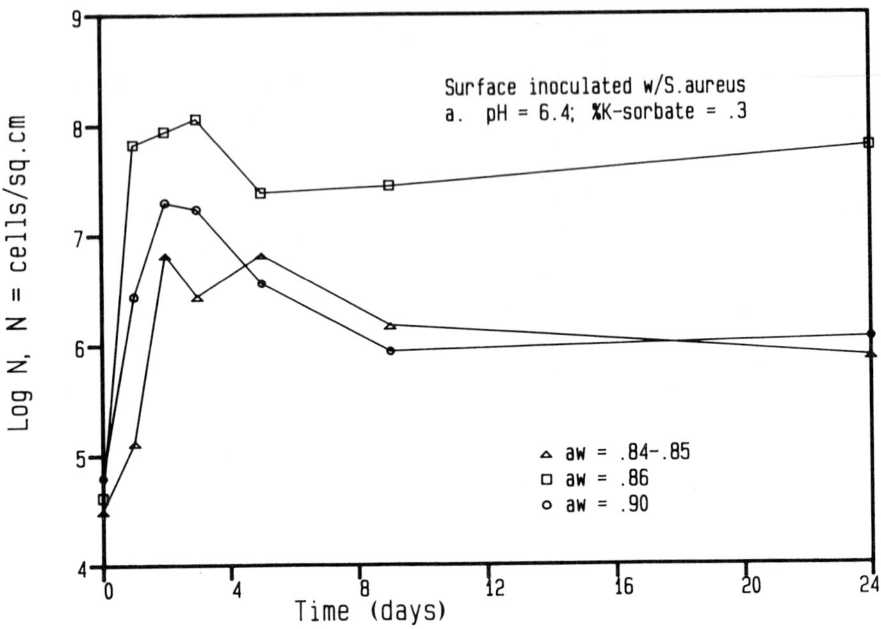
Surface inoculated w/S.aureus
a. pH = 6.4; %K-sorbate = .3
aw = .84-.85
aw = .86
aw = .90
Log N, N = cells/sq.cm
Time (days)
9
8
7
6
5
4
0
4
8
12
16
20
24

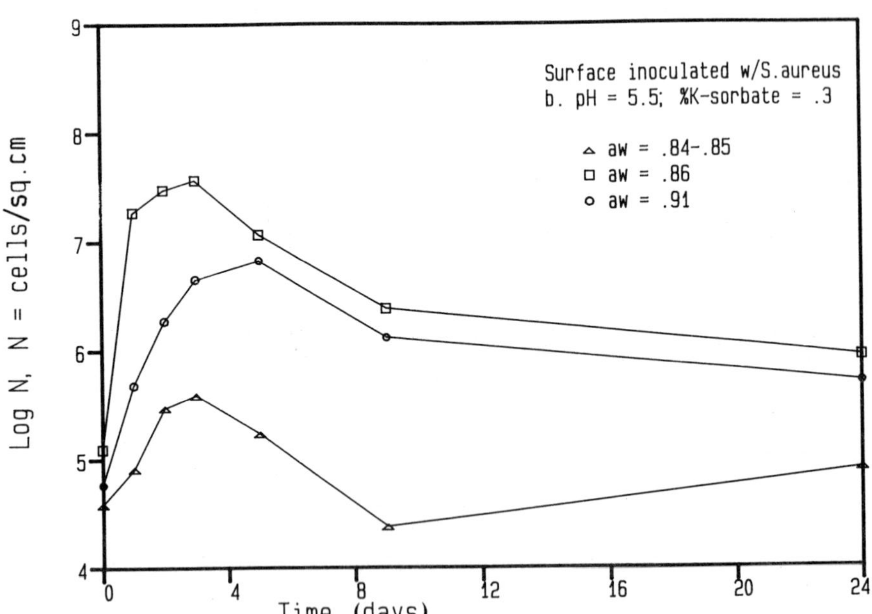
Surface inoculated w/S.aureus
b. pH = 5.5; %K-sorbate = .3
aw = .84-.85
aw = .86
aw = .91
Log N, N = cells/sq.cm
Time (days)
9
8
7
6
5
4
0
4
8
12
16
20
24

Microbial Surface Challenge Tests

The effectiveness of the reduced preservative diffusion approach has been demonstrated using the cheese analog described in Table 14.3 (Torres and Karel, 1985). As shown in Fig. 14.10, this model system allowed rapid outgrowth of *S. aureus* at pH above 6, a_w = 0.88 and a homogeneous 0.3% potassium sorbate concentration (Motoki et al., 1982). Based on this information two abuse testing conditions were selected. In the first one, samples with bulk a_w 0.88 were stored at 30°C and 88% relative humidity, and in the second, samples with bulk a_w 0.85 were stored at 30°C and exposed to 12-hr cycles of 85 and 88% relative humidity, i.e., over saturated $Sr(NO_3)_2$ and $BaCl_2$, respectively.

To allow for comparisons between different treatments a microbial stability limit was defined as the point when samples were no longer considered stable due to viable cell populations one log cycle above the inoculation level. At this point, it is important to mention that cell numbers, reported as time zero inoculation levels, were obtained on the average about 4 hr after inoculation.

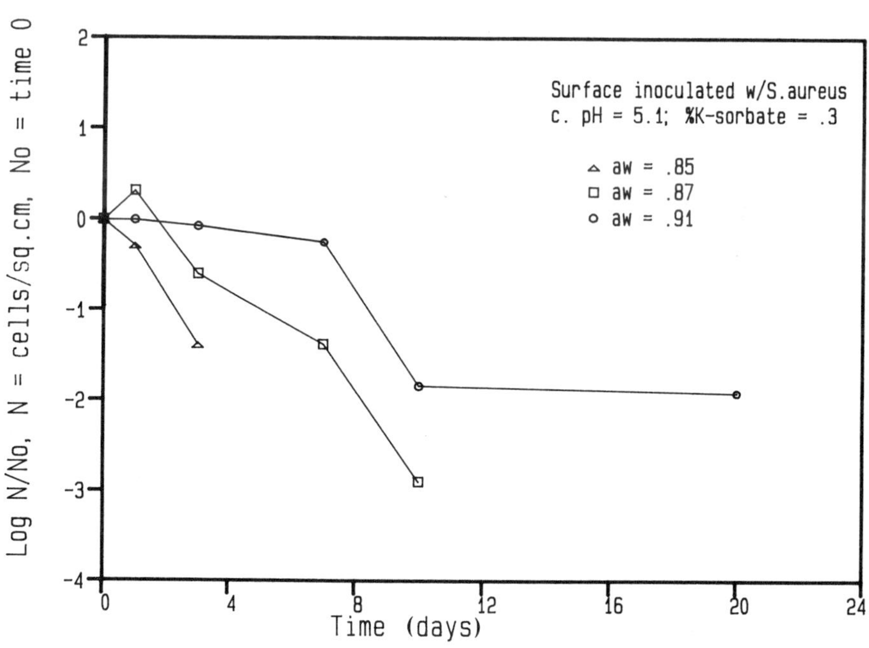

FIG. 14.10 Microbial stability of uncoated samples with an homogeneous 0.3% potassium sorbate and inoculated with *Staphylococcus aureus*: (a) pH = 6.4; (b) pH = 5.5; (c) pH = 5.1.

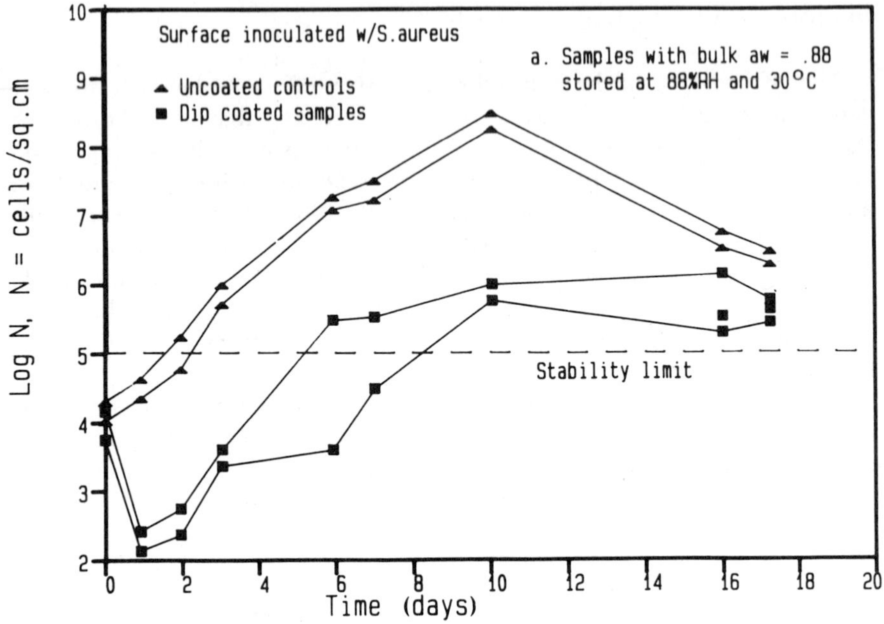

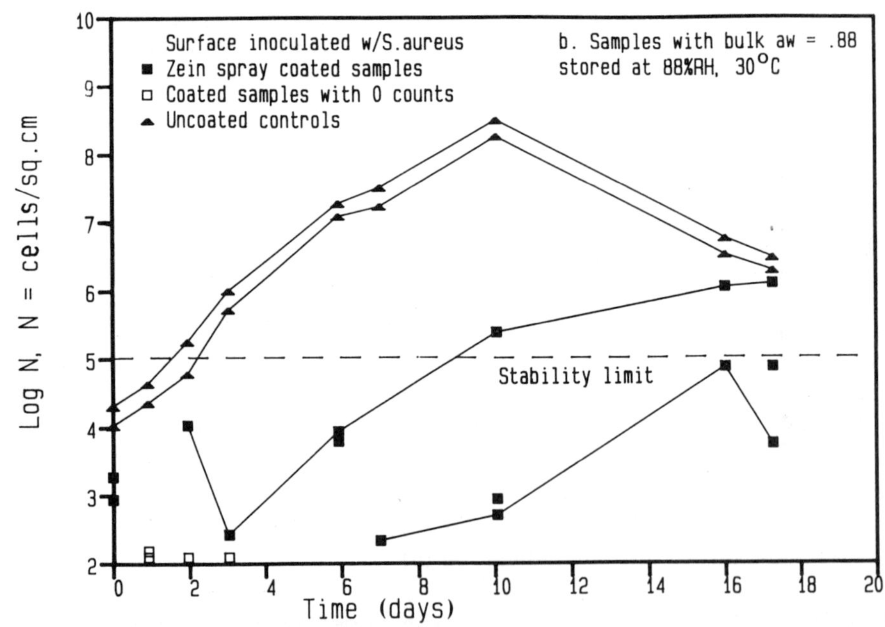

FIG. 14.11 Effect of sorbic acid surface treatment on populations of *Staphylococcus aureus* inoculated on surfaces: (a) Samples with bulk a_w 0.88, dip coated with zein and stored at 88% RH and 30°C; (b) Samples with bulk a_w 0.88, spray coated with zein and stored at 88% RH and 30°C; (c) Samples with bulk a_w 0.85, dip

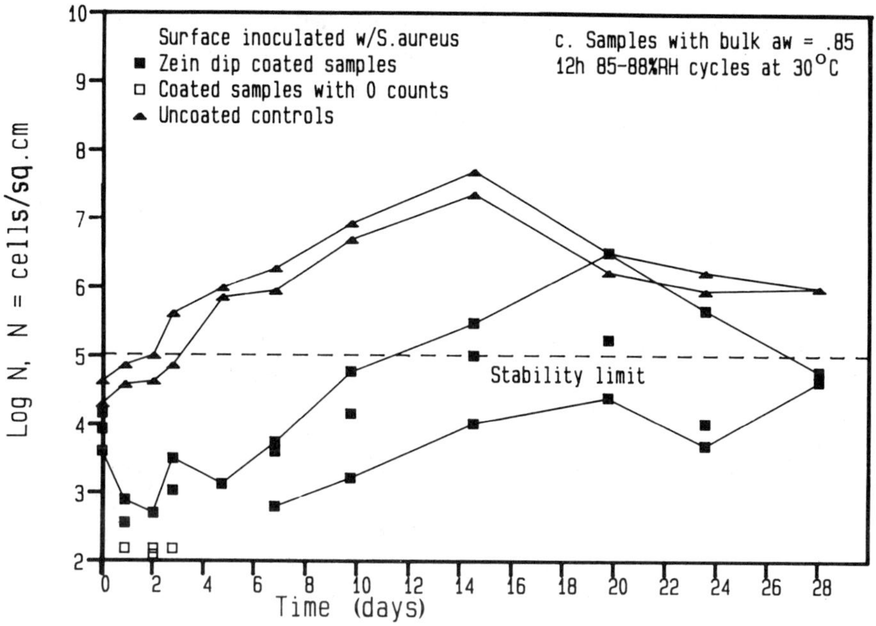

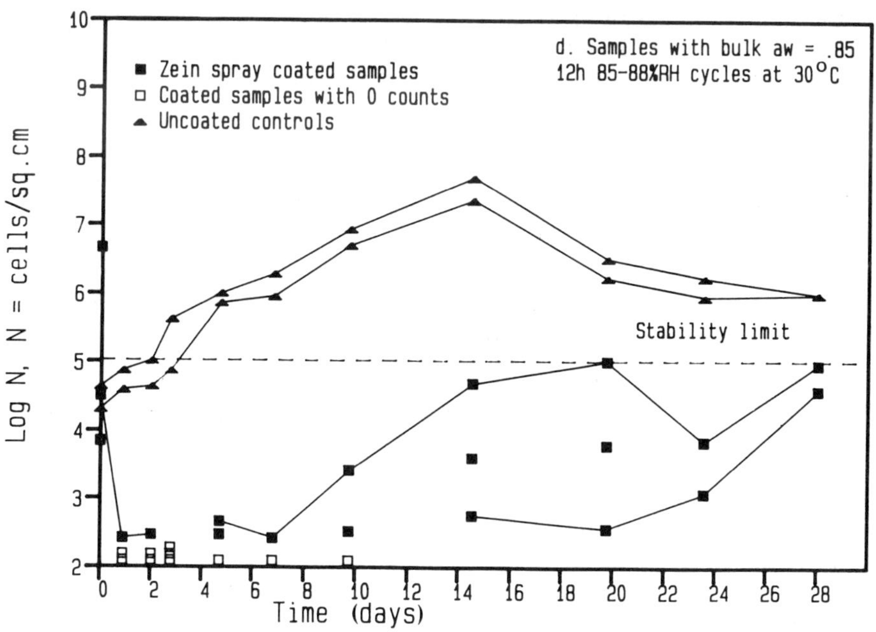

coated with zein and exposed to 12 hr 85–88% RH cycles at 30°C; (d) Samples with bulk a_w 0.85, spray coated with zein and exposed to 12 hr 85–88% RH cycles at 30°C.

That is why the inoculation levels for effective surface treatments, especially those showing strong bactericidal effects, seem to be lower than the values for positive controls.

Figure 14.11a shows the treatment effectiveness of samples with bulk a_w 0.88, stored at 30°C and constant 88% relative humidity, and coated by dipping in 20% zein solutions. The stability period can be estimated as being between 5 and 8 days. There is also a significant bactericidal effect during the initial storage period. Figure 14.11b shows results of stability tests of samples coated by spraying them three times with a 12% zein solution. The stability period was increased to about 10–16 days.

Coating effectiveness was also determined in tests with cycles of low and high relative humidity. Samples with bulk a_w 0.85, stored at 30°C, were exposed to cycles of 12 hr at 85% relative humidity and 12 hr at 88% relative humidity. Figure 14.11c shows the effectiveness of samples coated with 20% zein dips, a repetition of preparation conditions reported in Fig. 14.10a. The only difference was that storage conditions tested were less demanding. The change in bulk a_w and storage relative humidity conditions resulted in an increased stability period of 10–15 days. Finally, Fig. 14.11d shows results of stability tests of samples sprayed with 12% zein exposed to the relative humidity cycle. In this case, the stability period seems to be longer than or about the length of the testing period, 28 days.

Further analysis of Fig. 14.11 indicates that spraying of zein solutions on food samples is a more effective treatment than dipping. The exact nature of this difference has not been determined.

Samples with bulk a_w 0.88 and pH 6.4, stored at constant 88% relative humidity and 30°C, represent extreme testing conditions and were chosen to reduce testing time (Torres and Karel, 1985; Motoki et al., 1982). Samples with lower bulk a_w (0.86) challenged with cycles of low and high relative humidity (85 to 88%) and 12-hr cycles, had significantly longer stability periods. This cycling test was an approximation to situations found in the commercial distribution of IMF. Unfortunately, no model is available to extrapolate from these results the consequences of other less severe and more realistic abuse conditions. Only qualitative comments are possible.

The severity of the above tests should not be underestimated. Surface challenge conditions, in terms of a_w, pH, and temperature, were capable of supporting rapid outgrowth of *S. aureus* (Motoki et al., 1982). The only hurdle to microbial outgrowth was the high concentration of sorbic acid existing on the surface of coated samples.

REDUCED SURFACE pH

Lowering surface pH of foods will result in an increase in microbial stability of surfaces, for it increases the surface availability of the most effective form (undissociated acid) of the lipophilic acids commonly used as preservatives, e.g., sorbic acid (Eklund, 1983; Motoki et al., 1982; Akedo et al., 1977; Freese et al., 1973).

A nonpermanent pH difference between food bulk and food surface can be achieved by inclusion of low-molecular-weight acids in a zein-based coating as described by Torres et al. (1985b). Lactic acid was used in the example shown in Fig. 14.12. Coated samples and uncoated controls, with surfaces inoculated with *S. aureus* S-6, were stored at 37°C and 87% relative humidity. pH was measured with a surface pH electrode. These tests showed that, as soon as the surface-bulk pH difference disappeared, rapid growth occurred. Diffusion was assumed to be the mechanism for pH equilibration. These experiments illustrate the difficulties involved in establishing a significant and permanent pH difference, and the need for a theoretical approach to solve this problem.

At this point, it should be noted that bulk acidification is not always feasible. In most products it will adversely affect sensory acceptability (e.g., Motoki et al., 1982). This incompatibility between enhanced microbial stability and organoleptic quality, typical of IMF technology limitations, could be solved, in some cases, with surface pH reduction.

Model Development

Donnan equilibrium model. The Donnan model for semipermeable membranes (Hiemenz, 1977) can be used as a theoretical basis for the quantification of permanent pH differences between food surface and food bulk, and, most importantly, to identify the conditions for maximum pH difference. This model describes the concentration differentials between two solutions created by the presence of a separating membrane which is permeable to low-molecular-weight electrolytes but impermeable to a charged macromolecule (Grignon and Scallan, 1980; Scallan and Grignon, 1979; Donnan, 1934, 1924, 1911; Donnan and Guggenheim, 1932). Thus it can be used to represent the situation of a charged macromolecule (P^{-z}, z = number of charged groups) immobilized in the form of a food surface coating component. Other components, water and other solutes, particularly electrolytes (e.g., Na^+ and Cl^-), can be assumed to move freely from the food bulk to the surface, and vice versa (Fig. 14.13). This situation can be further schematized as indicated in Fig. 14.14, where M-M' represents the theoretical membrane, M^+ and X^- are electrolytes and sides 1 and 2 correspond to the coating and food model region, respectively. Using assumed molal concentration values, n, m, y, and x, and the equilibrium constant for water (K_w), the remaining expressions given in Fig. 14.14 can be derived from electroneutrality considerations.

Derivation of the pH differential expression. The pH differential expression can be derived starting from the six particular cases of equilibrium deviations

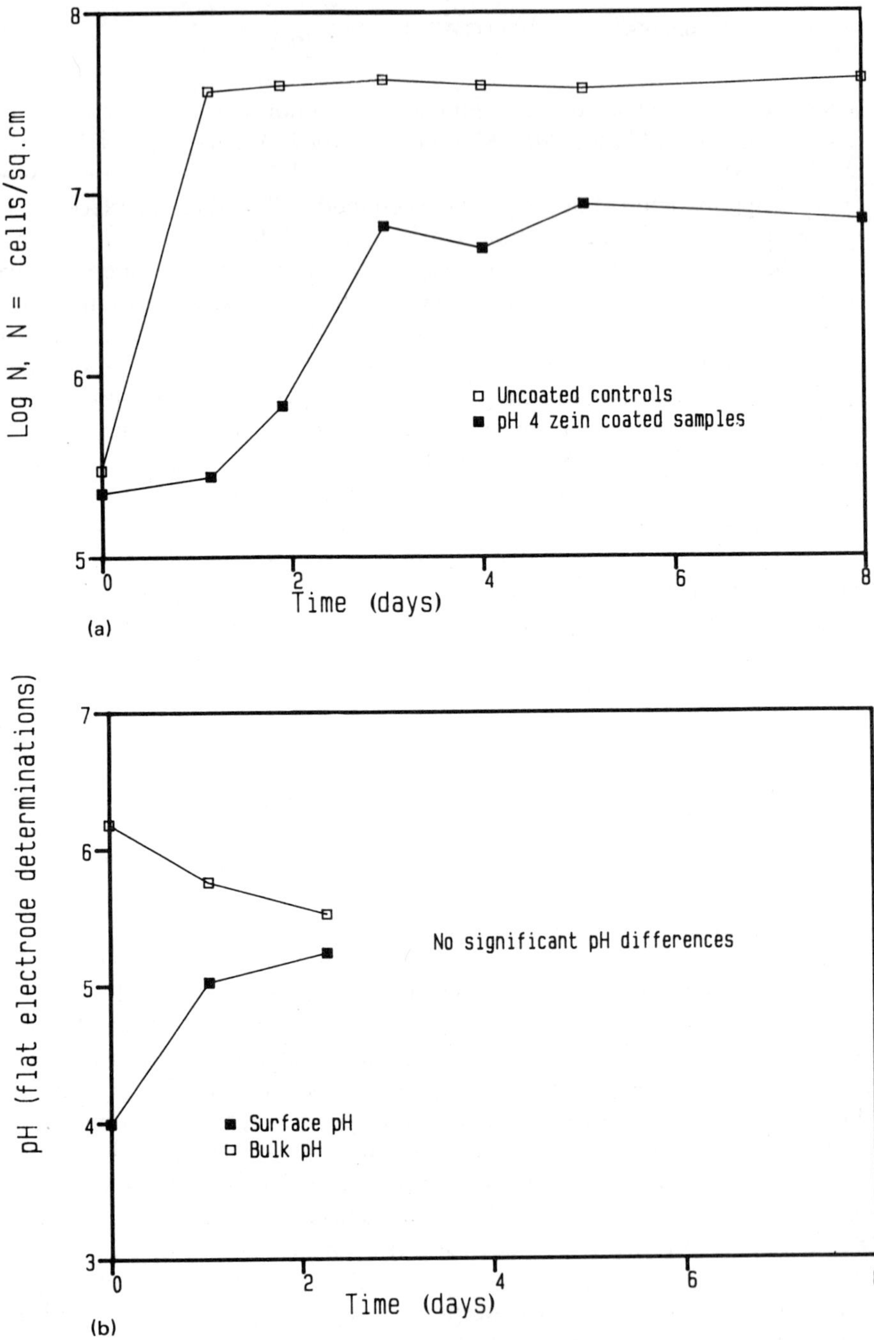

FIG. 14.12 Effects of temporary surface pH reduction (pH 4.0) on (a) *Staphylococcus aureus* populations; (b) Surface and bulk pH values.

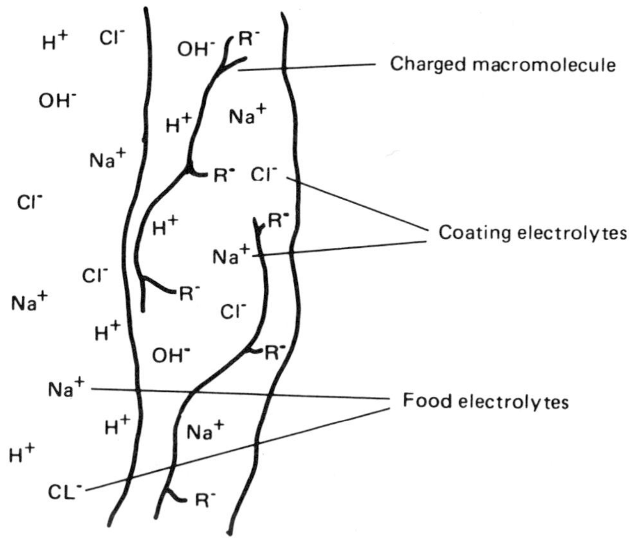

FIG. 14.13 Reduced surface pH upon coating with a charged macromolecule.

Component	Molal concentration Side 1 (Coating)	M	Molal concentration Side 2 (Food)
M^+	$K_w/y + n + zm_{p^{-z}} - y$		$K_w/x + m - x$
X^-	n		m
H^+	y		x
OH^-	K_w/y		K_w/x
p^{-z}	$m_{p^{-z}}$	M′	-----

FIG. 14.14 Charge distribution in the presence of a semipermeable membrane; see text for description of terms.

depicted in Fig. 14.14 (Hiemenz, 1977). All mirror image deviations were not considered since they would yield identical equations. We should also remember that:

$$a_{OH^-/1}\, a_{H^+/1} = K_w \tag{22}$$

$$a_{OH^-/2}\, a_{H^+/2} = K_w \tag{23}$$

where:

$a_{j/i}$ = activity coefficient of component j
i indicates side i, $i = 1,2$
$a_{j/i}$ = 1 for $j = H_2O$
K_w = equilibrium constant for water.

Applying the equilibrium condition to Case 1 (Hiemenz 1977)

$$\begin{aligned} dG = {} & \mu_{X^-/1}(-dn_{X^-/1}) + \mu_{OH^-/1}(dn_{OH^-/1}) \\ & + \mu_{X^-/2}(dn_{X^-/2}) + \mu_{OH^-/2}(-dn_{OH^-/2}) = 0 \end{aligned} \tag{24}$$

From mass and electroneutrality considerations,

$$dn_{X^-/1} = dn_{OH^-/1} = dn_{X^-/2} = dn_{OH^-/2} = dn \tag{25}$$

Therefore,

$$\mu_{OH^-/1} + \mu_{X^-/2} = \mu_{OH^-/2} + \mu_{X^-/1} \tag{26}$$

But:

$$\mu_i = \mu_i^\circ + RT \ln a_i \tag{27}$$

Substituting in Eq. (26):

$$a_{X^-/2}\, a_{OH^-/1} = a_{X^-/1}\, a_{OH^-/2} \tag{28}$$

Similarly for cases 2 through 6 described in Fig. 14.15:

$$a_{M^+/2}\, a_{H^+/1} = a_{H^+/2}\, a_{M^+/1} \tag{29}$$

$$a_{X^-/2}\, a_{M^+/2} = a_{X^-/1}\, a_{M^+/1} \tag{30}$$

$$a_{X^-/2}\, a_{H^+/2} = a_{X^-/1}\, a_{H^+/1} \tag{31}$$

$$a_{M^+/2}\, a_{OH^-/2} = a_{M^+/1}\, a_{OH^-/1} \tag{32}$$

$$a_{H^+/2}\, a_{OH^-/2} = a_{H^+/1}\, a_{OH^-/1} \tag{33}$$

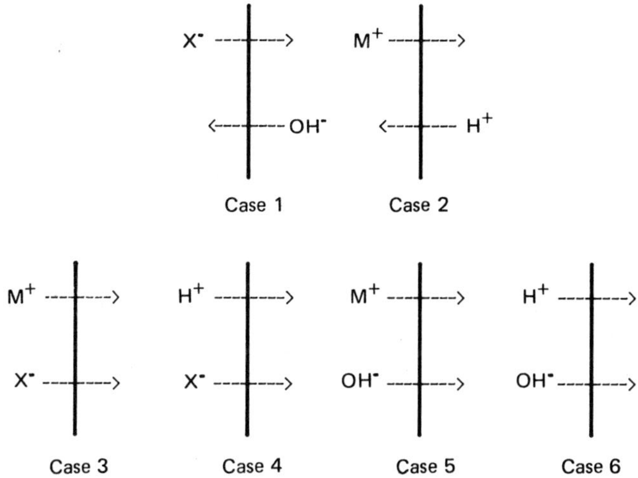

FIG. 14.15 Equilibrium analysis of semipermeable membranes; see text for description of terms.

Since there are more equations than the number of variables, four of them should be eliminated, e.g., Eqs. (28), (29), (32) and (33). For algebraic simplicity we shall substitute activities ($a_{j/i}$) by molal concentrations ($m_{j/i}$). From Eqs. (22), (23), (30) and (31) the following series of equalities was then obtained:

$$\frac{m_{M^+/1}}{m_{M^+/2}} = \frac{m_{X^-/2}}{m_{X^-/1}} = \frac{m_{H^+/1}}{m_{H^+/2}} = \frac{m_{OH^-/2}}{m_{OH^-/1}} \tag{34}$$

Substituting the expressions shown in Fig. 14.14 we obtain:

$$\frac{(K_w/y) + n + zm_{p-z} - y}{(K_w/x) + m - x} = \frac{m}{n} = \frac{y}{x} = \frac{K_w/x}{K_w/y} = \lambda \tag{35}$$

which can be rearranged to obtain:

$$\frac{(K_w/y) + n + zm_{p-z}}{(K_w/x) + m} = \lambda \tag{36}$$

Substituting in Eq. (36), the following expressions derived from Eq. (35):

$$y = \lambda x \tag{37}$$

$$m = \lambda n \tag{38}$$

we obtain:

$$\lambda = \sqrt{1 + \frac{zm_{p-z}}{(K_w/y) + n}} \simeq \sqrt{1 + \frac{zm_{p-z}}{n}} \tag{39}$$

This expression gives the distribution constant λ for all permeable solutes in terms of the number of charges and the concentration of the charged macromolecule and the proton and anion concentrations, all in side 1. The approximation is valid under food pH conditions and the electrolyte concentrations required for maximum pH difference. Finally, we should note that by definition in Eq. (35), log λ represents the pH difference between sides 1 and 2, i.e., between coated surface and food bulk in the case of our intended application. From Eq. (37) we then obtain:

$$\log \lambda = \log y - \log x = \text{pH (food bulk)} - \text{pH (coating)} = \Delta\text{pH} \tag{40}$$

Applications to an IMF Model

IMF model requirements. The key parameters in the expression for the pH difference between coated surface and food bulk are the electrolyte concentration and the number of immobilized charged groups on the surface. This is further emphasized in Table 14.6, which lists the calculated pH difference as

TABLE 14.6 Effect of Surface Charged Group Concentration (zm_{p-z}) and Food Electrolyte Concentration (n) on Surface pH Reduction

zm_{p-z}, [M][a]	n, [M]	$\log \lambda = \Delta\text{pH}$[b]
0.001	0.0100	0.02
	0.0010	0.21
	0.0001	1.00
0.010	0.0100	0.21
	0.0010	1.00
	0.0001	2.00

[a]Equivalent to a macromolecule with $z = -10$ and a concentration ranging from 0.0001 to 0.001M, i.e., a molecule with m.w. = 10,000 used in the 1 to 10% range.
[b]Eq. (39).
Source: Torres et al. (1985b).

affected by these two parameters and shows that significant ΔpH values can only be achieved in an IMF system with low electrolyte content.

The feasibility of the surface pH modification has been tested using the IM cheese analog with reduced total electrolyte concentration described in Table 14.7 (Torres et al., 1985b; Motoki et al., 1982). Electrolytes present in the isolated soybean protein (ISP) and the sodium and calcium caseinates were eliminated by dialysis. Electrolyte removal was followed by electrical conductivity measurements, using a NaCl standard curve to estimate electrolyte concentrations from conductivity measurements. Measurements on the protein solutions themselves are not possible because proteins are conductive. Permeate measurements have to be made instead.

Table 14.6 shows that significant pH differences between bulk food and treated surface are possible only when the total electrolyte concentration is kept below approximately 0.001M. This restriction was used to estimate protein dialysis requirements as follows. Protein concentration in the model system was 46.4g/100g water while in the dialysis experiment it was 7g/100g water. Assuming linearity between conductivity measurements and electrolyte concentration, an upper level of 0.001M represents a limit of 0.00015M in the dialysis experiment.

Dialysis experiments yielded fractions with characteristics summarized in Table 14.8. The three dialysis conditions yielded ISP fractions with similar electrolyte levels. Batch A was an ISP solution dialyzed as is for 7 days at 4°C.

TABLE 14.7 Formulas for a Low Electrolyte IMF Model

Ingredient	Amount (g)
Isolated soy protein	26.1
Caseinates	7.9
Hydrogenated vegetable oil	34.0
Emulsifer	1.6
Glycerol	5.9
Sorbitol	47.0
Sorbic acid	0.4
Water[a]	—

[a]The initial water content is 100 mL/100g solids; this amount is reduced by drying the mixture with warm air so as to achieve the desired final a_w, 0.88 (Motoki et al., 1982).
Source: Torres et al. (1985b).

TABLE 14.8 Final Permeate Conductivity Measurements

Ingredient[a]	Conductivity [micromho]	Equivalent NaCl[b] [M] $\times 10^{-3}$
ISP, batch A	30–40	3–4
batch B	20	2.0
batch C	18	1.8
Sodium caseinate	2	0.2
Calcium caseinate	2	0.2
λ-Carrageenan	⩽2	⩽0.2
Agarose	⩽2	⩽0.2

[a]See text for dialysis conditions.
[b]Calculated using NaCl calibration solutions.
Source: Torres et al. (1985b).

TABLE 14.9 Examples of Charged Macromolecules

Macromolecule	Molecular structure	Sulfate content[a] (%)
Furcelleran	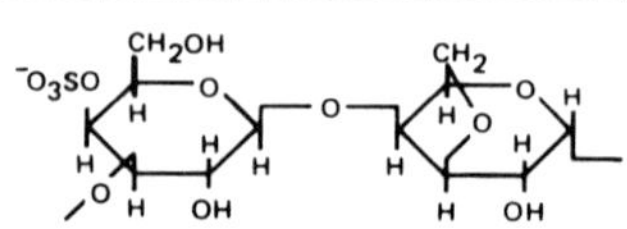	12–16
κ-Carrageenan	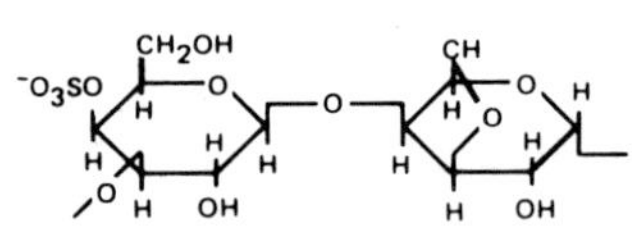	25
ι-Carrageenan	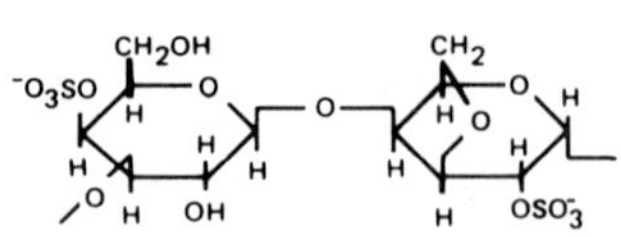	30
λ-Carrageenan	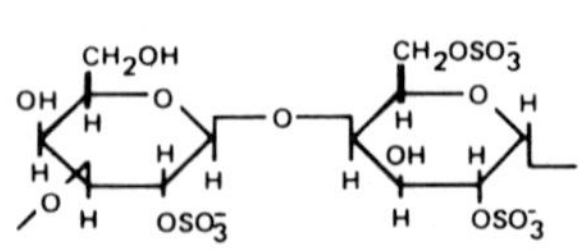	35

[a]Sulfate content percentage is given as an indicator of degree of substitution, e.g., λ-carrageenan has on the average about one sulfate group per repeating unit.

Batch B was an ISP solution adjusted to pH 8, dialyzed for 4 days at 4°C and then neutralized by dialysis against 0.0001M HCl. Batch C was an ISP solution dialyzed as is for 12 hr at 60°C. Sodium and calcium caseinate solutions were dialyzed as is against 6L deionized distilled water for 4 days at 4°C.

Selection of a charged macromolecule. The ideal polyelectrolyte should have a large number of strongly dissociated groups. Solubility and ease of application should also be considered. Several food grade polyelectrolytes satisfy these conditions, including pectic acid, xanthan gum, furcellaran, and carrageenans (Table 14.9).

The polyelectrolyte chosen to test the feasibility of the surface pH reduction approach was λ-carrageenan incorporated in an agarose gel as shown in Table 14.10. This polyelectrolyte was chosen for its solubility and high percentage of sulfate groups—a strongly dissociated group (sulfuric acid dissociation constants: K_1 = large, $K_2 = 1.2 \times 10^{-2}$ (Weast, 1972). Agarose was chosen as the coat-forming matrix because of its lack of charged groups, which explains why its gelling properties are independent of pH and salt concentration, important considerations in our intended application. Electrolyte removal changed the gelling properties of the protein oil emulsion used as a model system. Therefore, the sample preparation procedures used to test the reduced preservative diffusion approach were no longer possible. Instead, about 5g reduced electrolyte IMF samples were pressed into sterile 35-mm diameter disposable Petri dishes. About 0.5 mL of the λ-carrageenan/agarose mixture was poured on top.

Composition data, Tables 14.7 and 14.8, and Eq. (38) and (39), were used to estimate a food surface/food bulk pH difference of approximately 0.8 pH units. Experimentally the pH difference was observed to be 0.3–0.5 pH units, i.e., somewhat lower than the calculated value: 0.8 (Fig. 14.15). The difference

TABLE 14.10 Surface pH Reduction Coating

Ingredient	Amount (g)
Deionized agarose[a]	1.0
Deionized λ-carrageenan	1.0
Sorbic acid	1.2
Propylene glycol, 40% aq soln	96.8

[a]Deionized using dialysis.
Source: Torres et al. (1985b).

most probably was due to the assumption that the only electrolyte sources were the protein fractions.

In Fig. 14.16, the calculated percentages of undissociated sorbic acid ($pK_a = 4.8$) corresponding to the bulk and localized reduced surface pH conditions have also been indicated. The difference between these two curves visualizes the strong microbial stabilizing effect that can be achieved by reducing surface pH. The pH conditions established on the surface have more than doubled the concentration of the undissociated, i.e., the active form, of the preservative. Based on conductivity measurements, the similarity of the ΔpH values for batches A, B, and C was expected, and confirms the prediction power of the Donnan equilibrium model.

Microbial Surface Challenge Tests

After pH equilibration (about 4 days), samples were inoculated with *S. aureus* S-6. Inoculated, uncoated samples were used as positive controls. Figure 14.17 shows the results from batch A and visualizes the strong stabilizing effect of the reduced surface pH approach. In more extended microbial tests, growth occurred 21 days after inoculation, i.e., 25 days after surface treatment. Growth of unidentified bacteria and fungi was observed at the same time that the pH difference disappeared. Multiple samples taken at day 27 confirmed this finding. The observation could not be interpreted as a ΔpH disappearance caused by a "pH equilibration" between surface and food bulk (Torres et al., 1985b). Most likely, the explanation lies in unavoidable microbial contamination collected during the long sample preparation process. It should be stressed that at testing conditions food bulk is not stable; only the surface is affected by our treatment.

Note also that again extreme testing conditions were used. Samples with a_w 0.88, bulk pH 6.1, and sorbic acid 0.22% (w/w) were stored at 88% relative humidity and 30°C. The only hurdle to microbial growth was the reduced surface pH condition.

CONCLUSIONS

Surface treatments can improve the surface microbial stability of intermediate moisture foods. The effectiveness of a high surface preservative concentration, made possible by edible coatings acting as a diffusion barrier, has been demonstrated by extensive microbiological tests. It is important to note that these successful challenge studies utilized zein coatings only 0.03–0.04 mm thick. Since thicker films would be even more effective, it would be advisable to determine the maximum sensory acceptable coating thickness for each specific

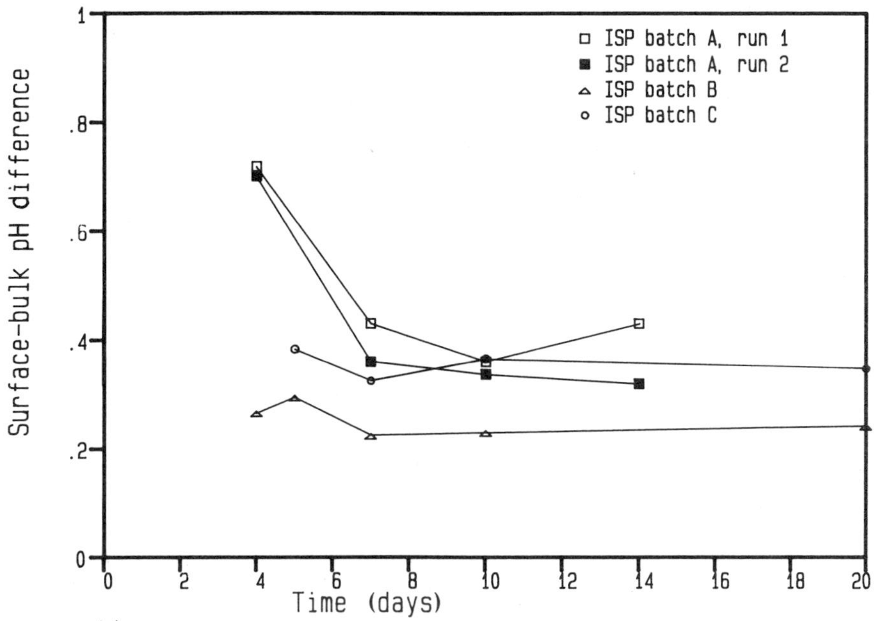

(a)

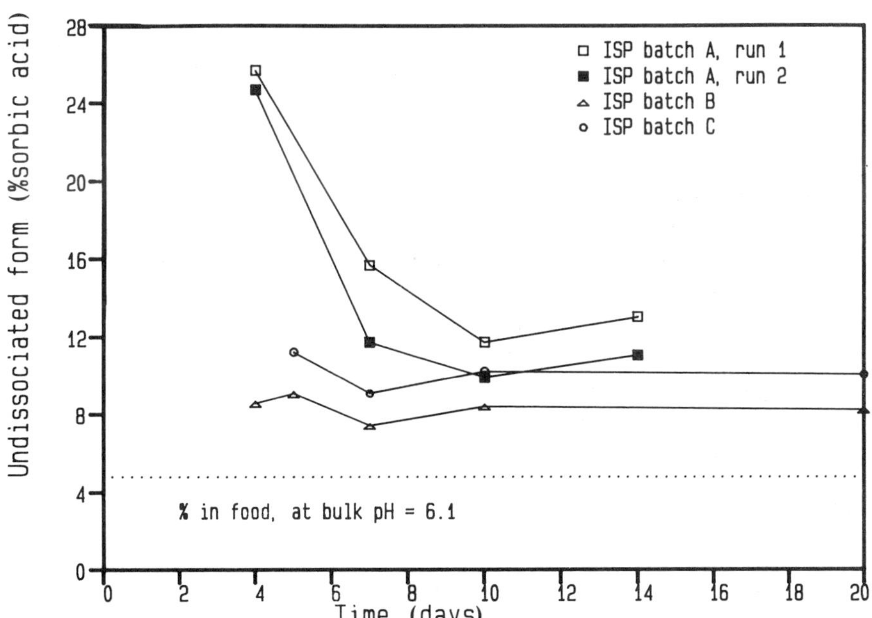

(b)

FIG. 14.16 Influence of surface pH reduction on (a) Surface-bulk pH difference; (b) Undissociated form, percent sorbic acid present.

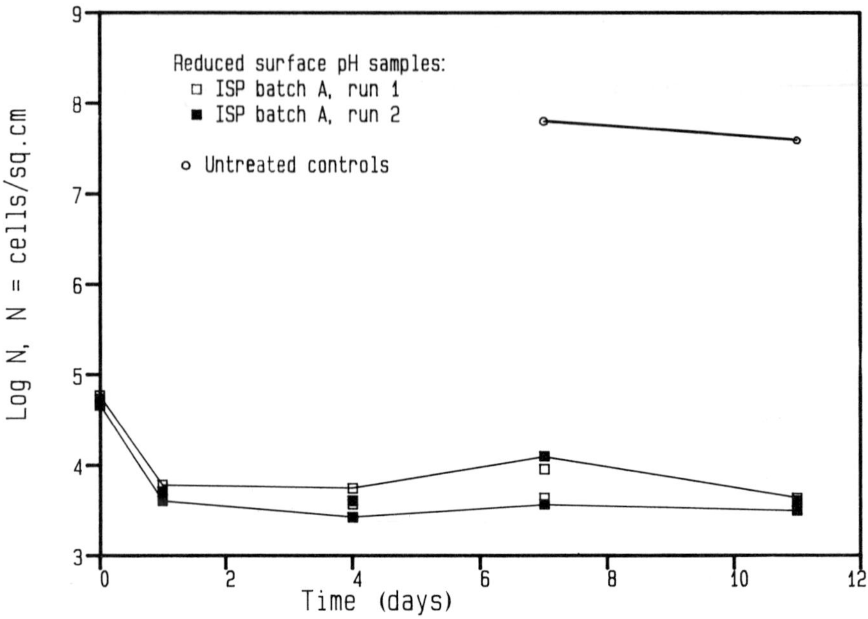

FIG. 14.17 Surface challenge test with *Staphylococcus aureus*; see text for description of terms.

food application. Current research in our laboratories has also been directed to the identification of other formulations as alternatives to zein coatings.

Coating a low electrolyte IMF model with a λ-carrageenan and agarose solution resulted in surface pH reduction and improved microbial stability. The commercial application of this approach is limited by the need to reduce the food electrolyte concentration to extremely low levels. The possibility of increasing the concentration of surface immobilized charged groups to overcome this limitation should be studied.

ACKNOWLEDGMENTS

Some of the original work described herein was conducted at the Department of Applied Biological Sciences, Massachusetts Institute of Technology, and supported by a grant from WESTRECO, Inc.

The author thanks the collaboration from Mr. Masao Motoki, Central Research Laboratories, Ajinomoto, Inc., Japan, and Dr. Jorge O. Bouzas, Molinos

Río de La Plata, S.A., Argentina, and very gratefully the encouragement from Dr. Marcus Karel, Massachusetts Institute of Technology.

O.S.U. Agricultural Experiment Station Technical Paper No. 7899.

REFERENCES

Adams, W. H. 1954. *Heat Transmission,* 3rd ed. McGraw-Hill Book Co., Inc., New York.

Akedo, M., Sinskey, A. J., and Gomez, R. 1977. Antimicrobial action of aliphatic diols and their esters. *J. Food Sci.* 42(3): 699.

Anderson, M. E., Segaugh, J. L., Marshall, R. T., and Stringer, W. C. 1980. A method for decreasing sampling variance in bacteriological analysis of meat surfaces. *J. Food Protection* 43: 21.

Anonymous. 1978a. Sorbic acid and potassium sorbate. Publication IC/FI-13, Monsanto Industrial Chemicals Co., St. Louis, MO.

Anonymous. 1978b. Here is the easy economical dependable way . . . to double, perhaps triple, the shelf life of your bakery products. Publication IC/FI-18, Monsanto Industrial Chemicals Co., St. Louis, MO.

Anonymous. 1977a. Potassium sorbate surface treatment for yeast raised bakery products. Publication IC/FI-21, Monsanto Chemicals Co., St. Louis, MO.

Anonymous. 1977b. Outlook expands for use of sorbate to preserve "natural freshness" of foods. *Food Process.* 46(11): 78.

Anonymous. 1976. Food Preservatives. Technical Information, Pfizer, Chemicals Division, New York.

Anonymous. 1972. Broad line of intermediate moisture foods approaching reality. Foods of Tomorrow, Autumn 1972. Supplement to *Food Process.* 33(10): 14.

Bhatia, B. S. and Mudahar, G. S. 1982. Preparation and storage studies of some intermediate moisture vegetables. *J. Food Sci. Technol.* (India) 19(1): 40.

Bolin, H. R., King, A. D., and Stafford, A. E. 1980. Sorbic acid loss from high moisture prunes. *J. Food Sci.* 45(6): 1434.

Bone, D. 1987. Practical applications. In *Water Activity: Theory and Applications.* Rockland, L. B. and Beuchat, L. R. (Ed.). IFT/Marcel Dekker, Inc., New York.

Crank, J. 1975. Methods of solution when the diffusion coefficient is constant. Ch. 2. In *The Mathematics of Diffusion.* 2nd ed., p. 11. Oxford University Press, Ely House, London.

Cunningham, F. E. 1979. Shelf-life and quality characteristics of poultry parts dipped in potassium sorbate. *J. Food Sci.* 44: 863.

Donnan, F. G. 1934. Die genaue Thermodynamik der Membrangleichgewichte. II. *Z. Physikal Chem.* Abt. A 168(516): 24.

Donnan, F. G. 1924. The theory of membrane equilibria. *Chemical Reviews* 1: 73.

Donnan, F. G. 1911. Theorie der Membrangleichgewichte und Membranpotentiale bei Vorhandesein von nicht dialysierenden Elektrolyten, ein Beitrage zur Physikalisch-Chemisch Physiologie. *Zeitschrift für Elektrochemie* 17: 572.

Donnan, F. G. and Guggenheim, E. A. 1932. Die genaue Thermodynamik der Membrangleichgewichte. I. *Z. Physical Chem.* Abt. A 162: 346.

Eklund, T. 1983. The antimicrobial effect of dissociated and undissociated sorbic acid at different pH levels. *J. Appl. Bacteriol.* 54: 383.

Erickson, L. E. 1982. Recent developments in intermediate moisture foods. *J. Food Protect.* 45(5): 484.

Flink, J. M. 1977. Intermediate moisture foods in the American marketplace. *J. Food Proc. Pres.* 1(4): 324.

Flora, L. F., Beuchat, L. R., and Rao, V. N. M. 1979. Preparation of a shelf stable food product from muscadine grape skins. *J. Food Sci.* 44(3): 854.

Freese, E., Sheu, C. W., and Galliers, E. 1973. Function of lipophilic acids as antimicrobial food additives. *Nature* 241: 321.

Gill, C. O. 1979. A review. Intrinsic bacteria in meat. *J. Appl. Bacteriol.* 47: 367.

Grignon, J. and Scallan, A. M. 1980. Effect of pH and neutral salts upon the swelling of cellulose gels. *J. Appl. Polymer Sci.* 25: 2829.

Grundke, G. and Kuklov, K. 1980. Das Kryptoklima in Lebensmittelverpackungen. *Lebensmittelindustrie* 27(1): 13.

Guilbert, S., Giannakopoulos, A., and Cheftel, J. C. 1983. Diffusivity of sorbic acid in food gels at high and intermediate water activities. Paper presented at the International Symposium on the Properties of Water, Beaume, France, Sept. 11–16.

Hanseman, J. Y., Guilbert, S., Richard, N. and Cheftel, J. C. 1980. Influence conjointe de l'activite de l'eau et du pH sur la croissance de *Staphylococcus aureus* dans un aliment canne a humidite intermediaire. *Lebensm.-Wiss. und-Technologie* 13: 269.

Hiemenz, P. C. 1977. Osmotic and Donnan equilibrium. Ch. 4. In *Principles of Colloid and Surface Chemistry.* 1st ed., p. 125. Marcel Dekker, Inc., New York.

Hsu, H. W., Deng, J. C., Koburger, J. A. and Cornell, J. A. 1983. Storage stability of intermediate moisture mullet roe. *J. Food Sci.* 48: 172.

Leistner, F. 1987. Shelf-stable products and intermediate moisture foods. In *Water Activity: Theory and Applications.* Rockland, L. B. and Beuchat, L. R. (Ed.). IFT/Marcel Dekker, Inc., New York.

Leistner, L. and Rodel, W. 1976. The stability of intermediate moisture foods with respect to microorganisms. In *Intermediate Moisture Foods.* Davies, R., Birch, G. G., and Parker, K. J. (Ed.), p. 20. Applied Science Publishers Ltd., London.

Mendoza, M. A. 1975. Preparation and physical properties of zein based films. Sc.M. thesis, University of Massachusetts, Amherst.

Miller, S. A. 1979. Personal interview with Director of Bureau of Foods, Food and Drug Administration, Washington, DC.

Motoki, M., Torres, J. A., and Karel, M. 1982. Development and stability of intermediate moisture cheese analogs from isolated soybean proteins. *J. Food Proc. Pres.* 6(1): 41.

Nury, F. S., Miller, M. W., and Brekke, J. E. 1960. Preservative effect of some antimicrobial agents on high-moisture dried fruits. *J. Food Technol.* 14(2): 113.

Park, G. L. and Nelson, D. B. 1981. HPLC analysis of sorbic acid in citrus fruit. *J. Food Sci.* 46(5): 1629.

Pavey, R. L. 1972. Controlling the amount of internal aqueous solution in intermediate moisture foods. Swift and Co., Research and Development Center, Oak Brook, Ill. Contract No. DAAG 17-70-CO-0077, Food Laboratory, U.S. Army Natick Laboratories, MA.

Peil, A. 1982. Development and characterization of fortified polymer coatings for rice. Ph.D. thesis, Massachusetts Inst. of Technology, Cambridge, MA.

Reid, D. S. 1976. Water activity concepts in intermediate moisture foods. In *Intermediate Moisture Foods.* Davies, R., Birch, G. G., and Parker, K. J. (Ed.), p. 54. Applied Science Publishers Ltd., London.

Robach, M. C. 1979. Extension of shelf-life of fresh, whole broilers, using a potassium sorbate dip. *J. Food Protect.* 42: 855.

Robach, M. C. and Ivey, F. J. 1978. Antimicrobial efficacy of a potassium sorbate dip on freshly processed poultry. *J. Food Protect.* 41: 284.

Robach, M. C., To, E. C., Mayday, S. and Cook, C. F. 1980. Effect of sorbates on microbiological growth in cooked turkey products. *J. Food Sci.* 45: 638.

Scallan, A. M. and Grignon, J. 1979. The effect of cations on pulp and paper properties. *Svensk papperstidning* 2: 40.

Scott, W. J. 1953. Water relations of *Staphylococcus aureus* at 30C. Austr. J. Biol. Sci. 6: 549.

Stiles, M. E. and Ng, L. K. 1979. Fate of pathogens inoculated onto vacuum-packaged sliced hams to simulate contamination during packaging. *J. Food Protect.* 42: 464.

Theron, D. P. and Prior, B. A. 1980. Effect of water activity and temperature on *Staphylococcus aureus* growth and thermonuclease production in smoked snoek. *J. Food Protect.* 43: 370.

To, E. C. and Robach, M. C. 1980. Potassium sorbate dip as a method of extending shelf life and inhibiting the growth of *Salmonella* and *Staphylococcus aureus* in fresh, whole broilers. *Poultry Sci.* 59: 726.

Torres, J. A. 1984. Exploration of the stability of intermediate moisture foods. Ph.D. thesis, Massachusetts Inst. of Technology, Cambridge.

Torres, J. A., Motoki, M., and Karel, M. 1985a. Microbial stabilization of intermediate moisture food surfaces. I. Control of surface preservative concentration. *J. Food Proc. Pres.* 9: 75.

Torres, J. A., Bouzas, J. O., and Karel, M. 1985b. Microbial stabilization of intermediate moisture food surfaces. II. Control of surface pH. *J. Food Proc. Pres.* 9: 93.

Torres, J. A. and Karel, M. 1985. Microbial stabilization of intermediate moisture food surfaces. III. Effects of surface pH control and surface preservative concentration on microbial stability of an intermediate moisture cheese analog. *J. Food Proc. Pres.* 9: 107.

Troller, J. A. and Christian, J. H. B. 1978. Microbial survival. Ch. 7. In *Water Activity and Food.* Academic Press, New York.

Weast, R. C. (Ed.). 1972. *Handbook of Chemistry and Physics.* 53rd ed. The Chemical Rubber Co., Cleveland, OH.

15

Practical Applications of Water Activity and Moisture Relations in Foods

David P. Bone

The Quaker Oats Company
Barrington, Illinois

INTRODUCTION

It was Scott (1957), an Australian bacteriologist, who made the key link between the thermodynamic concept of fugacity and the practical concept of relative vapor pressure, or a_w. He showed that the growth of most bacteria could be prevented by reducing the a_w of the growth media to values below about 0.90.

Soon after, a number of corporate and food research laboratories were researching practical applications of a_w. The first practical application of the new technology of a_w was made by General Foods Corp. when they marketed Gainesburgers in the late 1950s. Promoted as "the canned dog food without the can," the unique intermediate moisture dog food began to make serious inroads on sales of traditional canned dog foods.

REVIEW OF SELECTED U.S. PATENTS

General Foods Gainesburger

A U.S. patent on the new intermediate moisture product was assigned to General Foods (Burgess and Mellentin, 1965). A brief look at the patent provides clues as to why the novel dog food is a practical application of a_w. Oddly enough, a_w is not mentioned in the patent.

Following are some of the key features of the Gainesburger patent and the product itself:

a. Moist, meaty texture at 20%–30% moisture;
b. Neutral pH;
c. Shelf-stable at room temperature storage;
d. Bacterial inhibition caused by a high solute concentration in the moisture phase, estimated at about 5 molal;
e. Mold inhibition by potassium sorbate;
f. Pasteurized;
g. Simple low temperature, low pressure, continuous extrusion heat process or other means;
h. Postextrusion forming and handling steps at ambient septic conditions;
i. Relatively simple low-cost plastic film packaging;
j. Flexible formulation options—may contain animal by-products;
k. Anthropomorphic qualities—resembles a hamburger patty in form and appearance;
l. Advantageous or competitive cost per nutrition delivered to customer;
m. Highly palatable to dogs;
n. Convenient and easy to store and handle by the customer.

Table 15.1 lists a sample formulation from the Gainesburger patent. Without knowledge of a_w, one would be hard-pressed to creatively design practical new products by following the teachings of the Gainesburger patent.

The mixed ingredients are cooked at about 200°F (93°C) and formed into patties having about 25% moisture, a a_w about 0.90, and packaged.

During the early development of a_w applications, key information was provided by Dr. L. B. Rockland at the USDA Western Regional Research Laboratories. During the 1960s, information about a_w and the use of sorption isotherms began to appear rapidly in the literature.

TABLE 15.1 Gainesburger Patent

Example: Dog food formulation	%
Chopped meat by-products	32.40
Defatted soy flakes	31.40
Sucrose	21.98
Flaked soy hulls	3.04
Dicalcium phosphate	3.04
Nonfat dry milk	2.52
Propylene glycol	2.02
Tallow	1.01
Mono/diglycerides	1.01
Sodium chloride	1.01
Potassium sorbate	0.30
FD&C red dye	0.10
Garlic	0.20
Vitamin/mineral premix	0.06
Total	100.00

Source: Burgess and Mellentin (1965).

Quaker Oats Marbled Meat Pet Food

In the mid-1960s, The Quaker Oats Company introduced a shelf-stable, intermediate moisture, beeflike marbled meaty pet food. This highly successful dog food was reported to have generated more profit per square foot of display shelf space than any other product. In designing the marbled meaty pet food, as described in its patent (Bone, 1965), the known properties of sodium caseinate and starch were considered. Aside from their neutral color and excellent nutritional qualities, sodium caseinate slurries can be heated without denaturing at temperatures well above those that denature or "heat set" other commercial proteins. Caseinate compositions can form rubbery meatlike structures when cooled to room temperature. These features of caseinate enabled the formulation of a plastic melt that could be colored red or white to mimic the form, appearance, and texture of meat.

The a_w technology provided the means, following heat processing, to blend the red and white portions and form a marbled product resembling marbled meat. To conduct the marbling step, it was necessary to blend the red and white extrudates while they were soft and adhesive, at about 110–120°F (33–49°C) (Fig. 15.1).

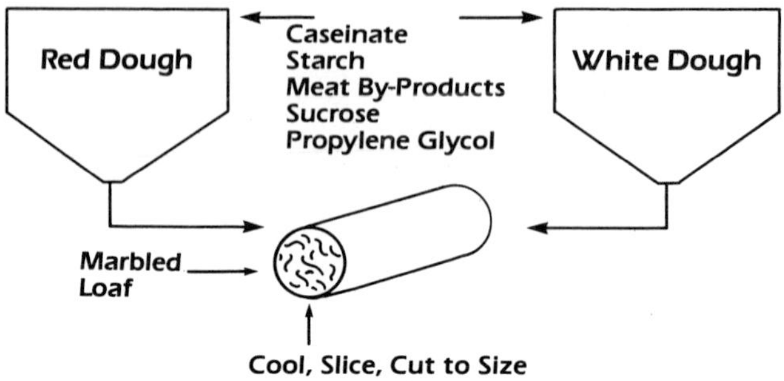

FIG. 15.1 Summary of process to prepare a shelf-stable intermediate moisture beeflike marbled meaty pet food.

When cooled to room temperature, it was necessary that the product retain an elastic meatlike quality without stickiness. These qualities are provided by the interaction of caseinate and starch. A final textural requirement, as explained in the patent, is that the product should not melt at temperatures up to 135°F (57°C). Such temperatures can occur in boxcars, etc., during distribution. This requirement was met by the interaction between starch and process temperature. It was observed that melting could be prevented at 135°F (57°C) and higher by inclusion of as little as 0.5% starch in the formulation, provided process temperatures greater than about 250°F (121°C) were employed. An example formulation from the patent is shown in Table 15.2.

TABLE 15.2 Marbled Meaty Pet Food

	Red (%)	White (%)	Total (%)
Beef by-products	42.50	42.50	42.50
Sugar	26.58	26.50	26.55
Caseinate	15.00	15.00	15.00
Starch	7.50	7.50	7.50
Propylene glycol	4.50	4.50	4.50
Dicalcium phosphate	2.50	2.50	2.50
Salt	1.20	1.20	1.20
Antimycotic	0.10	0.10	0.10
Vitamin supplements	0.12	—	0.09
Colorants	0.01	0.20	0.05
Total	100.00	100.00	100.00

Source: Bone (1965).

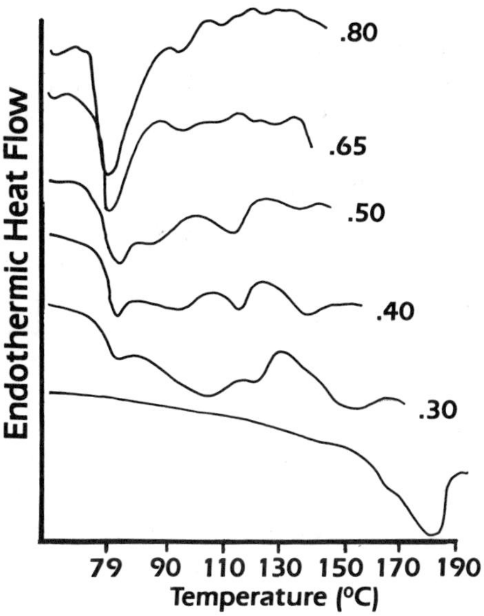

FIG. 15.3 DSC thermal curves of rice flour at various water contents (80–10%).

y dog food by uneven extrusion cooking. The objectives were to provide a oduct containing less than 15% moisture, with an a_w of about 0.75, re- ired nutritional qualities, and of course, high palatability.

To accomplish these objectives a dough was extruded which contained 20– % moisture at about a_w 0.85. Exit temperature of the extrudate was a odest 150–180°F (66–82°C) and back pressure at the exit die was less than psi. The combination of low shear back pressure and a low residence time the dough in the extruder produced a product having an uneven cook. The even cook resulted in a product that simulated a soft, meaty, marbled meat pearance and texture at a_w 0.75 and 14% moisture.

Table 15.3 lists a sample formulation from the patent. The Ross equation one et al., 1975) was used to estimate the dough a_w during calculation of the mula. By multiplying the grams of sucrose in 100 grams of dough (27.84g) the number of 24 gram units of dough moisture in a liter of water (1000g/ g) and then dividing the product of the multiplication by the gram molecu- weight of sucrose (342), the estimated molality of sucrose (3.39) in the ugh moisture was obtained. The a_w of a 3.39 molal sucrose solution was en estimated, using a table from Robinson and Stokes (1959).

Similar estimates were made for the a_w of sorbitol and sodium chloride in

The a_w of the raw doughs were about 0.88, dropping to abou final product due to moisture loss during processing. The requ processing temperature at least 250°F (121°C), to prevent me stored product is apparently due to the elevation of the gelatinizat ture caused by the reduced a_w of the doughs.

During development of the marbled meaty dog food, we w literature reports on the effects of a_w on the gelatinization ter starch. In Fig. 15.2, from a report by Baxter and Hester (1958), es have been added for the reported sugar concentrations in the doughs. Note that dough development was increasingly retardec 0.92 and was completely inhibited at a_w 0.84. The range of a_w inhibition of dough development is from 1.00 to 0.84, or 0.16 difference of only 0.10 a_w units encompasses at least 62% of t range. Thus relatively small differences in a_w, a dimensionless nu related to significant differences in process effects and product q

More recently, Maurice et al. (1985) reported scanning calori on the effects of moisture levels on the temperatures of starch rice flour (Fig. 15.3). The DSC endotherms for the gelatinization initiated at about 65°C and the endotherms for melting amylose c initiated at about 90–100°C, at a_w about 1.00. As flour moistu thus a_w are reduced, the temperature of the starch transition creasingly higher values. What is more, the shift is much gr amylose complexes than it is for gelatinization. These observ general interest, but especially with regard to the problem of the marbled meaty dog food discussed above.

Observations such as these, whereby seemingly small differen translate to large differences in process requirements and/or proc led to a patent (Bone et al., 1975) for making a soft, elastic, mar

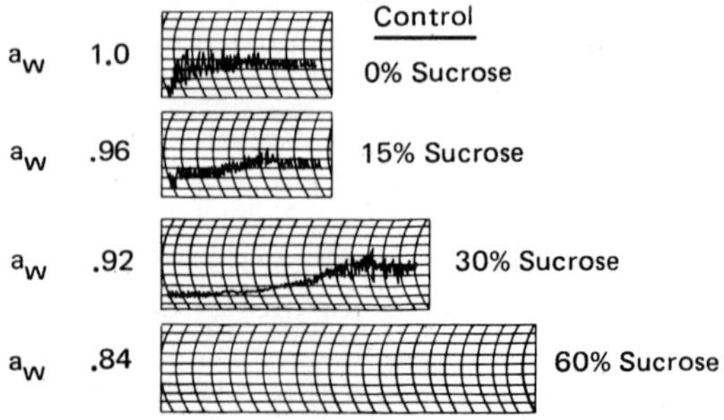

FIG. 15.2 Effect of sucrose on estimated a_w and mixograms of wheat flour dough.

TABLE 15.3 Patent Example Formula for a Soft, Elastic, Meatlike Dog Food by Uneven Extrusion Cooking

	%	Molality	a_w
Modified soy protein	19.54		
Defatted soy flour	4.88		
Meat and bone meal	4.88		
Sodium caseinate	4.88		
Sucrose	27.84	3.39	0.92
Beef trimmings	4.88		
Sorbitol, 70%	9.77	1.58	0.97
Prime steam lard	1.95		
Dicalcium phosphate dihydrate	4.79		
Sodium chloride	0.98	0.70	0.98
Potassium sorbate	0.29		
Trace minerals and vitamins	0.59		
Iron oxide	0.04		
Water	14.69		
Total	100.00		

[a]% Moisture = 24; a_w = 0.88; Est. $a_w = 0.92 \times 0.97 \times 0.98 = 0.87$.

the dough formula. The dough formula a_w was then calculated as the product of the three estimated a_w according to the Ross equation . . . $0.92 \times 0.968 \times 0.977 = 0.87$. This estimated a_w of 0.87 is close enough to the measured a_w of the dough (0.88) to be practical in designing formulas to a target a_w.

Quaker Oats Multi-Textured Pet Food

One of the most successful applications of a_w is a multi-textured pet food described in a U.S. patent (Bone and Shannon, 1977) assigned to the Quaker Oats Company. The objective was to combine the advantages of hard, dry-baked pet food with the advantages of soft–moist pet food while minimizing disadvantages of each type. The hard dry type component has the advantage of teeth-cleaning properties for dogs but has a disadvantage of being less palatable than a quality soft–moist type. While a soft–moist dog food can be highly palatable, it lacks the abrasive teeth cleaning properties of a hard dry type.

As shown in the process flow chart (Fig. 15.4), each component of the multi-textured pet food is processed separately. When the two components are mixed, they equilibrate to a common equilibrium a_w during storage. The common a_w must be a value that permits hardness to be maintained in the biscuit component and softness in the meaty portion.

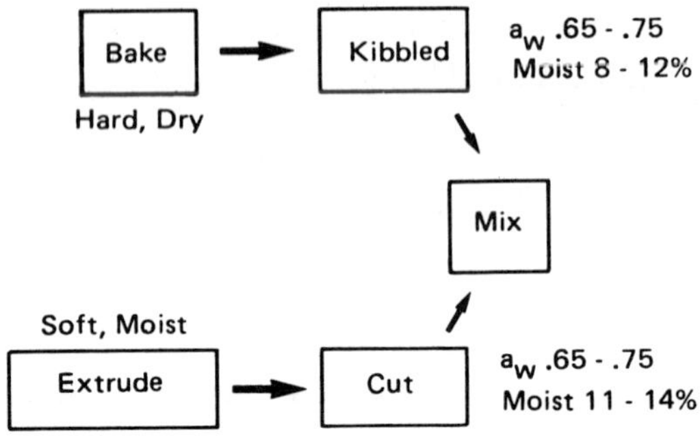

FIG. 15.4 Flow chart for components of a multi-textured pet food. (*From Bone and Shannonn, 1977.*)

There are a number of products on the market wherein texture contrasts are brought about by controlling moisture relations in the products. Examples are the Ralston Purina pet food "Bonz," a simulated bone filled with a simulated soft–moist marrow, and Quaker Oats "Moist Meals," consisting of a crunchy vitamin nugget in a soft–moist cat food. A new multi-textured shelf-stable product is Hostess "Pudding Pie," a flavored moist pudding in an eclair-type shell. The pudding pH is about 4.0, and a_w of the product is about 0.91–0.92. Ready-to-eat breakfast cereals with raisins are another example of multi-textured foods controlled by a_w.

APPLICATION OF THE SALWIN EQUATION

Salwin (1963) reported an equation (Fig. 15.5) based upon algebraic relationships of moisture and water activity of sorption isotherms. He showed how one could use the equation to make practical predictions of a_w and moisture content at equilibrium. The Salwin equation for a mixture containing two ingredients is shown in Fig. 15.5. However, the right side of the equation may have as many ingredient terms as desired. The equation is based on the assumption that no significant interactions occur that alter the isotherms of the mix ingredients. For practical purposes, the mix temperature should be reasonably consistent with the isotherm temperatures used in the prediction.

The components of the mixture may be simple solutes, polymers, such as starches and proteins, flours, or complex foods. By knowing the critical levels of moisture or a_w for textures, flavor, enzyme activity, and reactions such as browning and lipid oxidation, the equation becomes a very practical tool in designing food products.

$$a_w = \frac{(W_1 \cdot S_1 \cdot a_{w1|}) + (W_2 \cdot S_2 \cdot a_{w2|})}{(W_1 \cdot S_1) + (W_2 \cdot S_2)} \quad \ldots$$

W_1 = Grams Solids Ingredient 1
S_1 = Linear Slope Ingredient 1
$a_{w1|}$ = Initial a_w Ingredient 1

FIG. 15.5 Salwin equation for a mixture containing two ingredients to make practical predictions of a_w and moisture content at equilibrium.

Practicality of Salwin Equation

The practicality of the Salwin (1963) equation can be easily illustrated. First, as shown by Fig. 15.6, when two ingredients having different a_w and isotherms are mixed in a closed container, at least four changes may occur:

1. The lower a_w ingredient gains moisture while the higher a_w ingredient loses moisture.

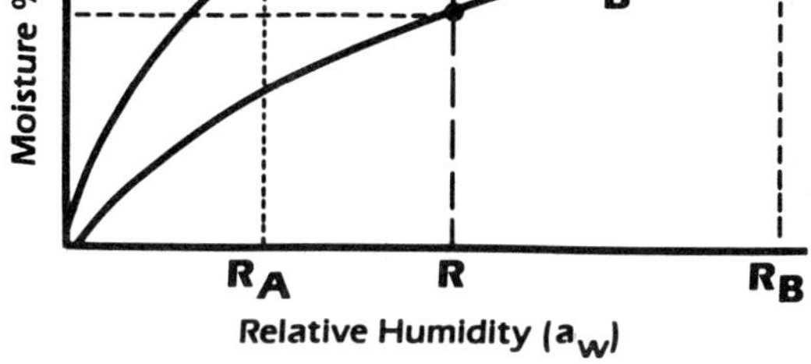

FIG. 15.6 Graphical representation of changes in moisture during equilibration.

2. Each ingredient changes its a_w upward or downward in the direction of its moisture change along its sorption isotherm curve.
3. Quality changes may occur within each ingredient.
4. All mixture components equilibrate to a common a_w value.

Estimating a_w of two-component mix. Table 15.4 shows an example of how the Salwin equation can be used to estimate the a_w of a mix containing only two components, each at specified levels, moistures, and a_w. The predicted a_w of the mixture is 0.41.

Predicting isotherms of simple/complex mixes. The Salwin equation is also very practical for predicting the isotherms of both simple and complex mixtures. As an example of predicting the isotherm of a simple mixture a 50:50 blend of dried egg albumen and soy protein was used. First, isotherms were developed for dried egg albumen and for soy protein (Fig. 15.7). Next the isotherm of a 50:50 mixture of egg albumen and soy protein was predicted followed by development of an actual isotherm for a 50:50 mixture of egg albumen and soy protein. A comparison of the predicted and the actual isotherms is shown in Fig. 15.8. There is good agreement between the predicted and the actual isotherm. Note that many prediction points were used to determine the curvature of the predicted isotherm, using different slope and initial a_w values obtained from each of the parent isotherms.

The next comparison shows that the Salwin equation can be used to accurately predict an isotherm segment of a complex mixture in the a_w range 0.75–0.85.

A model mixture consisted of: dried egg albumen; soy protein; Guar gum (Fig. 15.9); sorbitol, 70% solution (Fig. 15.10); corn oil; and water. Since water has no isotherm, water was entered into the equation by assigning it to the soy protein, using appropriate dry weight, slope, and initial a_w in reference

TABLE 15.4 Data for Predicting the Equilibrium a_w of a Mixture of 70% A and 30% B by the Salwin Equation

	A	B
% Moisture, dry basis	5.26	19.05
Initial a_w	0.20	0.65
Slope	0.25	0.60
Grams of solids	66.50	25.20

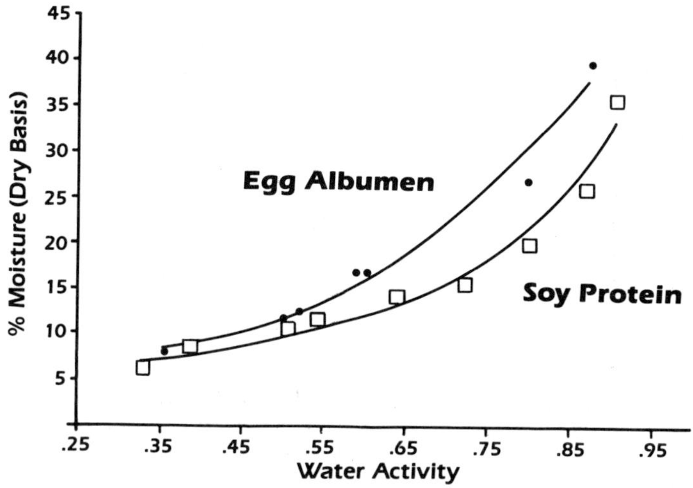

FIG. 15.7 Adsorption isotherms, 25°C, for dried egg albumen and soy protein.

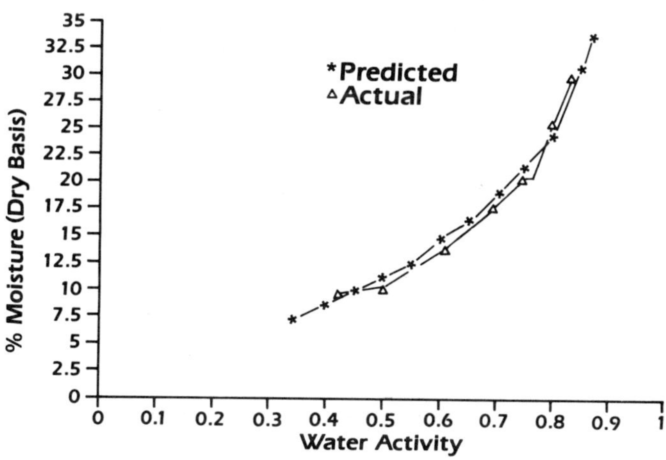

FIG. 15.8 Predicted and actual isotherms at 25°C for 50:50 mix of egg albumen and soy protein.

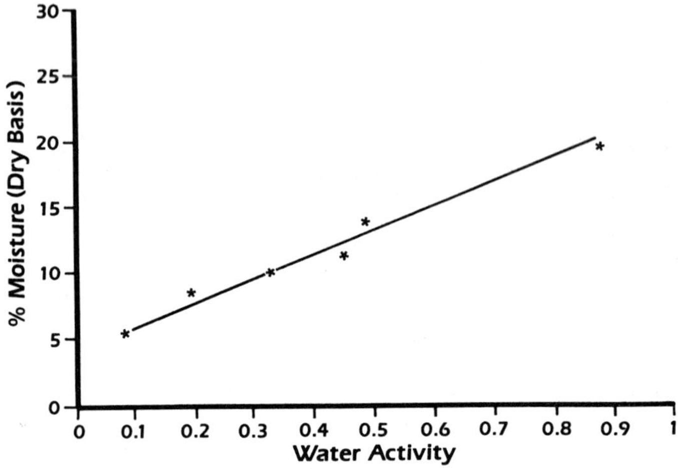

FIG. 15.9 Isotherm for guar gum, 25°C.

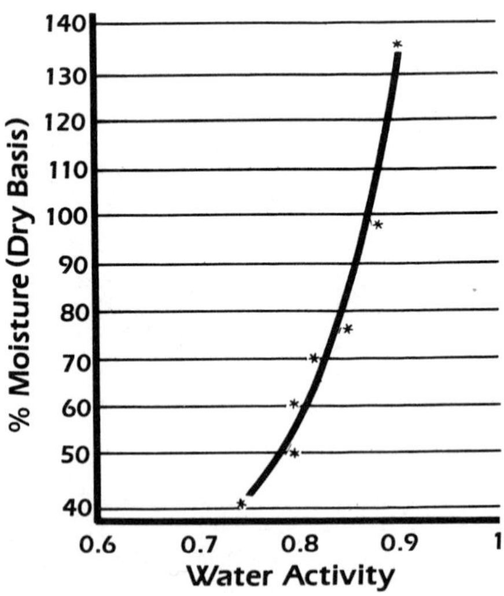

FIG. 15.10 Isotherm for sorbitol, 25°C.

to the soy protein isotherm. Table 15.5 contains actual and predicted values of a_w of a complex mixture in the targeted a_w range 0.75–0.85.

The practicality of the Salwin equation can be greatly increased by having a data base of critical a_w parameters for various food attributes. For example, Table 15.6 contains a set of critical values for the components of a model system. The set of critical values can be used to evaluate the practicality of a formulation and its predicted a_w.

PREDICTING PACKAGING SPECS RELATED TO SHELF LIFE

Packaged Product Moisture Gain/Loss

In addition to issues concerning moisture and a_w adjustments within a food, moisture transfer resulting in package moisture gain or loss is an important quality control and legal issue in foods. Generally the quality shelf life of a

TABLE 15.5 Actual and Predicted Moistures and a_w for a Complex Mixture in the Targeted a_w Range 0.75–0.85 at 25°C

% Moisture, dry basis		Water activity	
Predicted	Actual	Predicted	Actual
18.9	17.4	0.76	0.75
22.6	23.3	0.80	0.82
29.3	31.4	0.85	0.86

TABLE 15.6 A Set of Critical a_w Values for Ingredients in a Model Food Product

	Moistness	Crispness	Chewiness	Toughness
Cereal		<0.40		>0.50
Fruit	>0.30		<0.50	<0.30
Nuts		<0.65		

food packaged in an nonhermetically sealed container is dependent upon the rate of exchange of moisture through the package to the storage atmosphere plus the rate of change in a_w of the food toward critical upper or lower limits with respect to bacteria, fungi, texture, flavor, appearance, eating qualities, aroma, nutritional qualities, cooking qualities, and regulatory issues.

Current regulations on moisture loss from packaged foods are tied to good manufacturing practices. Most states have adopted a "model law" permitting moisture loss to below specified package weight in intrastate commerce only. Underweight food products, for whatever reason, are not legal for shipment into these states. Most companies overpack in order to counter moisture loss problems and carefully select packaging materials.

The technical problem is that foods packaged in nonhermetic containers are subject to moisture gain or loss dependent upon:

a. Storage relative humidity (a_w);
b. Sorption properties of the product;
c. The water activity gradient into or out of the product relative to storage atmosphere;
d. The permeability of the package material to water vapor.

Predicting Package Stability

Analogous to the use of the Salwin equation, sorption isotherms are combined with an equation based on Fick's law (Glasstone, 1946, p. 1257) and Henry's law (Glasstone, 1946, p. 696) to accomplish three main objectives:

1. Predict the storage time based on a critical a_w of the product for a particular package system under assumed storage conditions.
2. Predict packaging system specifications for obtaining acceptable shelf life with respect to moisture exchange.
3. Estimate packaging costs.

Typically, the prediction method involves three steps:

1. Obtain the isotherm (adsorption or desorption) of the product.
2. Identify the critical a_w for a product attribute, such as texture.
3. Solve the predicting equation.

As an example, the number of days required for a typical ready-to-eat cereal to reach its critical moisture level for loss of crispness when stored in a package in a humid atmosphere will be predicted. Figure 15.11 is a typical isotherm for a ready-to-eat cereal. For predicting the time for diffusion of the required amount of water vapor into the cereal, Fick's second law of diffusion takes the form shown in Fig. 15.12.

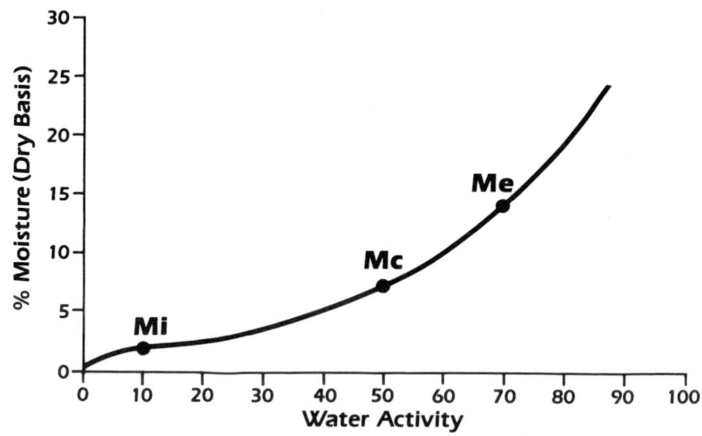

FIG. 15.11 Typical isotherm for a ready-to-eat cereal. *Mi* = Initial product moisture; *Mc* = Critical product moisture; *Me* = Equilibrium product moisture.

$$\vartheta_c = \text{Days to Mc} = \frac{\text{Ln}((\text{Me}-\text{Mi})/(\text{Me}-\text{Mc}))}{(\text{K/X})\ (\text{A/Ws})\ (\text{Po/B})}$$

Me = Equilibrium Product Moisture If Left in Contact with the Atmosphere Outside the Package (Depends on Temp., Relative Humidity, and the Product Isotherm)
Mc = Critical Product Moisture Level (Example: Crispness Is Lost at 10% Moisture)
Mi = Initial Product Moisture (Typically About 2%)
K/X = Permeability of Package to Moisture Vapor
A = Surface Area of the Package (Typically About 0.1303 Square Meters)
Po = Vapor Pressure in mm Hg at Storage Temperature
Ws = Grams of Dry Solids of Product in the Package
S = Slope of Product Isotherm (Assumed Linear) over the Range Between Mi and Mc
θ_c = Days It Takes for Product in the Package to go from Initial to the Critical Moisture Level

FIG. 15.12 Equation for predicting the time to diffuse the critical level of moisture into a packaged product.

The data in Table 15.7 are used in the equation to predict a crispness shelf life of only 46 days using typical values for package size, weight of product, moisture level, and storage conditions, and a package permeability factor (k/x) of 0.3.

A shelf life of 46 days is not nearly enough for a conventional ready-to-eat cereal. A typical option for increasing shelf life is to select a package material that has a lower permeability to water vapor, that is, a lower k/x value. By

TABLE 15.7 Equation for Predicting the Days to a Critical Moisture for Retention of Crispness in a Packaged Ready-to-Eat Cereal

$$\text{Days} = \theta_c = \frac{\ln\left(\frac{Me - Mi}{Me - Mc}\right)}{\left(\frac{K}{X}\right)\left(\frac{A}{Ws}\right)\left(\frac{Po}{B}\right)} = 46 \text{ days}$$

$Me = 0.06\text{g } H_2O/\text{g solids}$

$Mc = 0.05\text{g } H_2O/\text{g solids}$

$Mi = 0.02\text{g } H_2O/\text{g solids}$

$\frac{K}{X} = 0.3\text{g } H_2O/\text{day}/\text{m}^2/\text{mm Hg}$

$A = 0.1303 \text{ m}^2$

$Ws = 416.5\text{g}$

$B = 0.06154\text{g } H_2O/\text{g solids/unit } a_w$

$Po = 19.827 \text{ mm Hg}$

TABLE 15.8 Effect of Packaging Film k/x Values on Days to Critical Moisture (Mc)

k/x	Days to Mc
0.3	46
0.2	69
0.1	137
0.05	275

inserting increasingly smaller k/x values into the predicting equation, the days required to reach a predicted critical moisture content can be increased to a desired shelf life (Table 15.8).

Salwin Equation and "What If" Possibilities

The Salwin equation can be used to design "What if . . ." possibilities that could have significant advantages in packaging. For example, assume that two foods are identical in all respects except for their critical moisture levels for retention of flavor and in the shapes of their sorption isotherms (Fig. 15.13). The difference in predicted shelf life on storage in the same packaging material at 22°C and 60% relative humidity is shown in Table 15.9. Product B has a significantly longer shelf life than A, using the same packaging material at the same cost.

If 113 days is a sufficient shelf life for product B, a "what if . . ." possibility is to use a more permeable, and therefore cheaper packaging material for product B. The Salwin equation can be used to solve for the k/x factor for 113 days for product B. For product B, the k/x value is 0.217 for a packaging material that costs 35% less than the material for product A.

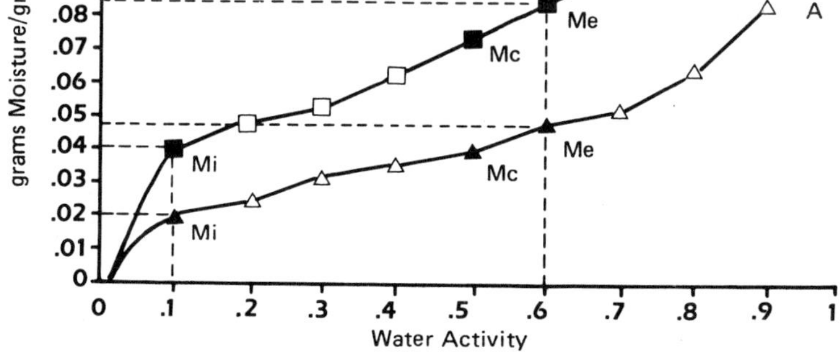

FIG. 15.13 Products A and B, identical in all respects except critical moisture levels for retention of flavor and shape of sorption isotherms, 22°C. (For difference in predicted shelf life on storage in the same packaging material, see Table 15.9.)

TABLE 15.9 Effect of Differences in Isotherms on the Time to Reach a Critical Moisture in a Packaged Food

	A	B
Me	0.048	0.085
Mi	0.020	0.040
Mc	0.040	0.080
k/x	0.100	0.100
A	0.1303	0.1303
Ws	416	408
Po	19.827	19.827
B	0.056	0.090
Predicted days to *Mc*	113	312

EFFECT OF a_w ON ENZYME ACTIVITY

A very important practical aspect of a_w is to control undesirable chemical and enzymatic reactions that reduce quality shelf life of foods. It is a well-known generality that rates of changes in food properties can be minimized or accelerated over widely different values of a_w. Figure 15.14 shows that different types of reactions in foods are accelerated or minimized as a function of a_w. Small differences in a_w can result in large differences in reaction rates. As a practical example, small differences in the a_w in a fudge chip oatmeal cookie mix can make large differences in lipase activity as measured by the rate of formation of free fatty acids (Webster, 1978).

Troller and Christian (1978) summarized the general behavior of enzyme activity as a function of a_w in Fig. 15.15. Near or below the BET monolayer a_w value, enzyme activities are generally minimized or cease. Above the monolayer value of a_w the enzyme activities would be expected to be increased.

Lipase Activity and Sorption Isotherms in Defatted Ground Oats

Acker and Beutler (1965) reported on the activity of lipase in defatted ground oats mixed with monoolein. The activity curve for lipase closely matches the sorption isotherm for the mixture of oats and monoolein (Fig. 15.16), thus lipase activity appears to be quite sensitive to differences in a_w.

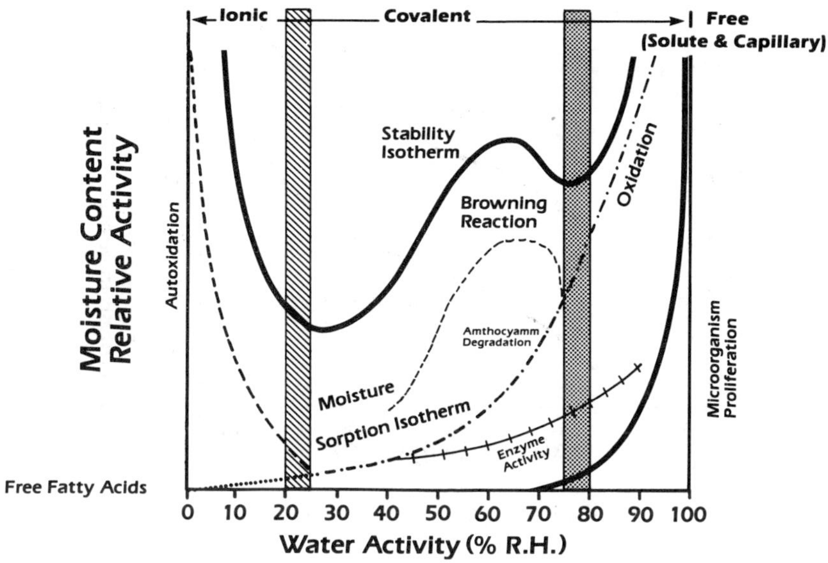

FIG. 15.14 Different types of reactions in foods are accelerated or minimized as a function of a_w.

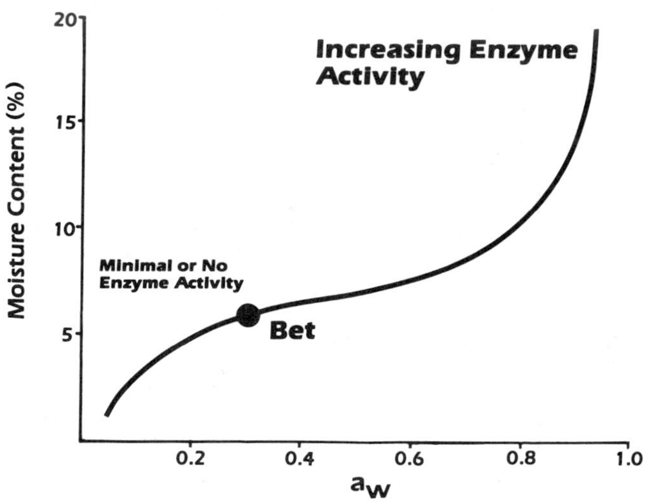

FIG. 15.15 General behavior of enzyme activity as a function of a_w. (*From Troller and Christian, 1978.*)

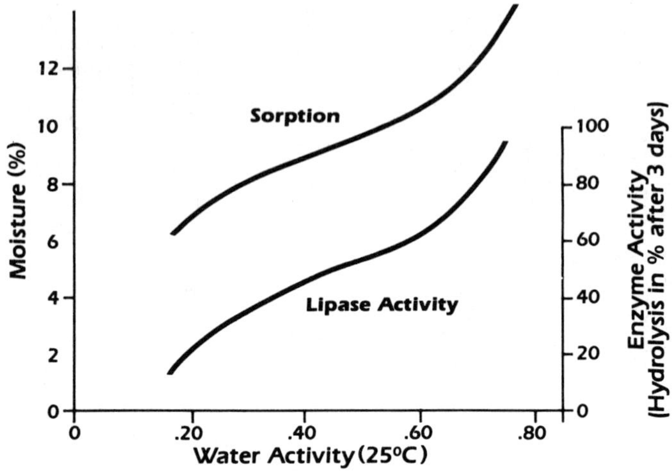

FIG. 15.16 Lipase activity in defatted ground oats + 4% monoolein. (*From Acker and Beutler, 1965.*)

Effect of a_w on Reaction Rates in a Fudge Chip Cookie Mix

In addition to the sensitivity of enzymes to moisture, the rates of enzyme reactions may also be a function of the mobility of the enzyme on the substrate and/or on the mobility of the substrate itself (Acker, 1962; Dapron, 1985). In a

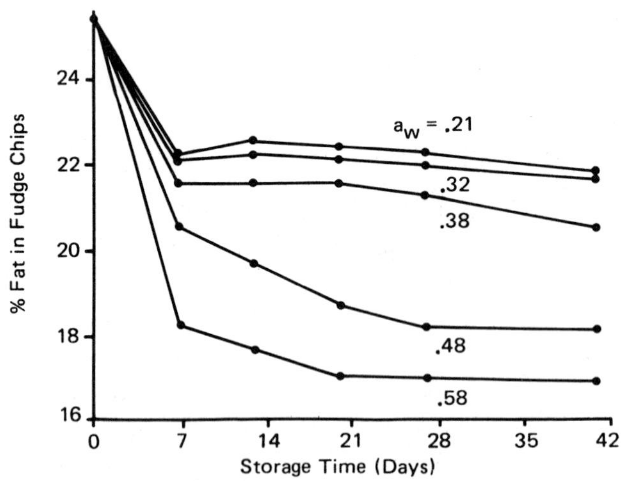

FIG. 15.17 Mobility of substrate diffusing from chips (in a cookie mix) as a function of a_w.

cookie mix, both oatmeal and fudge chips are sources of substrate for the lipase in flours. In Fig. 15.17, the mobility of the substrate diffusing from the chips as a function of a_w is evident. In a related graph (Fig. 15.18), the interaction of a_w and days stored at 100°F (38°C) on the formation of free fatty acids in the cookie mix is shown.

A plot of reaction rates vs. a_w is given in Fig. 15.19. The minimum data

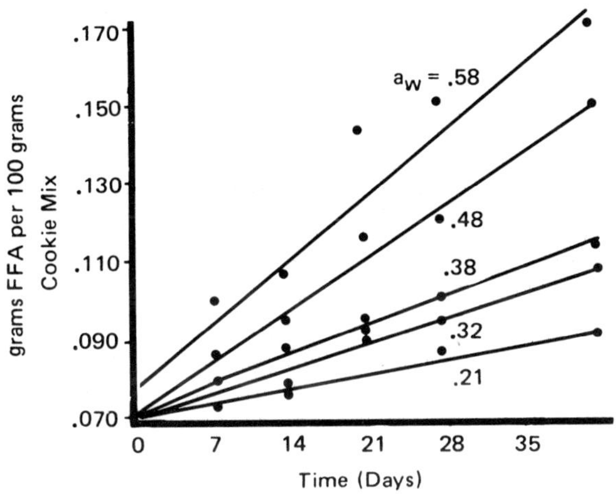

FIG. 15.18 Effect of storage time at various a_w on free fatty acid concentration in fudge chip cookie mix stored at 100°F.

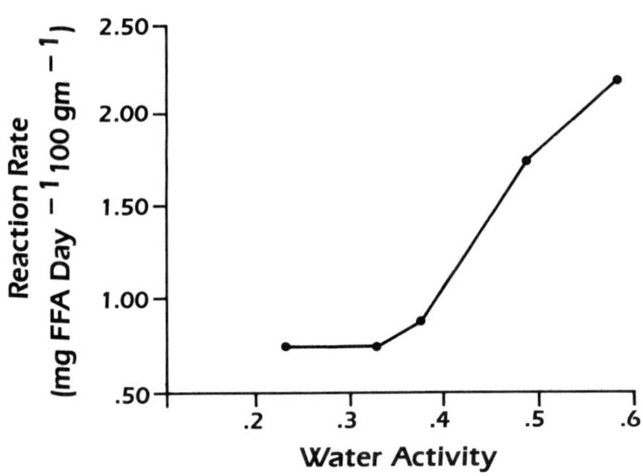

FIG. 15.19 Effect of a_w upon lipase reaction rates in fudge chip cookie mix stored at 100°F.

points for lipase activity are in the range of a_w 0.22–0.33. As a generalization, then, one would formulate to a_w of 0.3 and less in order to minimize lipase activity.

PRACTICAL APPLICATIONS OF a_w IN FOODS

Since the early 1960s, a large number of patents describing practical applications of a_w in foods have been issued in the U.S. and abroad. A few recent patents have been selected here to illustrate the various ways the food industry is using a_w today.

Table 15.10 (Staley and Pelaez, 1985) (McCormick & Co.) shows the key features of a patented process for producing intermediate moisture quick-cooking beans. Packed in a plastic film container, the beans are claimed to be quickly reconstituted to a cooked state having excellent texture and the flavor of home-cooked beans.

Table 15.11 relates to a patent assigned to the Procter & Gamble Co. (1985) for a shelf-stable liquid tea concentrate with added flavors. Teas are prepared by simple mixing of the concentrate with water.

TABLE 15.10 Process for Producing Intermediate Moisture Quick-Cooking Beans

1. Cook dry beans under pressure in a humectant solution.
2. Raise moisture to 50–60%.
3. Dry to 30–40% moisture.
4. a_w 0.75–0.85

Source: Staley and Pelaez (1985).

TABLE 15.11 Stable Liquid Tea Concentrate

a_w 0.75–0.85
pH $\leq$3
Total solids $\geq$55%
No preservatives

Source: Procter & Gamble (1985).

Sun-Diamond Growers (1985) were assigned a U.S. patent for preparing intermediate moisture fruit pastes and purees (Table 15.12).

The Pillsbury Co. has been assigned a patented method (Durst, 1985) for producing shelf-stable ready-to-eat baked products (Table 15.13).

Nabisco Brands Inc. (1985a, b, c) is the assignor of three 1985 patents: see Table 15.14 on intermediate moisture fruit fillings; Table 15.15 describes a gel for use in soft–moist cookies that have an extended shelf life; and Table 15.16 lists a Nabisco patent for an intermediate moisture dog food that is either soft or hard and contains meat particles. The last recent patent is perhaps the most interesting because it clearly illustrates the kind of opportunities that can be found for practical applications of a_w.

TABLE 15.12 Produce Pastes or Purees of Apricots, Peaches, Nectarines, Mangoes, or Papayas

1. Treat with SO_2.
2. Dehydrate whole fruit to $a_w \leq 0.70$.
3. Extrusion separates a substantially solid-free mix.
4. Use in candy, cookies, ice cream, etc.

Source: Sun-Diamond Growers (1985).

TABLE 15.13 Ready-to-Eat Cakes and Breads Shelf Stable Up to One Year

$\geq$5% Polyhydric alcohol
$a_w \leq 0.85$
Pound cake, carrot cake, sponge dough bread, brownies

Source: Durst (1985).

TABLE 15.14 Intermediate Moisture Fruit Fillings Patent

a. $a_w \leq 0.70$
b. Moisture 22–25%
c. Shelf stable
d. Heat stable (as used in toaster pastries)

Source: Nabisco Brands Inc. (1985a).

TABLE 15.15 Gel for Use in Soft Cookies Patent

High fructose syrup	100
Glycerol	3
Sodium alginate	1.25
Calcium sulfate	1.25
Propylene glycol	1
Gel formed on ordered mixing	

Source: Nabisco Brands Inc. (1985b).

TABLE 15.16 Soft or Hard Dog Biscuits with Meat Inclusions Patents

a. Prepare spiced cured meat particles.
b. Mix into a dough ± plasticizing humectant for soft or hard texture.
c. Bake to $a_w \leq 0.70$.

Source: Nabisco Brands, Inc. (1985c).

There is a consensus that emulsions have little effect on raising or lowering a_w at temperatures above freezing. However, it has recently been reported that certain emulsions can lower the freezing point of aqueous systems by 10°C (Arai and Watanabe, 1985). Apparently certain emulsions can interfere with the nucleation of ice crystals and thereby lower freezing points. Arai and Watanabe used the plastein-type reaction to link gelatin with an amino acid

TABLE 15.17 Observation of the Presence or Absence of the Antifreeze Effect at −10°C in Emulsions of Various Surfactants

	Frozen
EMG-12	No
PGS	No
Tween-80	Yes
Sodium caseinate	Yes
Sodium soy proteinate	Yes

TABLE 15.18 Antifreeze Agent for Aqueous Media

- Commercial proteins (casein, gelatin, soy, albumen, etc.) modified with an amino acid ester containing an aliphatic alcohol.
- Emulsions are nonfreezable at −10°C.
- Use in freezer-stored foods.

Source: Kanegafuchi Chemical KK (1985).

ester of dodecyl acid to form a surfactant that has an unusual affinity for ice nuclei. As shown in Table 15.17, the plastein product, EMG-12, as well as the surfactant PGS (polyglycerol stearate), formed emulsions that did not freeze at −10°C, whereas freezing did occur in three other emulsions made with different surfactants.

Related to this new technology, a patent has been assigned (Kanegafuchi Chemical KK, 1985) for producing an antifreeze agent for use in the manufacture of foods stored at freezer temperatures (Table 15.18). A stable, nonfreezing emulsion system could replace the high solute load that is typically required to achieve nonfreezing in a food system at −10°C.

CONCLUSIONS

In conclusion, a few interrelated examples of practical applications of a_w in foods have been presented. The practical applications of a_w throughout the food industry have generated creative new food products worth many billions of dollars since the 1960s.

Perhaps of even greater potential value is the pool of basic information about a_w and its connection to moisture relations and food qualities that has been generated since the 1960s. There are tremendous opportunities for food scientists and technologists to creatively translate this pool of basic information into practical applications of great value.

REFERENCES

Acker, L. 1962. Enzymatic reactions in foods of low moisture content. *Adv. Food Res.* 11: 263.

Acker, L. and Beutler, H. O. 1965. Über die enzymatische Fettspalltung in wasserarmen Lebensmitteln. *Fette, Seifen, Anstrichm.* 67: 430. [In *Food Technol.* (1969) 23(10): 31.]

Arai, S. and Watanabe, M. 1985. Modified protein as a new surfactant. In *Properties of Water in Foods.* Simatos, D. and Multon, J. L. (Ed.). Martinus Nijhoff, Dordrecht, The Netherlands.

Baxter, E. J. and Hester, E. E. 1958. The effect of sucrose on gluten development and the solubility of the proteins of a soft wheat flour. *Cereal Chem.* 35: 366.

Bone, D. P. 1965. Method of preparing a solid semi-moist marbled meat pet food. U.S. patent 3,380,832. April 30.

Bone, D. P. and Shannon, E. L. 1975. Method of making a dry type pet food having a meat-like texture and composition thereof. U.S. patent 3,883,672. May 13.

Bone, D. P. and Shannon, E. L. 1977. Process for making a dry pet food having a hard component and a soft component. U.S. patent 4,006,266. Feb. 1.

Bone, D. P., Shannon, E. L., and Ross, K. D. 1975. The lowering of water activity by order of mixing in concentrated solutions. In *Water Relations of Foods.* Duckworth, R. B. (Ed.), p. 613. Academic Press Inc., London.

Burgess, H. M. and Mellentin, R. W. 1965. Animal food and method of making the same. U.S. patent 3,202,514. Aug. 24.

Dapron, R. 1985. Enzyme activity as a function of water activity. In *Properties of Water in Foods.* Simatos, D. and Multon, J. L. (Ed.), p. 171. Martinus Nijhoff, Dordrecht, The Netherlands.

Durst, J. R. 1985. Storage stable, ready-to-eat baked goods. U.S. patent 4,511,585. April 16.

Glasstone, S. 1946. *Textbook of Physical Chemistry,* 2nd ed. D. Van Nostrand Company, Inc., Princeton, NJ.

Maurice, T. J., Slade, L., Sirett, R. R., and Page, C. M. 1985. Polysaccharide-water interactions—thermal behavior of rice starch. In *Properties of Water in Foods.* Simatos, D. and Multon, J. L. (Ed.), p. 211. Martinus Nijhoff Publishers. Dordrecht, The Netherlands.

Kanegafuchi Chemical KK. 1985. Japanese patent 60-235,880.

Nabisco Brands, Inc. 1985a. Heat stable fruit filling for pastry doughs-comprising high fructose corn syrup, apple powder, tapioca starch, and concentrated fruit juice or puree. U.S. patent 4,562,080. Dec. 31.

Nabisco Brands, Inc. 1985b. Method and composition for soft edible baked products having an improved shelf-life and an edible firm gel for this use. EP patent 155,203. Sept. 18.

Nabisco Brands, Inc. 1985c. Production of soft or hard dry canine biscuits containing meat particles by stepwise blending of dough containing the particles, baking, and drying. GB patent 2,115,265. Dec. 18.

Procter & Gamble Co. 1985. Stable liquid tea concentrates-containing tea, acid, and sweetener. EP patent 162,526. Nov. 27.

Robinson, R. A. and Stokes, R. H. 1959. *Electrolyte Solutions,* 2nd ed. Butterworths, London.

Rockland, L. B. and Nishi, S. K. 1980. Influence of water activity on food product quality and stability. *Food Technol.* 34(4): 42.

Salwin, H. 1963. Moisture levels required for stability in dehydrated foods. *Food Technol.* 17(9): 34.

Scott, W. J. 1957. Water relations of food spoilage microorganisms. *Adv. Food Res.* 7: 83.

Staley, L. L. and Pelaez, J. 1985. Process for producing intermediate moisture quick-cooking beans. U.S. patent 4,510,164. April 9.

Sun-Diamond Growers. 1985. Producing fruit pastes and purees from drupacious fruit by partial dehydration followed by compression of the whole fruit. U.S. patent 4,554,168. Nov. 19.

Troller, J. A. and Christian, J. H. B. 1978. *Water Activity and Food,* p. 50. Academic Press, New York.

Webster, F. 1978. Unpublished data. Quaker Oats Co., Barrington, IL.

Index